Recent Trends in
Water Research
Hydrogeochemical and Hydrological Perspectives

Recent Trends in

Water Research

Hydrogeochemical and Hydrological Perspectives

Editors

S. Chidambaram

AL. Ramanathan

K. Shivanna

R. Arthur James

I.K. International Publishing House Pvt. Ltd.

NEW DELHI • BANGALORE

Published by

I.K. International Publishing House Pvt. Ltd.
S-25, Green Park Extension, Uphaar Cinema Market
New Delhi – 110 016 (India)
E-mail: info@ikinternational.com
Website: www.ikbooks.com

ISBN 978-93-80026-91-6

Published by Krishan Makhijani for I.K. International Publishing House Pvt. Ltd., S-25, Green Park Extension, Uphaar Cinema Market, New Delhi – 110 016
Printed by Avantika Printers Private Limited New Delhi - 110 065.

Prof. Dr. M. Ramanathan,
M.S. (G.S.) F.R.C.S.(Glasg)
Vice-Chancellor

Phone : Office : 04144-238283
Res. : 04144-238214
Fax : 04144-238925
04144-238080
E-mail : auvc_mr@hotmail.com
ANNAMALAI UNIVERSITY,
ANNAMALAINAGAR P.O. 608 002
INDIA

27-07-2009

MESSAGE

I feel happy to learn that the Department of Earth Sciences, Annamalai University has successfully brought out a book on "The Recent Trends in Water Research: Hydrogeochemical and Hydrological process". The demand of freshwater continues to grow in manifold times during the recent years. The diversion of freshwater supply to agricultural, industrial, domestic and municipal needs results in stretching hydrological systems both natural and man made limits within the country. A continuous supply of freshwater may already be in scarce and may vary both seasonally and geographically. The study of quantity of water alone is not sufficient to solve the water management problem because their uses for various purposes depend on its quality.

In this scenario, it is interesting that this book is being brought out with nearly twenty three papers incorporated with different aspects of recent trends in water research, which was discussed as issues of interest in the National Seminar on Recent Trends in Water Research (RETWAR-2009). It consists of key issues like, isotopic studies, rain water chemistry, pollution, geophysical techniques for quantification, identification of good groundwater potential regions, instrumentations, etc., I do hope that this book will be a cluster of good scientific contributions. It will serve to fill up the gaps and enlighten the new avenues in this area of research.

(Dr. M. Ramanathan)

Message

Prof. M. Arumugam
The Dean,
Faculty of Science
Annamalai University

28.07.2009
Annamalai Nagar

The book on Hydrogeological and Hydrogeochemical process, the outcome of the National Seminar on Recent Trends in Water Research by Dr. S. Chidambaram can be singled out for one special feature — it has got included various principles in its substance and style. These include comprehensive coverage of various themes, a balanced treatment of any topic and a lucid simplicity in expression. The book tries to encompass the latest development and the subject, spells out basics clearly and makes liberal use of relevant illustrations for the desired effect.

The collections of papers included in the edited volume are of great value in pursuing understanding and application in future studies. Among the papers; the paper by Shahul Hameed and Gowtham are found to be quiet useful to researchers in Groundwater studies with the application of stable isotopes. The paper on rainwater status of Puducherry is found to be quite useful in coastal groundwater studies. Kushagara et al., have distinguished indigenous bacterial stains from tannery effluent sediments and suggested the same can be utilized for pollution abatement.

The paper on speciation analyses of trace elements in water brings out the applicability of AAS in determining trace metal concentration. The papers on defluoridation of groundwater by Aquatic Macrophyte *Hydrilla verticillata* enlighten the role by *Hydrilla verticillata* in removing fluoride which is the most dangerous pollutant of water.

Thus, this volume provides an invoking impact to the researchers.

M. Arumugam

Preface

The specific role of water in socio-economic development has been widely recognized for the past few decades. The availability of freshwater is one of the major issues facing mankind today in many ways. These problems are very closely related with lives of many millions of people. The rapid growth of population coupled to steady increase in water requirements as industrial, agricultural and domestic. Due to climate change and over exploitation and other anthropogenic impact, surface water resource is under stress and are depleted or abused and hence the focus of the planners is now more on the groundwater resources management. The specific advantage of groundwater is its ubiquity, and this ubiquity presents particular problems in its management. The spatial and temporal distribution of groundwater occurrences has presented certain opportunities for the development of rural communities, urban centers and irrigated agriculture. It is the pattern of groundwater use that serve as a starting point for considering management options (United Nations, 1960 and 2000).

Long-term, systematic measurements of groundwater levels provide essential data needed to evaluate changes in groundwater storage over time. Using these data, the assessment would include two types of indicators to describe the nationwide status and trends of groundwater availability: groundwater-level indices and periodic assessments of changes in groundwater storage. No nationwide, systematic groundwater level monitoring programme exists. At present, there are few long-term observation wells record are available for any meaningful interpretation. The density of existing monitoring wells, however, varies considerably from a combination of analytical methods and delineation techniques in order to differentiate between various methods of assessment of groundwater resource. Pollution threatens surface water supplies and the availability of the precious resource–groundwater. More and better water cannot be acquired without the continuation and extension of water research. The analytical programme should include the determination of trace elements, mineralogy, organic components and micro-organisms, which underlines the necessity for interdisciplinary collaboration in groundwater studies.

This book is a conglomeration of twenty-three multi-disciplinary field papers covering various aspects of water resources, viz., Geophysical Techniques applications, Hydrogeochemical interpretations, Isotopic signature, speciation of trace elements, etc. The case studies from different parts of India are well illustrated with detailed analysis. Impacts of anthropogenic activities and the resultant deterioration of water quality in different parts of India were thoroughly introspected in different case studies. The case studies on anthropogenic aspect gives an insight into the impact from dye industries of Tiruppur, Industrial pollutants along the coastal aquifer of Cuddalore region, Impact of fertilizer, nitrates and ore mining/exploitation impact on coastal groundwater. The hydrogeochemical changes due to natural geological process from Salem, Kalpakkam, Kanniyakumari and Chennai are discussed in detail. Some of the key issues like targeting the availability of potential fresh groundwater regimes in hard rocks and sedimentary terrains with the aid of Geophysical Resistivity survey methods used in conjunction with borehole geophysical techniques were discussed as well. Hydraulic continuity determination in different aquifers was attempted by different authors using Stable Isotopes and Tritium Isotopes techniques. Phytoremediation to remove excess fluoride from drinking groundwater, bacterial treatment of the tannery effluent is also dealt in length in this book. Trace metal speciation and their significance in water analysis is one of the chapter in this book which gives an overall outlook of metal species in water quality control. Hence, this book is an agglomeration of different aspects of water research, covering surface, rain and groundwater. This book is useful for teaching, research and for management of water and its resources in India. We thank Ministry of Earth Science, India, for their financial support to bring out this volume. We thank sincerely all the contributors to this book for sparing time to spare their findings and experience which will help to manage the water resource of our country.

S. Chidambaram
AL. Ramanathan
K. Shivanna
R. Arthur James

Contents

Editors

S. Chidambaram, Reader, Department of Earth Sciences, Annamalai University is an expert in the field of groundwater resource management especially on its quality, sources and evolution (isotope hydrogeology). He is a university gold medallist in the UG and PG courses. He received DST young scientist project and best paper award on drinking water supply in SAARC countries. He has got several ongoing and has completed numerous projects on hydrogeochemistry, F contamination and Isotope from various governmental agencies like DST, MOEN and F, UGC, BRNS, etc. He has produced 3 Phds and 12 publications in reputed referred journals in his field of expertise. He has visited and worked in Russia for his groundwater research collaborative projects. He is doing numerous collaborative research with institutes like BARC, NIH, NGRI, JNU and other universities and institutions in India and abroad.

AL. Ramanathan is a Professor in the School of Environmental Sciences, Jawaharlal Nehru University, and New Delhi, India. His research area is in the field of hydrogeochemistry from inland and coastal surface and groundwater and their resource management. He has got 18 yrs of teaching experience (PG) and research experience in this subject. He has taught in numerous universities in India and abroad. He has guided quite a number of Ph.D. research students on groundwater quality and modelling aspects. He has widely travelled all over the world for the advanced research work in groundwater. He is also a recipient of various international and national scholarship and had collaborations with institutes and universities reputed in groundwater research in India and abroad. He has published two dozen articles in reputed referred journals and authored five books in these aspects. He has completed and continuing his research work on groundwater got from Indian and international agencies.

K. Shivanna, Head, Isotope Hydrology Section, Isotope Applications Division of Bhabha Atomic Research Centre (BARC) and Associate Professor, Homi Bhabha National Institute, BARC, has made outstanding contribution in the field of isotope hydrology for the last 30 years. He is a specialist in the field of isotope hydrology, hydrogeology and hydrochemistry for water resources development and management and internationally known in isotope hydrology. He served as a UNDP/IAEA regional expert in the field of water resources development in African countries and national co-coordinator for geothermal systems under IAEA/RCA programme. He has authored over 80 scientific publications.

R. Arthur James, Senior Assistant Professor, Department of Marine Science, Bharathidasan University, Tiruchirapalli has 15 years experience in the field of Environmental Impact Assessment (EIA), Quality Control (QC) analysis and Environmental Audit. He has conducted detailed geological surveys and environmental audit for various projects. He has published nearly 35 papers in reputed national and international journals. Dr. James is the recipient of Young Scientist (2006), Young Investigator (2007) and Indian Academy's summer fellowship (2008) awards. He has been serving as board of studies member and resource person for various academic bodies.

1

Geophysical Investigation in the Different Litho Units of Gadilam River Basin, Tamil Nadu, India

*PRASANNA M.V.[1], CHIDAMBARAM S.[2], NAGARAJAN R.[1], RAJALINGAM S.[1] AND ELAYARAJA A.[3]
[1]School of Engineering and Science, Curtin University of Technology, Miri, Sarawak Malaysia–98009
[2]Department of Earth Sciences, Annamalai University, Annamalai Nagar–608002
[3]School of Business Curtin University of Technology, Miri, Sarawak, Malaysia–98009
*E-mail: prasanna@curtin.edu.my

ABSTRACT

Resistivity survey was conducted in 20 locations to cover different lithologies in the study area. Three major zones were delineated as weathered, fractured and massive in hard rock region and top soil, sandy layer and clayey layer in sedimentary region. Interpreted master curves are mostly three-layered curves, namely A type followed by H, K and Q type. This study helps us to identify the weathered zone in the hard rock and contaminated zone by anthropogenic impact in the downstream of river Gadilam.

Keywords: Resistivity, thickness, Gadilam river basin.

INTRODUCTION

Geophysical resistivity surveys are regularly used for studies related to groundwater investigations. Resistivity profiling delineates the lateral changes in resistivity that can be correlated with steeply dipping interfaces between two geological formations in the subsurface. Resistivity sounding determines the thickness and resistivity of different horizontal or low dipping subsurface layers, including the aquifer zone (Kalpan Choudhury and Saha, 2004). The electrical resistivity method is widely used in groundwater exploration studies (Todd, 1959) because it's least expensive of all geophysical methods requiring no specially trained technicians to operate the instrument. Water barren formations can be identified based on the contrast in electrical resistivity (Zohdy *et al.* 1974).

The study area Gadilam river basin is located in Cuddalore and Villupuram district, Tamil Nadu, India. The area is bounded between latitudes 11°30'N – 11°55'N and longitudes 79°0'E – 79°47'E. It covers a total area of about 1,394 sq. km (Fig. 1). The location of Neyveli mining region and Veeranam Lake from which groundwater extraction is at its maximum falls in this environment. There are group of industries sparsely distributed which generate multi-faceted chemicals and raw materials is located along the downstream of the river Gadilam. The Gadilam river basin covers different stratigraphic units, viz. Archaean, Cretaceous, Tertiary to Recent alluvium (Fig. 2). Archaean complex consists of garnetic gneiss, charnockite, leptinites

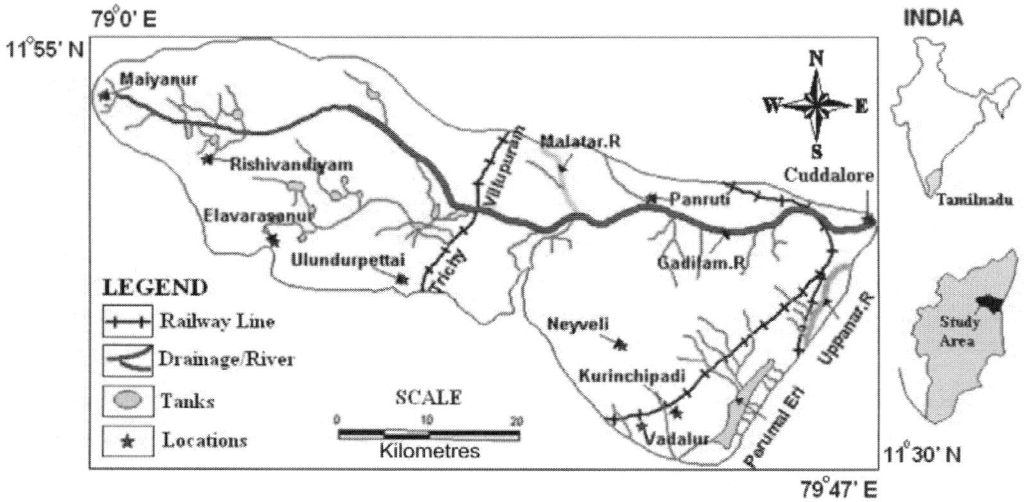

Fig. 1. Location map of the study area.

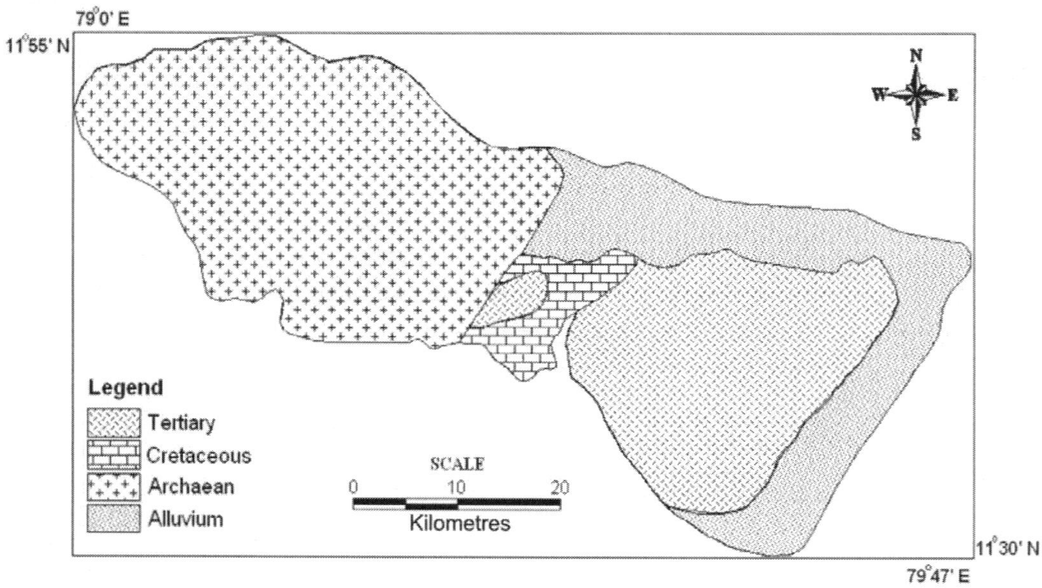

Fig. 2. Geology map of the study area.

and schists. Limestone constitutes the cretaceous formation and Cuddalore sandstone of the tertiary age are the litho units of the area. The western part of the study area is composed of hard rock and the eastern part by sedimentary formation. It is inferred as a faulted contact between the hard rock and the sedimentary formation (Aravindan *et al.*, 2004). The basin lies between the Ponnaiyar in the north and Vellar in the south. The open wells of the study area show a range of water levels from 3.10 m BGL (meter below ground level) to 48.85 m BGL with an average of 15.25 m BGL in sedimentary terrain and 5.23 m BGL to 7.48 m BGL with an average of 6.21 m BGL in hard rock terrain.

MATERIALS AND METHODS

The resistivity survey was done with the help of SSR-MP-AT Resistivity meter, ±1 accuracy (IGIS, Hyderabad, India) adopting Schlumberger electrode configuration and vertical electrical sounding (VES) method. The apparent resistivity of the subsurface formation was determined. The results were manually plotted by curve matching method and cross-checked with IPI2Win (v.2.0) software (Geoscan-M Ltd, 2002). The location map of resistivity sounding is given in the Fig. 3. The resistivity sounding was carried out in the study area which covering major litho units.

RESULT AND DISCUSSION

The Vertical Electrical Sounding (VES) soundings were carried to cover different lithologies, 6 soundings falls in Alluvium formation, 6 in Tertiary and 8 in Archaean formation. The interpretation was done through the following procedures and it was found that 3-4 layers were identified in this region.

The geophysical survey was conducted and the curves were manually plotted and then checked with IPI2Win software and the RMS error was less than ±2%. Most of the values of

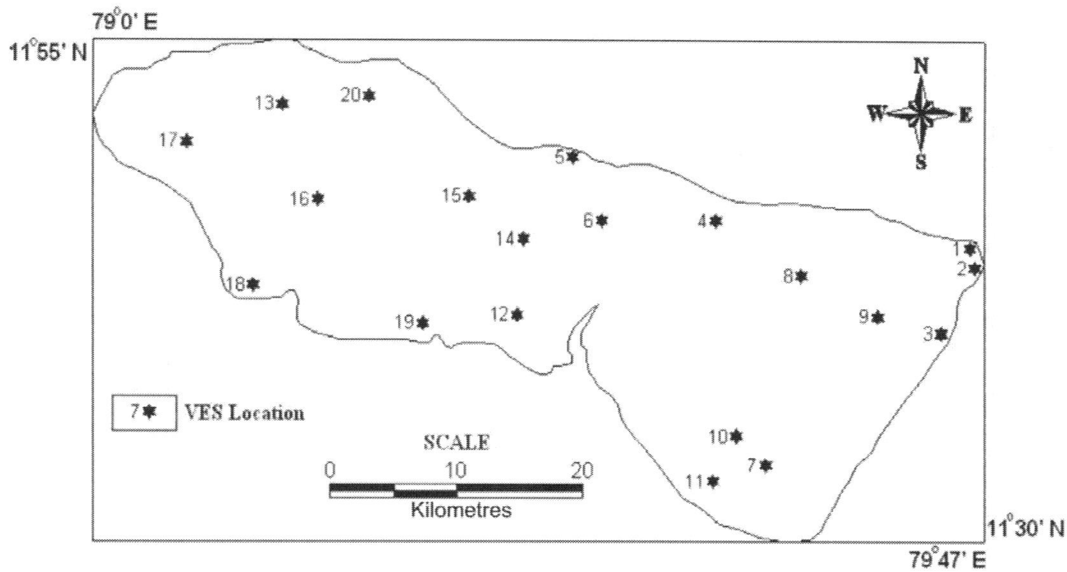

Fig. 3. Resistivity sounding (VES) location of the study area.

manual matched curves and that of IPI2Win v.2.0 software were same. Thickness of weathered zone estimation and depth of bedrock studies were carried out by using inverse slope and curve matching techniques in comparison to lithology data and the curve matching was found to give fruitful results (Sankaranarayana and Ramanujachary, 1967). Curve matching is adopted here with the objective to delineate the thickness of the aquifer zones. The data analyzed by curve matching techniques and IPI2Win for 20 locations have been compared. The layer parameters like apparent resistivity (ρ_a) and thickness (h) of different layers are arrived. Minor deviations due to subsurface discrepancies and the error in the field data is corrected by smoothening the field curve to have easy interpretation. More importance was given for thickness and resistivities of various layers in hydrogeochemical studies. The values obtained by the manual curve matching were used for further studies since, the deviation was lesser. A maximum of 4 layers was identified in few regions, but major part of the study area is dominated mainly by 3 layers (Table 1).

Interpreted master curves are mostly three-layered curves, namely A type followed by H, K and Q type. VES location numbers 2, 11, 13, 16, 19 and 20 are of A type, whereas the VES No. 4, 7, 12, 17 and 18 is of H type. Four VES curves (5, 8, 14 and 15) are K type, three belong to HQ combination type (3, 9 and 10) and two are of Q type (1 and 16). Generally, A type curve is dominant in Archaean terrain indicates the resistivity values are gradually increased ($\rho1 > \rho2 > \rho3$). Minor representation of K and H type curves are also observed in this region indicating few weaker zones. In Tertiary formation, HQ and H are dominant curve types indicate that a low resistivity layer is sandwiched between the top soil and sand layer of high resistivity. In alluvium formation, the aquifer consists of alternate bands of sand, clayey sand and fine sand with clay with varying thickness. Hence, a specific dominant layer is not evident in this terrain.

In general, three major zones were delineated as weathered, fractured and massive in hard rock region and top soil, sandy layer and clayey layer in sedimentary region. It is observed that in alluvium terrain, the resistivity value ranges from 0.17 Ωm to 4999 Ωm with an average of 420.08 Ωm. The thickness ranges from 0.62 m to 16.63 m with an average of 6.16 m. In tertiary formation, resistivity values range from 0.18 Ωm to 7156 Ωm with an average of 468.15 Ωm. The thickness ranges from 0.55 m to 40.18 m with an average of 12.89 m. In Archaean terrain, resistivity ranges from 0.03 Ωm to 2952 Ωm with an average of 350.55 Ωm. The thickness ranges from 0.85 m to 42.12 m with an average of 7 m respectively. Generally, the thickness of the first layer shows decreasing trend from Tertiary, Alluvium and Archaean but in the second layer, decreasing trend from Archaean, Tertiary and Alluvium was observed. The area of low apparent resistivity values indicates the occurrence of relatively good conductors while those with high values indicate poor conductors (Srinivasamoorthy, 2004).

The apparent resistivity values and the thickness have been distributed spatially to develop a resistivity map to get an overall perspective of the study area. First two layers were selected to prepare the isoresistivity map by using Surfer software. The first layer isoresistivity map indicates that in the SE side shows higher resistivity values indicating the presence of top soil, away from the coast values are get decreased (Fig. 4). Higher thickness value is noted in Panruti of Alluvium formation due to the deeper water level by over exploitation of groundwater in this region. In Archaean formation, both the resistivity and the thickness values are lesser. In the second layer isoresistivity map (Fig. 4), lower resistivity values are noted along the coast indicates the influence of saline water into the coastal aquifer. The resistivity values get increased away

Table 1: Resistivity and thickness of geoelectrical sections.

Sounding no.	Lithology	Location	Resistivity (in ohm.m)				Thickness (in m)				Curve Type
			Layer 1	Layer 2	Layer 3	Layer 4	Layer 1	Layer 2	Layer 3	Layer 4	
1	Alluvium	Cuddalore (NT)	50.63	50.63	18.3	–	1.36	2.04	∞		Q
2		Cuddalore (OT)	0.411	41.96	929.6	–	1.2	1.5	∞		A
3		Karaikadu	2894	79.22	191.6	1.33	0.73	3.4	16.63	∞	HQ
4		Panruti	17.44	4.589	4999	–	3.89	5.89	∞		H
5		Arasur	36.54	456.2	43.68	0.1709	1.04	2.08	16.17	∞	K
6		Veeraperumanallur	253.8	7.311	1.52	1.371	0.62	1.93	9.84	∞	Q
7	Tertiary	Kurinchipadi	35.78	5.57	2.717	5.543	0.85	3.03	14.73	∞	H
8		Naduveerapattu	32.09	2621	177.4	–	7.2	26.8	∞		K
9		Ramapuram	7156	39.74	90.83	0.4023	0.55	13.42	21.92	∞	HQ
10		Vadakuttu	159.3	5.26	9.16	3.97	0.72	2.79	40.18	∞	HQ
11		Vadalur	0.2272	30.33	849.5	–	1.2	1.68	∞		A
12		K.Kallakurichi	0.5159	0.1873	0.486	–	0.6711	19.52	∞		H
13	Archaean	Ariyur	1.625	5.14	1928	–	0.85	9.83	∞		A
14		Gadilam	0.0309	39.06	3.41	–	1.2	22.5	∞		K
15		Kalamaruthur	0.247	29.96	0.105	–	1.02	9.78	∞		K
16		K. Palayam	0.5249	817	848.3	–	1.2	10.52	∞		A
17		Pasar	25.28	3.33	2952	–	0.86	5.12	∞		H
18		Sembimadevi	9.77	0.9	126.6	–	0.88	1.3	∞		H
19		Ulundurpettai	0.1145	11.11	17.52	–	1.2	1.33	∞		A
20		Villivalam	2.84	41.61	1549	–	2.4	42.12	∞		A

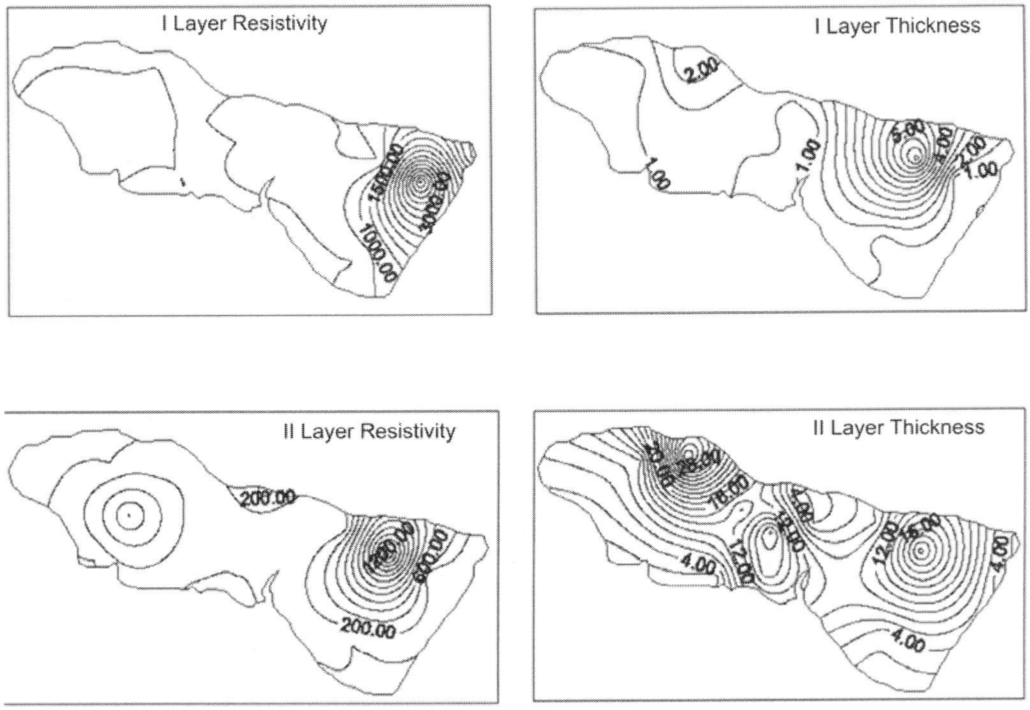

Fig. 4. Spatial distribution of resistivity and thickness of I and Ii layers.

from the coast and higher resistivity value is also noted at K. Palayam of Archaean formation considered as a weathered layer has a thickness of 10 m. In the NW side of the Archaean formation, thickness of second layer is around 40 m with lesser resistivity values.

CONCLUSION

The resistivity studies indicate that three major zones were delineated as weathered, fractured and massive in hard rock region and top soil, sandy layer and clayey layer in sedimentary region. Interpreted master curves are mostly three-layered curves, namely A type followed by H, K and Q type. A type curve is dominant in Archaean terrain indicates the resistivity values are gradually increased ($\rho_1 > \rho_2 > \rho_3$). Minor representation of K and H type curves are also observed in this region indicating few weaker zones. In Tertiary formation, HQ and H are dominant curve types indicate that a low resistivity layer is sandwiched between the top soil and sand layer of high resistivity. In Alluvium formation, the aquifer consists of alternate bands of sand, clayey sand and fine sand with clay with varying thickness. Hence, a specific dominant curve type is not evident in this terrain. The isoresistivity maps indicate that contaminated water/saline water is noted near the coast with low resistivity values. Good potential water was observed away from the coast in the study area.

ACKNOWLEDGEMENTS

The authors wish to express thank to DST-SERC for provided the necessary financial support to carry out this study. We are very much thankful to the CWRDM, Kozhikode for their cooperation.

REFERENCES

Aravindan, S., Manivel, M. and Chandrasekar, S.V.N. (2004a). Groundwater quality in the hard rock area of the Gadilam river basin, Tamilnadu. *J. Geological Soc.* India. **63:** 625–635.

Geoscan-M, Ltd. (2002). IPI2Win (v.2.0) User's manual. Moscow State University, Geological faculty, Department of Geophysics, pp. 15.

Kalpan Choudhury and Shah, D.K. (2004). Integrated Geophysical and chemical study of saline water intrusion. Groundwater. **42(5):** 671–677.

Sankaranarayanan, P.R. and Ramanujachary, K.R. (1967). An inverse slope method of determining absolute resistivities. *Geophysics.* **32(6):** 1036–1040.

Srinivasamoorthy, K. (2004). Hydrogeochemistry of groundwater in Salem District, Tamilnadu, India. Unpublished Ph.D. Thesis, Annamalai University.

Todd, D.K. (1959). Groundwater Hydrology. New York. John Wiley & Sons.

Zohdy, A.A.R., Eaton, G.P. and Mabey, D.R. (1974). Applications of surface geophysics to groundwater investigations. Techniques of water resource investigation of the U.S Geological Survey Chapter D Book. **2:** pp.116.

2

Delineation and Correlation of Fresh and Salt Water Interface of Shallow Coastal Pheratic Aquifer by Surface and Subsurface Geoelectrical Investigations

SENTHIL KUMAR, G.R.,*[1] BADRINARAYANAN, T.S.[2] AND AHILAN, J.[1]

[1]*Dept. of Earth Sciences, Annamalai University, Annamalainagar–608002*
[2]*Consultant Hydrogeologist, Kollidam–609102*
**Email: gr_senthilkumar@yahoo.com*

ABSTRACT

Seawater intrusion into the freshwater aquifers is a common scenario in the coastal aquifer system. Since fresh potable water is one of the basic needs of human being, salt water intrusion hazard which pollutes the drinking water is a global problem, causing much concern to the groundwater scientists. The role of science and technology, particularly the hydrogeological investigation techniques help us to assess the seawater intrusion. Among the various geophysical techniques one of the method, Electrical Resistivity Technique (ERT) is useful to delineate the subsurface lithology, quality and groundwater potentiality. Under ERT both surface and subsurface exploration methods were conducted at Vadakkupoigainallur (N10° 738313 & E 79° 846343) which is one of the coastal village of Nagapattinam district of Tamil Nadu. The purpose of the study is to ascertain the depth to fresh water – salt water interface. For this study a detailed hydrogeological investigation and both surface and subsurface ERT have been conducted to correlate the fresh water – salt water interface. The TDS parameter of shallow tube well water samples has been collected. The study reveals that the depth to freshwater – salt water interface is about 7 to 8 m bgl.

Keywords: Pheratic aquifer, seawater intrusion, freshwater – salt water interface, electrical resistivity techniques, well logging.

INTRODUCTION

Seawater intrusion is the movement of seawater inland into fresh groundwater aquifers as a result of higher seawater density than freshwater and heavy withdrawal of freshwater. A natural equilibrium exists between discharging fresh groundwater and seawater in a coastal aquifer. This interface is dynamic and depends on geological formations, hydraulic gradient, topography and the quantity of freshwater moving through the aquifer system. The density contrast between fresh groundwater and salt water leads to mixing and convective circulation at the saline interface. The depth to interface will vary from place to place depending upon the geomorphology, hydrogeology and topography of the area. Electrical resistivity methods have been widely used to investigate groundwater occurrences, groundwater contamination, lithology, etc. with the help of electrical resistivity techniques we made an attempt to study the fresh water–salt water interface and to delineate the subsurface lithology. The coastal aquifers that are prone to saline water intrusion are identified by relatively low resistivity values indicating salt water intrusion.

METHODOLOGY

- TDS parameter of water samples of shallow tube wells obtained by TDS pen.
- Conducted Vertical Electrical Sounding (VES) using DC resistivity meter to acquire resistivity data.
- Acquired resistivity data interpreted and analyzed both qualitative & quantitative methods using softwares.
- Obtained the geoelectrical parameters and prepared geoelectrical sections.
- Electrical well logging conducted and correlated with the litholog of bore holes.
- Delineated the subsurface lithology and freshwater – salt water interface.

STUDY AREA

The study area Vadakkupoigainallur falls in Nagapattinam district, which is one of the coastal districts of Tamil Nadu. The study area lies between latitude N 10°738 and longitude E 79°846. The district headquarters Nagapattinam is about 8 km away north of the study area (Fig. 1).

PHYSIOGRAPHY AND DRAINAGE

The entire district is a peneplained terrain with a gentle slope towards east and south-east. The area is drained by the rivers Cauvery, Vennar and their distributaries which are ephemeral in nature. The drainage pattern is subdentritic.

GEOMORPHOLOGY

The geomorphic setup is the result of action of the major rivers, fluctuation in the sea level, tidal effects of Bay of Bengal and forces of wind. The depositional regime comprises a coastal plain under marine influence, a flood plain of fluvial regime with an intermixing section of both fluvial and marine influences. Sand dunes and beach ridges are very common along the sea coast. Deltaic plains, alluvial plains and flood plains are the geomorphic units. Major part of

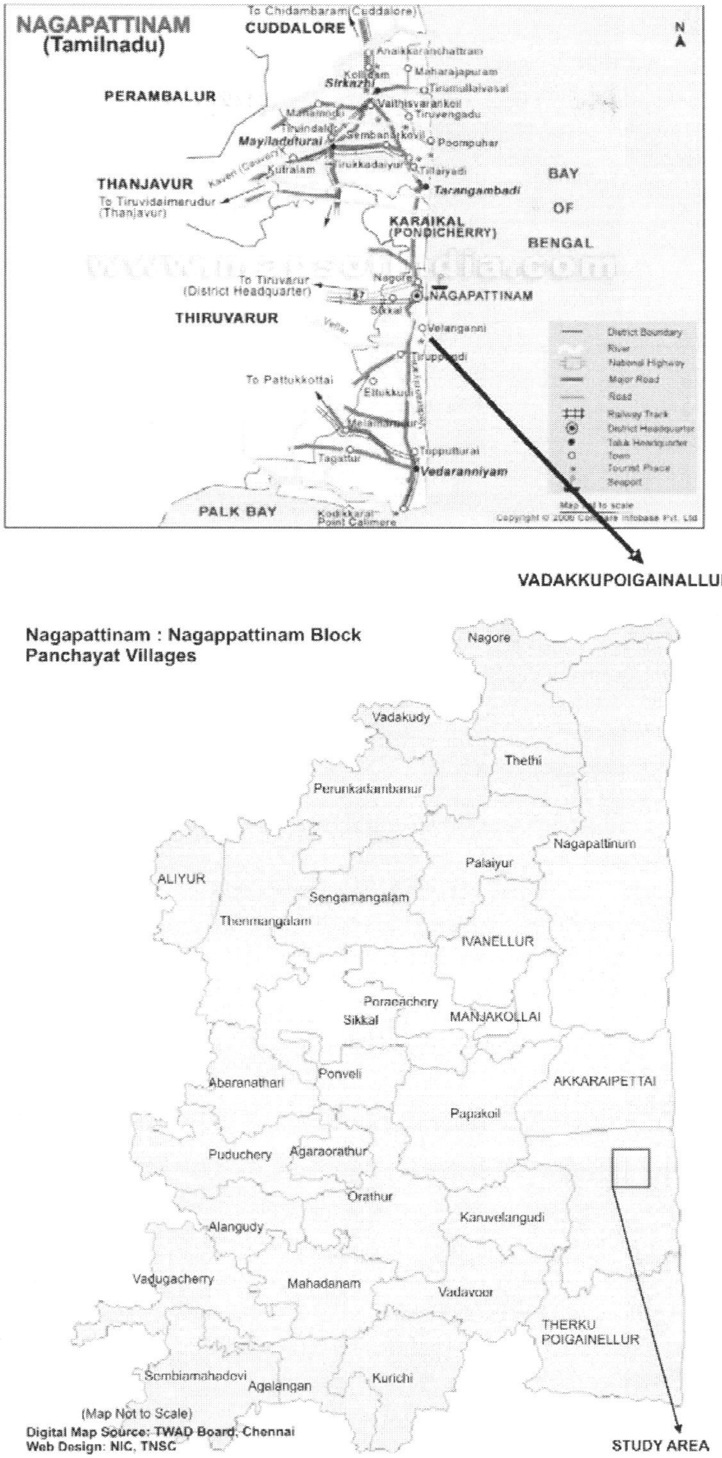

Fig. 1. Location map of the study area.

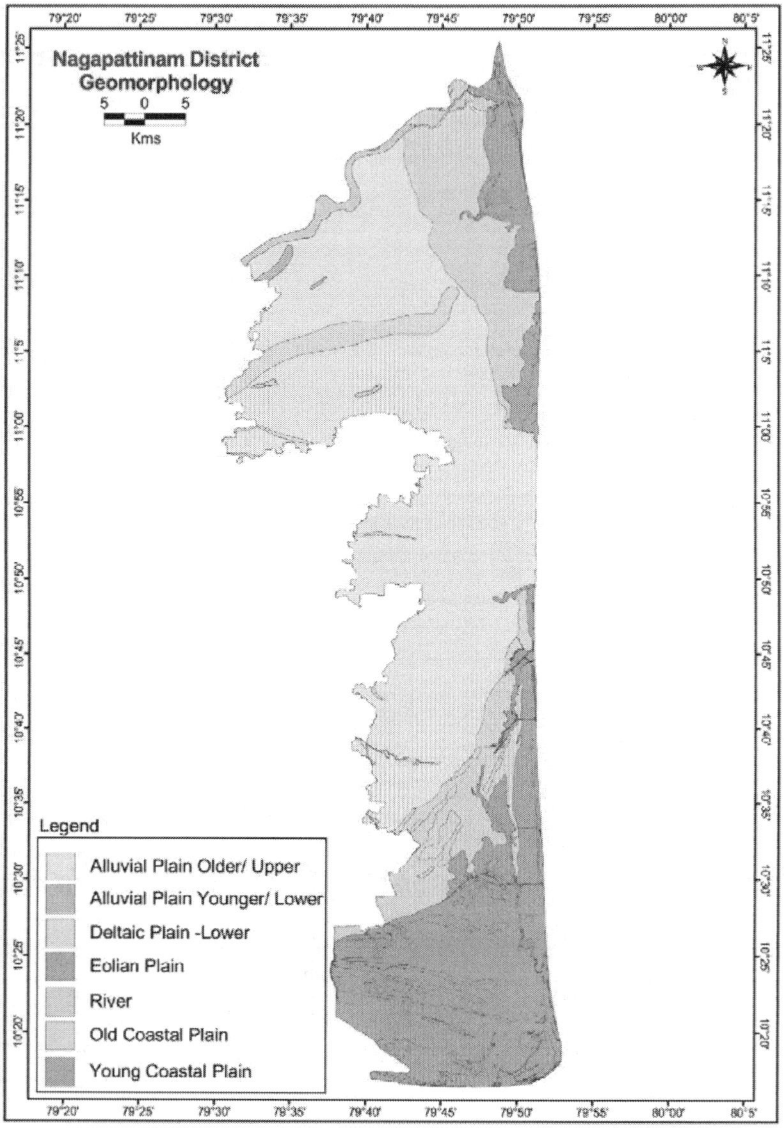

Fig. 2. Geomorphology map of Nagapattinam district.

the district is covered by black clayey soils. Some patches of arenaceous soils are found along the coastline. The tail ends of the Cauvery and Kollidam rivers are rich with fertile alluvial soils. Geomorphology of Nagapattinam district shown in Fig. 2.

GEOLOGY

The study area is a sedimentary terrain underlain by various geological formations ranging in age from Pliocene to the Recent alluviums. The main lithological units of Pliocene deposits are sands, sandy clay, grey and variegated clay with gravel beds. Pliocene formations underlie the Quaternary deposits in the Cauvery and Vennar sub basins.

HYDROGEOLOGY

The occurrence and movement of groundwater are controlled by climate, topography, geomorphology, geology and structures, etc. Groundwater occurs in all geological formations under unconfined, semiconfined and confined conditions. In major parts of the district, the depth to water table lies within 4.0 metres below ground level. In the north eastern part, the depth to water table is shallow i.e. less than 2.0 m. The general post-monsoon water level ranges between 0.2 m and 4.0 m bgl. The water level fluctuations in the district are related to recharge from different sources, hydrogeological setting, topography and status of groundwater developments. The study area receives maximum rainfall during North-east monsoon period, i.e. October to December. The normal annual rainfall of the district is 1372 mm. Precipitation is the main source of groundwater recharge. The impact of surface water irrigation is very significant. The lateral surface water flow through the river channels and canals form the principal source of groundwater recharge in the area. The principal and potential aquifers are the sands, gravel and alluviums.

THE COASTAL AQUIFER SYSTEM

The entire coastal belt of Nagapattinam district is occupied by the Quaternary formations. In the coastal areas, the groundwater occurs in the alluvial, flood and deltaic plains under pheratic conditions. The principal and potential aquifers are alluvium and sands. The quality of groundwater of shallow aquifer is generally good along northern, north-eastern and central parts. The quality deteriorates in the south and south-eastern parts, i.e. along Vedaranyam coast, which may be attributed to the marine and semi-marine origin sediments. The groundwater occurring in the sand dunes are generally fresh in nature. Seawater intrusion in coastal aquifers occurs when permeable formation outcrops into the body of seawater and the freshwater is underlined by salt water. A landward slopping freshwater – salt water interface will be formed with the depth governed by Ghyben-Herzberg relationship.

GROUNDWATER EXPLORATION

The groundwater exploration may be broadly classified as aerial, surface, sub surface and esoteric methods. Groundwater exploration comprises a number of stages. The first stage is identification of suitable well sites by integrated hydrogeological and geophysical method while the next stage is drilling and the last stage is development. Geophysical technique is one of the common techniques used for groundwater exploration. One of the geophysical technique is electrical method wherein the electrical conductivity of the subsurface formations is measured. The ERT is a non-invasive method and cost-effective. The inverse of conductivity is resistivity. In the ERT the conductivity is measured in terms of apparent resistivity. The ERT for groundwater exploration is of two types. One is surface and the other is subsurface exploration. In the surface method the resistivity of the subsurface formations is measured from surface itself. In the subsurface method the resistivty of the different geological formations encountered in the bore hole is measured through the probe traversing the borehole.

ELECTRICAL RESISTIVITY TECHNIQUES (ERT)

Surface Exploration

In the Electrical Resistivity Technique (ERT) of surface method, vertical electrical soundings have been conducted in the study area using DC resistivity meter adopting Schlumberger configuration. The depth of investigation AB/2 is 30 m. The VES curves have been qualitatively and quantitatively interpreted, analyzed and correlated by software packages IPI2WIN version 3.0 and INVERSE SLOPE version 1.0. The VES curve is a typical 'Q' type curve with six layered geoelctrical section. The geoelectrical parameters were obtained (Figs. 3 and 4).

Subsurface Methods

Well logging is the subsurface exploration method. There are several methods of well logging. The common methods employed in groundwater exploration, are the self potential (SP) log and resistivity logs – lateral. Short normal and long normal logs. In this study electrical well logging in pilot bore hole drilled by calyx rig to a depth of 75 m, has been carried out using portable spot logger with DC resistivity meter. The logging modes measured are SP and LN 64". SP and resistivity measurements have been recorded for 2 m interval. SP measurements are highly useful in deciphering the saline and clay predominant zones. Saline water saturated zones produce sharp – SP and low resistivity. Normal resistivity logs are four electrode system widely used in groundwater hydrogeology which measures the apparent resistivity. There are two types of normal resistivity logs SN 16" & LN 64". The normal resistivity logs clearly distinguish the clay, saline & freshwater zones. The interpretation of electro logs have been done in correlation with the litholog of boreholes (Figs. 5 and 6).

RESULTS AND DISCUSSION

The VES curve is a typical 'Q' type curve with six layered earth. The top three layers with the true resistivities of 543, 317 and 126 Ωm; indicate the shallow pheratic sandy aquifers. The sudden drop of resistivity from 126 to 12 Ωm may be attributable to the salinity of aquifer. The litho units of the pilot borehole very well coincide with that of geoelctrical section. The electro log curves of normal resistivity log LN-64" exhibit an increasing trend up to a depth of 8 m and below which the curve shows a decreasing trend with a steep fall of resistivity of 6 Wm, indicating the saline nature of aquifers. Below 10 m depth though the normal resistivity log curve shows an increasing trend, aquifers are saline. Again below 30 m depth the resistivity falls to 4 Wm which can be attributed to the brine water. Generally, in Nagapattinam areas brine waters are encountered in the depth range of 40-60 m. The TDS of water samples of the shallow tube well in Vadakku & Therkkupoigainallur are within the ranges of 654 to 1240 ppm. reflecting the fresh water aquifers.

SUMMARY AND CONCLUSION

Vadakkupoigainallur of Nagapattinam district in Tamil Nadu is a coastal sedimentary terrain with unconsolidated formation ranging in age from Pliocene to the recent alluvium. Groundwater exploration of both surface and subsurface geoelectrical investigations has been conducted to

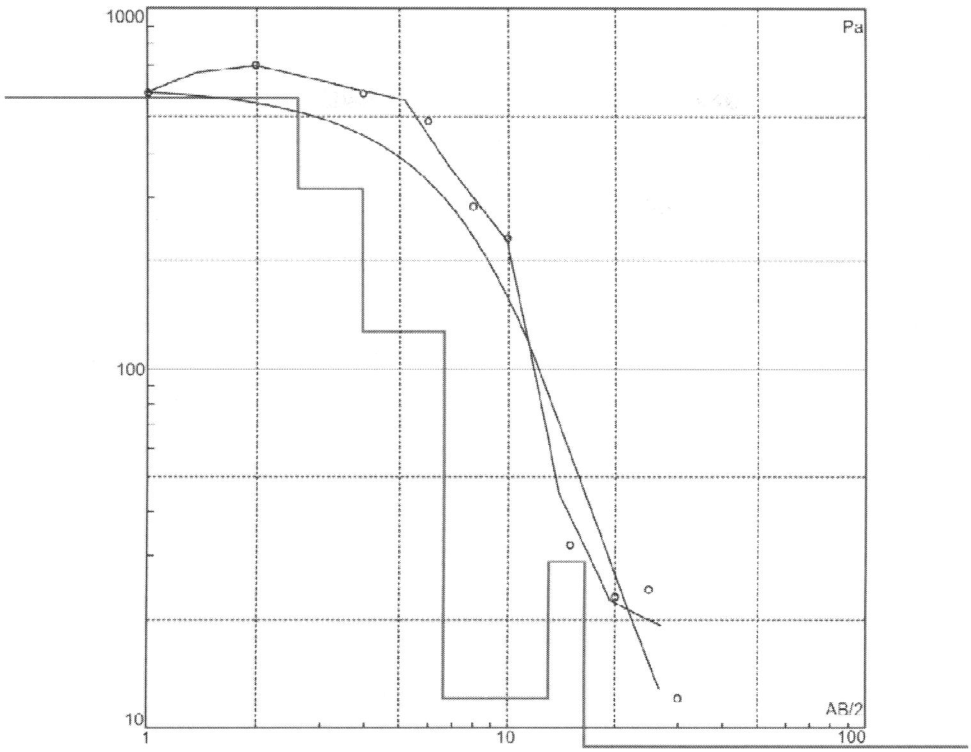

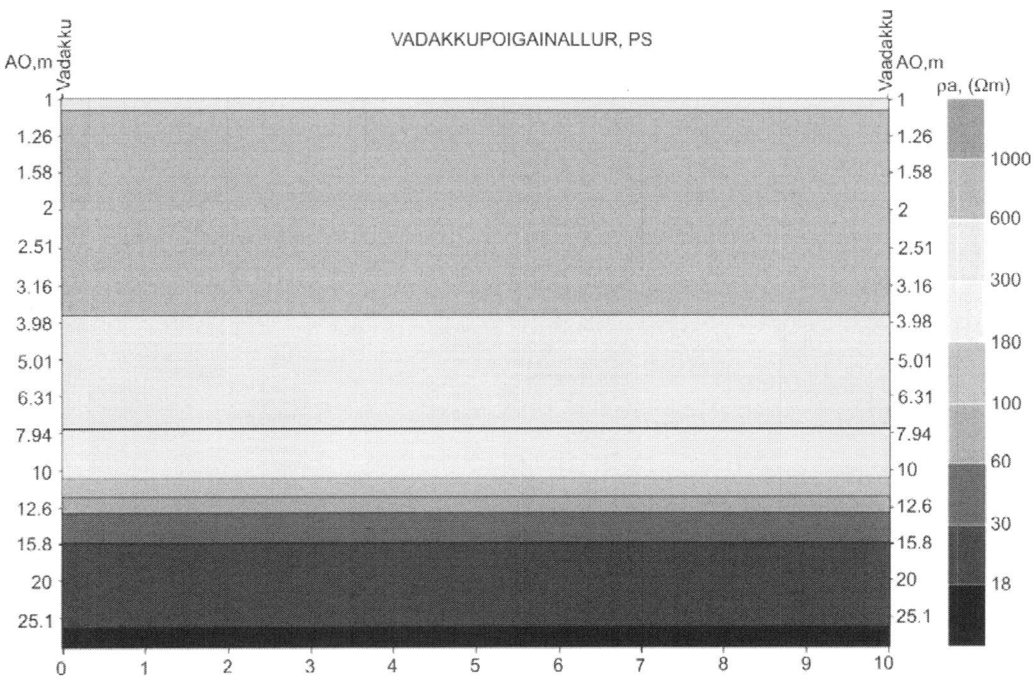

Fig. 3. Vadakkupoigainallur curve type and pseudo section.

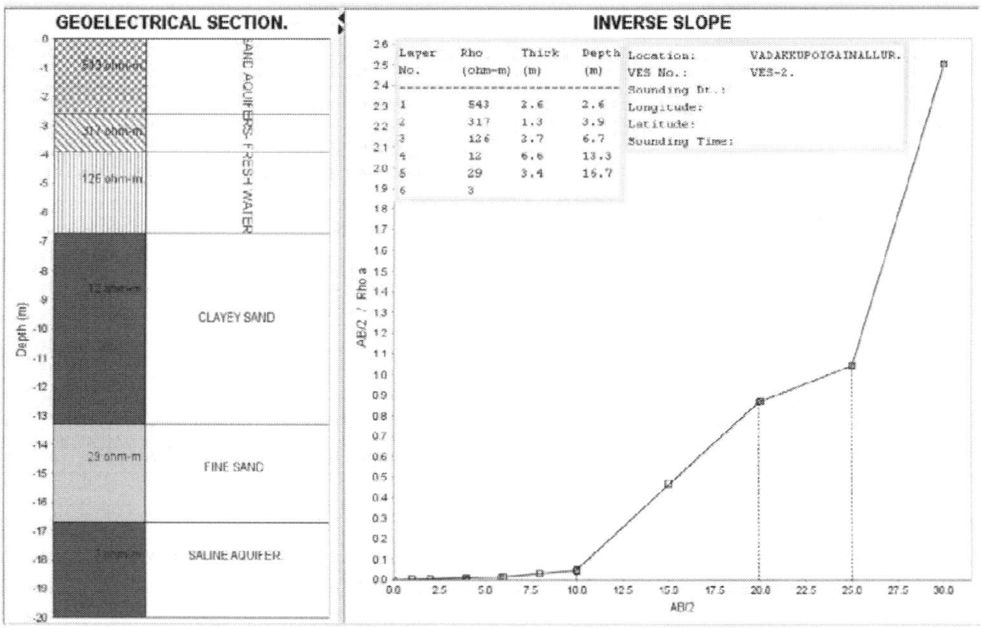

Fig. 4. Vadakkupoigainallur geoelectrical section – inverse slope.

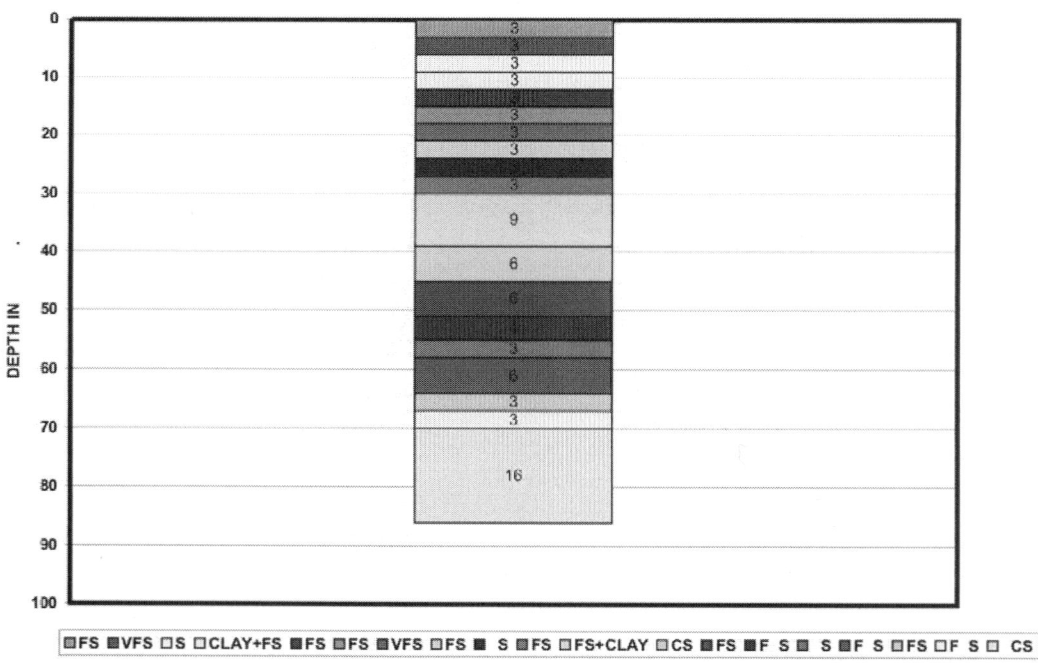

Fig. 5. Vadakkupoigainallur lithology section.

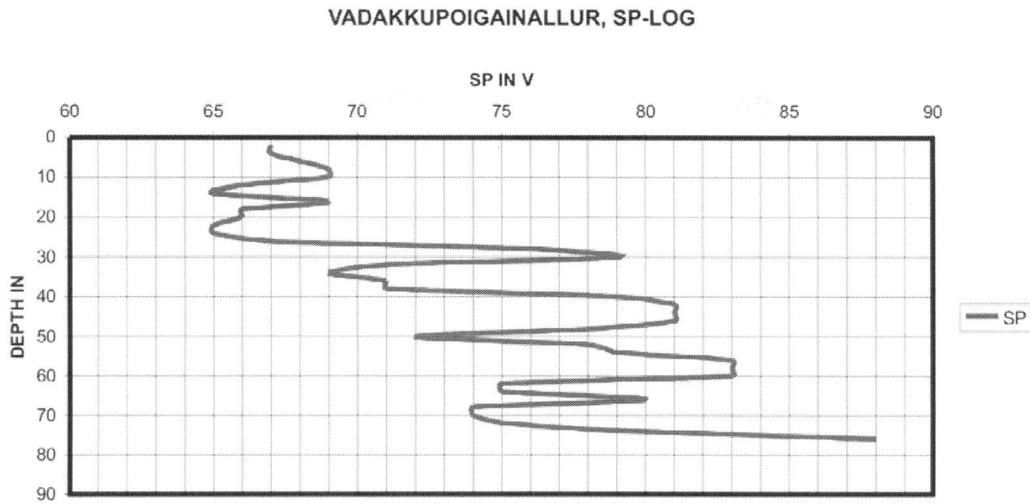

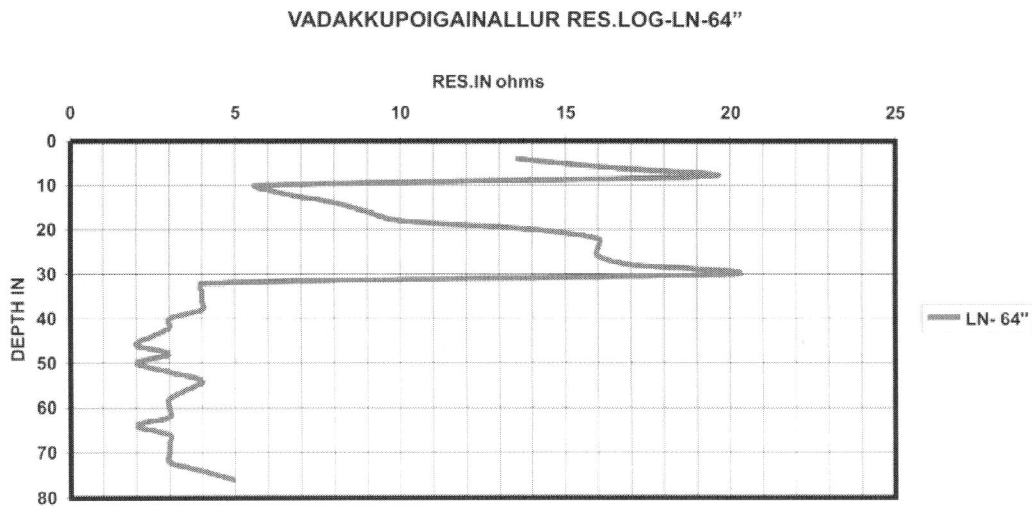

Fig. 6. Vadakkupoigainallur SP-LOG and RES.LOG-LN-64.

delineate the fresh – salt water interfaces and the subsurface lithology. The VES curve is a 'Q' type curve with six layered earth. The top 3 layers with high resistivities indicate the fresh water sand aquifer and the very low resistivity is because of the salinity of aquifers. From the VES curve of geoelectrical section, it may be interpreted that the freshwater – salt water interface is about 6.7 m bgl. A pilot bore hole has been drilled and electrically well logged. The normal resistivity log LN-64" electro log pattern exhibits a increasing trend up to a depth of 8 m and below which shows a very deep decreasing trend. From the electric log curve pattern, it is interpreted that the top shallow aquifers are fresh up to a depth of 8 m and below which saline. The fresh water – salt water interface may be about 8 m bgl. The TDS of shallow tube well samples are within the range of 654 to 1240 ppm indicating the freshwater nature. From the hydrogeological set up, geoelectrical investigation and TDS parameters, it may be inferred that the shallow pheratic aquifers up to a depth of 7 to 8 m are fresh in nature and below 8 m depth it is brackish. The depth to freshwater – saline water interface is 7 to 8 m bgl. Both the surface and subsurface geoelectrical investigation findings reveal that it very well coincides. Thus, the surface and subsurface geoelectrical investigations are good tools to ascertain and delineate the fresh water – saline water interface.

REFERENCES

Central Ground Water Board (CGWB), (1997). Groundwater resources and development prospects in Nagapattinam Quiad-e-Millath district, Tamil Nadu.

Carg, S.P. (1998). Ground water and tube wells, 3rd edition.

Raghunath, H.M. (1987). Groundwater, 2nd edition.

Ramanujachari, K.R. and Gangadhar Rao, T. (2005). A monograph on groundwater exploration by resistivity, SP and well logging techniques.

www.tnmaps.tn.nic.in

www.geophys.geol.mou.ru

3

Hydrogeological Studies of Coastal Zone from Triplicane to Tiruvanmiur, Chennai, Tamil Nadu

KUMAR, R.S.,* BALACHANDER, V., PRABHAKARAN, R., SELVARAJ, B., CHIDAMBARAM, S., ASAIMANI, S. AND RAMASAMY, V.

Department of Earth Sciences, Annamalai University
**e-mail: rskgeo@gmail.com*

ABSTRACT

Chennai city is located in the north-eastern corner of Tamil Nadu State. It is bound by Bay of Bengal in the east, Thiruvallur district in the north and west and Kancheepuram district in the south. The meteorological study point out the higher amount of rainfall is during the north-east monsoon. In the present study it has been tried to establish the various relations between the pre-monsoon and post-monsoon features of the coastal groundwater characteristics. The classification of the water based on SAR (Richards 1964) most of the samples are falling in the excellent water class. According to classification of water based on EC values most of samples fall in the permissible to doubtful category. The groundwater nature is explained by the Piper Trilinear diagram, which indicates that the groundwater samples come under Ca-Mg-Cl and Na-Cl types. The quality of the water for irrigation was estimated by Wilcox diagram classification and Na%, indicates that 5 samples fall under doubtful category during pre-monsoon whereas almost all the samples fall under the Excellent to Good category during post-monsoon. The main sources of water supply for city are surface water and groundwater. The groundwater sources for the city are six well fields namely Poondi, Tamaraipakkam. Flood plains, Kannigaiper, Panjetty, and Minjur in Araniyar-Korattalaiyar basin. Further the southern coastal aquifer and brackish water Reverse Osmosis plants are also the sources of groundwater.

Keywords: Coastal zone, groundwater, coastal aquifer, alluvium, salinity hazard.

INTRODUCTION

Ever increasing population of the Chennai metropolitan is threatening the Government and policy makers to meet the demand especially water. A report says that Chennai city people purchase water to fulfil their needs to the tune of Rs. 50 crores per month. Chennai city is constantly plagued by water shortage. In spite of the best efforts by the Governments, the water demand and supply positions are precarious. In this context the state Ground and Surface Water Resources Data Centre (SG&SWRDS) of Public Works Department has published a bulletin containing the results of groundwater investigations carried out in Chennai city, status of water supply, demand, availability, hydrogeological conditions, etc. In the present study it has been attempted to study the hydrogeological characteristics of the coastal zone from Tirplicane to Tiruvanmiyur.

Chennai city is one of the 30 districts of Tamil Nadu and it is located in the north-eastern corner of Tamil Nadu. Chennai is a plain land surface with a very gentle slope towards the Bay of Bengal. The land elevation is within 10 m and the maximum elevated lands are on the western part of the city. The almost even topography of the land throughout the district renders subdivision into natural regions rather difficult.

The geological set up of the district remains as that of other coastal districts, i.e., river alluvium and coastal alluvium. But the thickness of the alluvium is around 25 m.

The extraction of groundwater is only for domestic purpose and due to excess pumping, the water level has gone down considerably in entire Chennai district.

LOCATION

Chennai city is located in the north-eastern corner of Tamil Nadu State. It is bound by Bay of Bengal in the east, Thiruvallur district in the north and west and Kancheepuram district in the south. Chennai district is bound by North latitudes 12°59'10" and 13°08'50" and East longitudes 80°12'10" and 80°18'20" and it lies in Survey of India Toposheets No. 66 C/4 and 66 D/1&5. Chennai district's geographical area is 170.98 sq.km. or 17098 ha. The forest cover is only around 275 ha. The location map of the study area is shown in Figure 1.

DRAINAGES

Two rivers, the Cooum and the Adyar, intersect the city. Cooum runs through the heart of the city and enters the sea in between the university building and Fort. St. George underneath the Napier Bridge. The Adyar river flows through the southern part of the city. These two rivers are almost stagnant and do not carry enough water except during rainy seasons. Cooum river starts from Kesavaram anaicut in Kesavaram village of Thiruvallur district. This Kesavaram anaicut is built across the river Korattalaiyar. The surplus water from Cooum tank joins this course at about 8 kms lower down and this point is actually the head of Cooum river which is located at 48 kms west of Chennai. The river receives a sizeable quantity of sewage from its surroundings for disposal. The average flow measured at Korattur is 40.2 MCM/year. The river Adyar can be traced to a point near Guduvancheri, it assumes the appearance of a stream only after it receives the surplus water from the Chembarambakkam tank as well as the drainage of the areas in the southwest of Chennai the length of the river is 42 kms and has a catchment area of 800 sq. km.

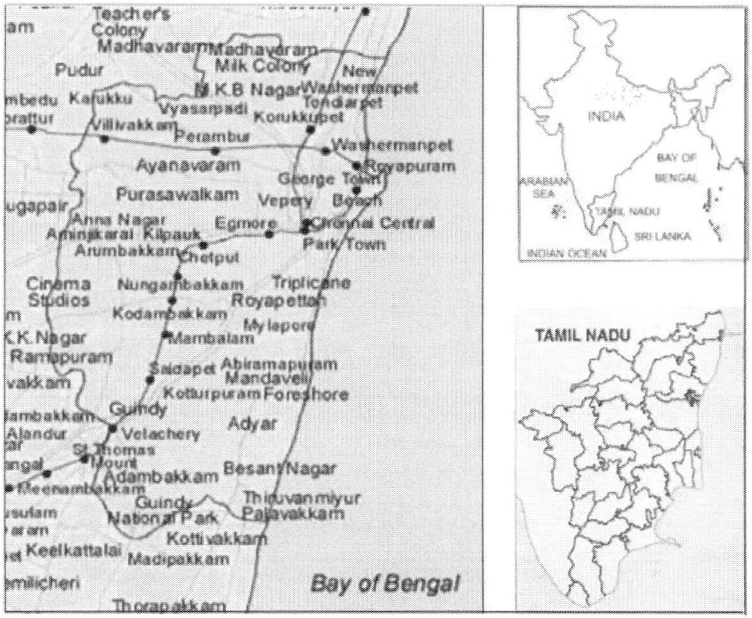

Fig. 1.

its average discharge is 89.43 MCM. During the high tides the backwaters from the Bay of Bengal enter inland up to 3-4 kms.

MATERIALS AND METHODS

In this present study about 10 groundwater samples have been collected from bore wells during pre-monsoon and post-monsoon season (2007) as shown in the Fig. 2. The samples were collected in a clean polythene bottles as prescribed by Hem (1959), Brown et al. (1974). The analysis has been carried out for pH, Electrical Conductivity (EC), major cations such as Ca, Mg, Na, K and anions namely HCO_3, Cl, SO_4, NO_3. Besides, other parameters like TDS, Total Hardness (TH) and SAR were calculated. EC values are expressed in µ mohs/cm at 25°C and except pH other parameters are expressed in mg/L. The chemical composition of the groundwater analyzed for different parameters are given in the Tables 1 and 2.

Table 1 Geochemical analysis results of groundwater pre-monsoon season Triplicane to Tiruvanmiyur, Chennai ionic concentration in mg/L

Well Loc.	Location	TDS	Ca	Mg	Na + K	HCO_3	CO_3	Cl	SO_4
1	Secretariate	1918	67.2	84.3	432.6	331.6	6	792	101.6
2	Chepauk Stadium	1685	89	76	360	483	12	574	68
3	Triplicane V.P.Street	1782	76	81	378	430	7	584	88
4	St.Thom	1883	74	83	386	396	6	582	96
5	Chettinad Palace	1826	84	77	369	468	6	571	76
6	Foreshore Estate	2162	3	152	466	583	10	682	208
7	Besant Nagar	3942	189	16	1129	442	6	1721	188
8	Tiruvanmiyur	646	29	48	71	242	0	101	74.2
9	Tharamani	2743	171.6	169	421	862	0	892	118.2
10	Kottivakkam	1272	41.6	192.3	66.2	529	0	232	218

Fig. 2. Sample location map of the study area.

Table 2: Geochemical analysis results of groundwater during post-monsoon Triplicane to Tiruvanmiyur, Chennai ionic concentration in mg/L

Well Loc.	Location	TDS	Ca	Mg	Na+K	HCO_3	CO_3	Cl	SO_4
1	Secretariate	992	90	77	106	226	24	369	50.6
2	Chepauk Stadium	1659	156	35.3	341	372	24	639	48.3
3	Triplicane V.P.Street	1058	160.3	74	126	253	22	390	51.3
4	St. Thom	1126	157	72	132	236	12	382	50
5	Chettinad Palace	1176	121	62	152	291	12	361	39
6	Foreshore Estate	1906	262	152	72.5	172	0	789	178
7	Besant Nagar	1016	80	71	93	391	0	151	188
8	Tiruvanmiyur	380	172	3	51	401	6	31	176
9	Tharamani	1906	132	212	92	751	0	455	198
10	Kottivakkam	1201	191	68	83	272	0	351	189

GEOLOGICAL SETTING OF CHENNAI

The major part of the district is covered by recent alluvium and only a small part is covered by crystalline rocks of Archaean age. The geological map of the study area is shown in Fig. 3. The Tertiary and Gondwana rocks encountered at depth. Cooum alluvium varies from 10 m to 28 m thickess and is more granular in Kilpauk-Perambur area. Beach ridges and sand dunes constitute good freshwater aquifer in Adyar-Besant Nagar area.

Fluvial marine and fluvio marine landforms are noted in the coastal area. Sand bars have scattered along the course of the river Adyar.

Though the alluvium deposits are major geological formation encountered, Tertiary and Gondwana deposits are encountered on the boreholes drilled on the western fringes of city by residents of Kulathur, Villivakkam, Anna Nagar, etc. Due to more clayey nature of these formations the yield is reported to be poor.

Charnockites are the crystalline formations encountered in and around Guindy and Velachery (south of Adyar river.) in Tharamani old lagoonal deposits are encountered and the presence of thick shell bed below clay proves this fact.

Geological succession in Chennai city is given below.

GROUP	AGE	LITHOLOGY
Quaternary	Sub recent	Soils, alluvium (sand and silt)
Tertiary	Pleistocene	Black clay
	Eocene to Pliocene	Sand stone and shale fossiliferous
	UNCONFORMITY	
Mesozoic	Lower Cretaceous To Lower Jurassic	Brown sandstone, Siltstone, grey and black shale
	UNCONFORMITY	
Azoic	Pre Cambrian	Charnockite, Granitic gneiss

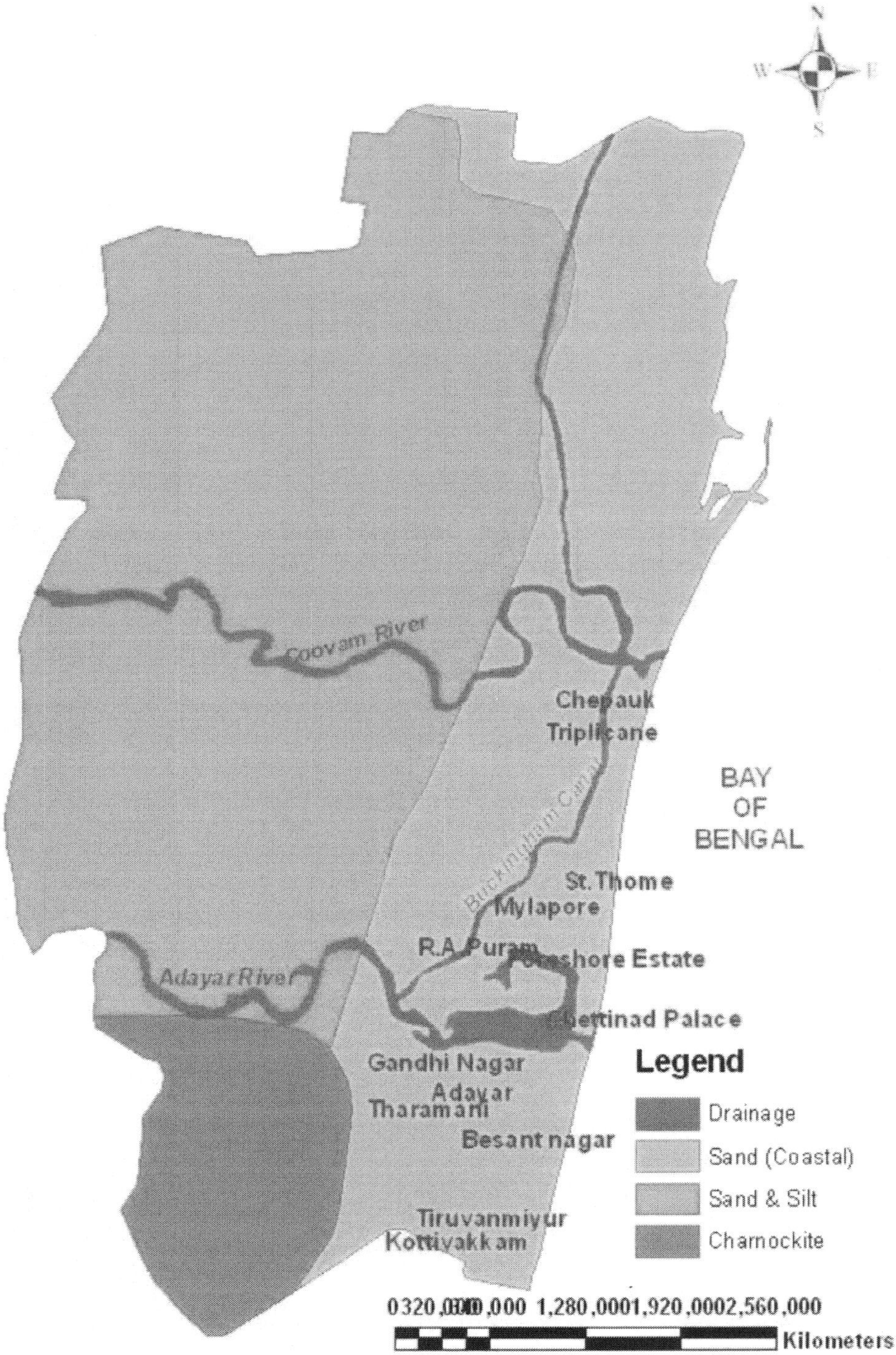

Fig. 3. Geological map of Chennai.

HYDROGEOLOGY

Groundwater occurs in all the geological formations of the city. But it's occurrence and quantity depends upon the percentage of granular zone available in the aquifers of Chennai city are phreatic. The groundwater extraction in the city is only for domestic purpose.

Alluvium

River alluvium and coastal alluvium are mostly seen in the city. Generally, in most part of the city the occurrence of groundwater is limited to thin granular zones in the alluvium. The thickness of alluvium is highly variable in space. The variation is about 10 m to 28 m. in most parts of the city, initially (up to 1970) groundwater extraction was done only by open well and the mode of extraction was hand bailing. Then the tradition of boreholes was introduced during early seventies in the city. Then began the theory, of groundwater mining. Instead of open wells of 8.00 m to 10.00 m depths, boreholes to a depth of 20.00 m to 25.00 m have been drilled. Population also increased with city developments. Net result is indiscriminate pumping of groundwater that causes groundwater depletion. Now in some places like Saidapet, Alwarpet, Nungambakkam, Egmore and Royapettah people gone down further to tap groundwater from the underlain crystalline basement (fractures & joints) rocks.

The beach sands and Adyar river alluvium aquifers in Adyar and Besant nagar were a good potential zone in the past. But due to the same indiscriminate mining of groundwater from this rich potential aquifer, the water level has gone down considerably in these areas also. The groundwater level has gone below Mean Sea Level and faces serious threat of seawater intrusion.

Tertiary and Gondwana

Groundwater in sandstone of tertiary and Gondwana that occur below alluvium especially on the western fringes of Chennai city is moderate to poor in quantity. This is due to more clayey nature of sandstone. (In the western fringes of the city the thickness of alluvium is around 10 m)

Archean (Crystalline)

This hard rock formation is seen the southern side of Adayar river in places like Guindy, Raj Bhavan, Velacherry and Tharamani. Open wells and boreholes are common groundwater structures in these areas. Nowadays people prefer only borehole as the open wells have gone dry. Boreholes have been drilled to a depth of 35 m to 40 m. But the yield is reported to be poor.

Total Wells in the City

It is estimated that there are 4,20,000 wells with a density of 2400 weklls per sq. km. There has been a squrt in the number of bore wells after year 1980 and the depth is also increasing day by day.

RESULTS AND DISCUSSION

The use of groundwater for domestic, agricultural and industrial needs are determined by chemical, physical and bacterial characteristics. In the study area about 10 samples were analysed for their chemical composition of the groundwater.

Hydrogen ion concentration (pH)

The pH of water is very important indication of its quality and provides important piece of information in many types of geochemical equilibrium or solubility calculations (Hem, 1985). The pH of the groundwater in the study area is varying from 7.07 to 8.36 with the average value of 7.56.

ELECTRICAL CONDUCTIVITY (EC)

The electrical conductivity measurements provide an indication of ionic concentrations. It depends upon temperature, concentration and types of ions present (Hem, 1985). The Electrical conductivity of the groundwater is varying from 250 to 5000 microsiemens/cm at 25°C with the average value of 2100 microsiemens/cm. Out of 10 groundwater sample locations, 6 samples have been identified as above the permissible limit of 750 microsiemens/cm (WHO,1993).The conductance has exceeded the maximum permissible level of 2000 microsiemens/cm in 3 sample locations. The high conductance was observed in certain groundwater samples may be attributed to high chloride concentrations in groundwater (Davis and Dewist, 1966).

TOTAL DISSOLVED SOLIDS (TDS)

The principal ions contributing to TDS are bicarbonate, carbonate, chloride, sulphate, nitrate, sodium, potassium, calcium and magnesium. The TDS of the groundwater varies from 646(8) mg/L to 3942(7) mg/L during pre monsoon and during post monsoon it varies from 380 mg/L to 1906 mg/L.

CATIONS

In this study, the major cations like Ca, Na, Mg, and K were analysed. The Ca concentration is varying from 3 mg/L to 189 mg/L during pre-monsoon and during post-monsoon it varies from 80 mg/L to 262 mg/L. The limit of Ca for drinking water is specified as 100 mg/L (WHO, 1993).The analysis result shows that almost all the 10 samples are exceeding the desirable limit.

The magnesium concentration is varying from 16 mg/L to 192.3 mg/L during pre-monsoon whreas it is found to range from 3 mg/L to 212 mg/L during post-monsoon. The limit of Mg for drinking water is 30 mg/L (WHO,1993). In the study area, about 8 samples are exceeding the maximum permissible limit of 50mg/L.

Similarly, the Sodium and Potassium concentration is varying from 71 mg/L to 1129 mg/L during pre-monsoon and 51 mg/L to 341 mg/L during post-monsoon season. The limit for drinking water is specified as 30 mg/L (WHO, 1993). However, nearly 8 samples were exceeding the maximum permissible limit of 175 mg/L.

ANIONS

The major anions like Cl, SO_4, HCO_3 and NO_3 were analysed in this study. The Cl concentration is varying from 101 mg/L to 1721 mg/L during pre-monsoon and it varies from 31 mg/L to 789 mg/L during post-monsoon. The limit of chloride concentration for drinking water is specified as 25 mg/L (WHO, 1993). All the sample locations have exceeded the desirable limit. However, 4 samples are exceeding the maximum permissible limit of 600 mg/L.

The sulphate concentration is varying from 68 mg/L to 218 mg/L during pre-monsoon and it varies from 39 mg/L to 198 mg/L during post-monsoon. The limit of sulphide concentration for drinking water is specified as 25 mg/L (WHO, 1993). Almost all the samples have exceeded the desirable limit.

CHEMICAL RELATIONSHIP

Among the various Trilinear methods of plotting Palmer (1911), Hill (1940) and Piper (1953), Piper's diagram has been extensively used for understanding problems about the geochemical evolution of groundwater (Karanth, 1987). The Trilinear diagrams of Piper are very useful in bringing out the chemical relationships among groundwater in more definite terms. The present study area groundwater samples have been plotted with a piper (1953) Trilinear diagram (Figs. 3 and 4).

The geochemical evolution of groundwater can be understood by plotting the concentration of major cations and anions in the piper Trilinear diagram. The plot shows that most of the samples fall in the field of 'Ca-Mg-Cl' type. The groundwater samples are distributed in the 'Ca-Mg-Cl' and Na-Cl facies.

SODIUM ABSORPTION RATIO (SAR)

The SAR is useful for judging the quality of groundwater for the use of agricultural purposes (Todd, 1980, Balasubramanian, 1986; Sastri and Lawrence, 1988). Richards (1954) classified

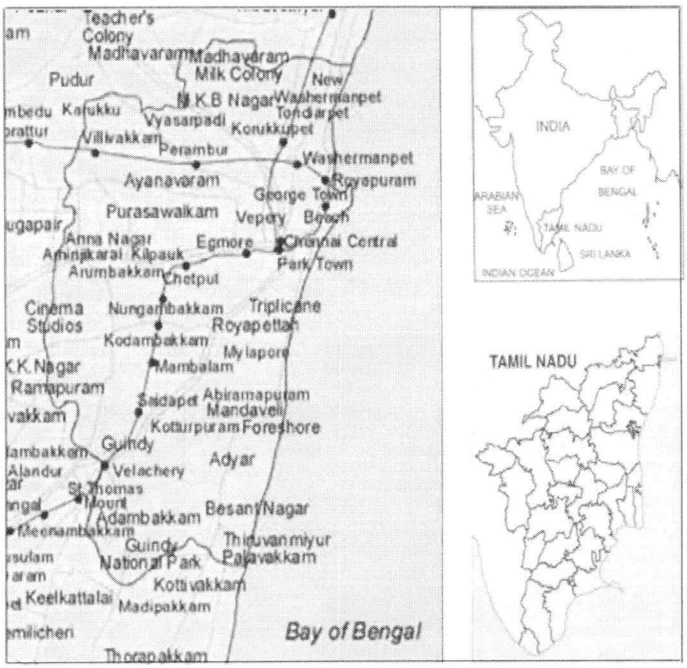

Fig. 4.

the waters in relation to irrigation based on the ranges of SAR values. According to Richards's classification all the samples of the study area have been classified as excellent for irrigation.

A more refined way of expressing the alkali hazard in irrigation water is shown by the Wilcox (1955) diagram which takes into the relative activity of sodium in the exchange reaction in soil as expressed in terms known as SAR. The Wilcox diagram has been used to understand the alkali hazard of the groundwater samples for the study area, because this interpretation is very much useful for judging the quality of groundwater for the use of agricultural purposes (Todd, 1980). The Wilcox diagram prepared for the groundwater samples of the study area is shown in the Figure 5, where the sodium adsorption ratio is plotted against Salinity Hazard.

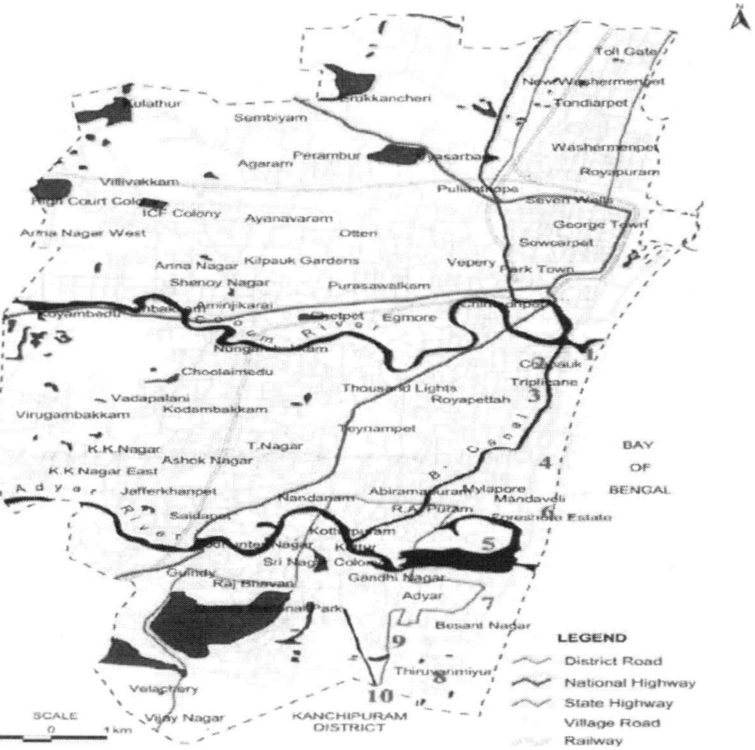

Fig. 5. Wilcox Diagram for the groundwater samples of the study area.

A better measure of the sodium hazard for irrigation is the Sodium Adsorption Ratio (SAR) which is used to express reactions with the soil. The SAR is computed as:

$$\frac{Na}{\sqrt{\frac{Ca + Mg}{2}}}$$

where all the ionic concentrations' are expressed in epm. When the SAR and salinity hazard of water are known, the classification of the water for irrigation can be determined by graphically plotting these values on the Wilcox diagram. The significance and interpretations of quality ranges on the Wilcox diagram can be summarized as follows:

The study area samples were identified with the field of C2S1, C3S1, C4S1 and C4S2. Majority of premonsoon samples fall in C4S2 indicating very high salinity and medium sodium hazard. Majority of post-monsoon samples fall in C3S1 indicating high salinity and low Na hazard, satisfactory for plants having moderate salt tolerance on soils.

SUMMARY AND CONCLUSION

Sodium and potassium is the dominant cation in both pre-monsoon and post-monsoon whereas bicarbonate is the dominant anion in the study area in both the season. Based on hardness, the groundwater samples are very hard in post-monsoon except one sample which falls in the moderately hard in pre-monsoon. In all locations the pH value is in the acceptable range of 6.9 - 8.2. The quality of groundwater is verified with WHO and ISI standards which shows that most of the samples are falling in the permissible limit for drinking purpose except few samples. The classification of the water based on SAR (Richards 1964) most of the samples are falling in the excellent water class. According to classification of water based on EC values most of samples fall in the permissible to doubtful category. The groundwater nature is explained by the Piper Trilinear diagram, which indicates that the groundwater samples come under Ca-Mg-Cl and Na-Cl types. The quality of the water for irrigation was estimated by Wilcox diagram classification and Na%, indicates that 5 samples fall under doubtful category during pre-monsoon whereas almost all the samples fall under the excellent and Good category during post-monsoon.

REFERENCES

Aravindan, S. *et al.* (2004). Ground water Quality in the Hard Rock area of Gadilam River basin, Tamil Nadu, *J. Geological Soc.* of India, **Vol. 63:** pp. 625-635.

Balasubramanian, A., (1986). Hydrogeological investigations of Tambaraparni River Basin, Tamil Nadu, unpublished Ph.D. Thesis, University of Mysore. pp. 349.

Brown, E., Skovgestad, M.W. and Fishman, M.J. (1974). Methods for collection and Analysis of Water samples for Dissolved Minerals and Gases, U.S Department of Interior. **Vol 5:** pp.160.

Chandra Sekhar. M. (1996). Water Quality criteria for Irrigation, *Indian J. Environmental Protection*, **Vol. 16**: and No. 3.

Dhenbare, A.J., Pondhe, G.M. and Singh, C.R. (1998) Ground water characteristics and their significance with Special Reference to public health in pravara area, Maharashtra, poll. *Res.* **17(1):** 87-90.

Doneen. L.D. (1963). Water Quality for Agriculture Department of Irrigation, University of California, Davis, 48.

Davis. C.D and Dewiest. R.J. (1966). Hydrogeology John Wiley, New York. 463.

Durov, S.A. (1948). Natural waters and graphic representation of their composition, Dokl. Akad. Nauk, SSSR. **59:** 87-90.

Eation, F.M. (1950). Significance of CO_3 in irrigation water, solid *Sci.*, **69:**128-133.

EPA Report, (1976). National interim primary drinking water regulation. EPA Publication NO. EPA570, 9-76-003.

Garrels, R.M. and Christ, C.L. (1965). Solutions, minerals and equilibria, Harper and Row.

Gibbs. R.J. (1970). Mechanisms controlling world water: Science, **Vol. 170**, pp.1088-1090.

Handa, B.K. (1967). Geochemical characteristics of ground water in the Cuddalore sand stone of Madras State, Indian, *J.Geochem.Soc.* India, **2:** 29-44.

Hem, J.D. (1959), Study and interpretation of the chemical characteristics of natural waters. US GS water supply paper 1935 - C, 17 p.

Hill. R.A., (1940), Geochemical patterns in Coachella valley, California, *Am. Geophy, Union Trans.*, **21:** pp. 46-53.

Jain, C.K., Sharma, M.K., Omkar, (1977). Ground water quality, *Indian J. Environmental Protection*, **Vol. 17:** No. 6.

Johnson, J.H. (1975). Hydogeochemistry in Groundwater exploration; Groundwater, Symposium, Bulawano.

Karanth, K.R. 1987. Ground water Assessment, Development and Management, Tata Mc Graw-Hills Publications Company Limited, New Delhi.

Kelly, W.P. (1957). Adsorbed sodium, cation exchange capacity and percentage sodium sorption in alkali soils. *Sci.*, **84:** 473-477.

Manade, P.N., Shrivastava, V.S., (1988). Nitrate studies of ground water in belt of Sat Valley, *Indian J. Envronmental Location* **Vol. 17, No. 6**.

Mehrota, C.L. (1969). Water quality and use of saline water for crop growth in U.P., *J. Indian Soc. Soil Sci.*, **17**: 442-46.

Meiri, A. and Shal Haret, J. (1973). Physical aspects of soil water and salts in Ecosystem, Springer Verlag, New York.

Naidu, N. (1988). Urban water quality in North Coastal Andhra Pradesh. Indian *J. Environmental Protection*, **Vol. 18:** No.4.

Palmer, Chase. (1911). The geochemical interpretation of Water analysis. USGS Boll. No 479. p. 31.

Piper, A.M. (1944). A graphic procedure in the geochemical interpretation of water analyses. *Trans. Am. Geophys. Union*, **25(9):** 4-23.

Piper, A.M. (1953). A graphic Procedure in the geochemical interpretation of water analysis. USGS Ground Water Note, No. **12:** pp. 63.

Porchelva, P. and Indrasensn, N. (1999). Environmental Hazards of Palar River Basin, *J. Indian Water Resources Society*, **Vol. 19(5):** No.1.

Prasad, B.V. and Ramesh Chandre P. (1997). Ground water Quality in an industrial Zone, Poll Res., **16 (2):** 105-107.

Rahul Mohan, and Abhay Kumar Singh, (2000). Hydrochemistry and Quality assessment of GW in Naini Industrial area, Allahabad District.Up *J. Geological Soc.* India, pp. 77-89.

Ragunath, H.M. (1992). Ground Water, 2nd edition, Wiley Eastern Limited, New Delhi, p. 563.

Rajmohan, N., Elango, L. and Elampooranan. T. (1997). Seasonal and spatial variation in Mg and Chloride concentration in groundwater of Nagai-Quaid-e-millrth dist. Tamil Nadu, *Indian J.Environmental Protection*, **Vol.17, No.6**.

Sambasiva Rao, T. (1997). Sustainability of Groundwater for Irrigation in Chandragiri block, Chittor District., (A.P), Ecology, **11(10):** 1-9.

Sastri, J.C.V. and Lawrence. J.F (1988). Hydrogeochemical evaluation of Karaikudi, Environs, Tamil Nadu Geophys *Res. Bull.*, **Vol. 26**. **No. 1**. pp. 39-44.

Sathi Babu, Y., Suryanarayana, K. and Srinivasa Rao, B. (1989). Chemical Quality of groundwater and its mechanism. – A Case study at Muvvalavanipalem, Vishakapainam, *Indian J. Environ. Protection*, **Vol. 9, No. 10,** October 1989.

Singanan, M. and Somasekharan Rao, K. (1996). Evaluvation of Rameswaram Island groundwater Resources for Irrigation and Industrial Purposes, *Indian J. Environ. Protection*, **Vol. 16, No. 3.**

Singh, K.P. (1998). Significance of Unsaturated zone in Understanding Pollution of groundwater. – A Case study of Ludhiana Area, *Indian J. Environ. Protection*, **Vol. 18**, **No. 5.**

Srinivasa Rao, K. and Sambassiva Rao, T. (1997). Hydrogeochemical Evaluvation of Erpedu Mandal Aquifers Chittoor District, *Indian J. Environ. Protection*, **Vol. 17**, **No. 5.**

Subbarao, C., Subbara, N.V., Priyadarshini, J.P. and Anila Vijay Nair, (1977). Temporal variation of groundwater corduation in two industrial localities of Vishakapainam, *Indian J. Environ. Protection*, **Vol. 17, No. 6**.

Todd, D.K. (1980). Groundwater Hydrology, 2nd edition, John Wiley & Sons, New York, pp. 535.

WHO (1984). Guidelines for drinking water quality. Geneva. **1**: 53-73.

Wilcox, L.V. (1955). Classification and use of irrigation water, U.S. Dept. of Agriculture, Circular 969.

4

Hydraulic Continuity between Groundwater Sources in the Malabar Region of Kerala—Evidence from Stable Isotopes and Ion Chemistry

SHAHUL HAMEED A.,* UNNIKRISHNAN WARRIER C., VELAYUDHAN K.T. AND SURAJ S.

Centre for Water Resources Development and Management Kozhikode, Kerala
**Email: hameedsha@rediffmail.com*

ABSTRACT

Better understanding of groundwater sources in terms of its behaviour and quality is of prime concern for judicious management of the resources in view of the fast pace of groundwater development to meet the growing demand. A study on water chemistry including stable isotope composition of water (Deuterium and Oxygen-18) was undertaken by selecting 30 shallow dug wells in the midlands of Malabar region of Kerala. The predominant lithological unit of the study area is laterite. Water samples were examined for different characteristics like pH, electrical conductance, alkalinity, fluoride, chloride, nitrate, sulphate, sodium, potassium, calcium and magnesium as well as for stable isotope ratios, δD and $\delta^{18}O$. Most of the waters in the region are represented by Na-Ca-Mg-HCO_3 and Ca-Na-Mg-Cl-HCO_3 types. Groundwater level fluctuations in different seasons indicate the direct influence of monsoon waters. The total dissolved ions in water range from 0.837 epm to 9.608 epm and 0.484 epm to 4.438 epm during pre-monsoon and post-monsoon periods, respectively. δD and $\delta^{18}O$ values of groundwater during dry months (March-April) range from –16.58 to –5.02% and –3.86 to –1.80%, respectively. The observed values during post-monsoon (November-December) season vary from –26.69 to –10.45% and –4.32 to –2.13%, respectively. Spatial and temporal trends in the water composition are found to be primarily influenced by local precipitation, as evidenced from the ion chemistry and stable isotope ratios. Data point distribution on composition diagrams of ion chemistry and stable isotope ratios establish hydraulic interconnection between the groundwater sources in the study area.

Keywords: Ion chemistry, stable isotopes, hydraulic connectivity, malabar.

INTRODUCTION

Groundwater, a vital resource for the sustenance of human life is diminishing at an alarming rate due to both natural and anthropogenic activities. In order to make use of the available groundwater sources in an optimal way, the precious resource has to be well understood in terms of its occurrence, behaviour and quality. Quality of water is a function of its physical and chemical characteristics which depend upon many factors like site characteristics, type of geologic formations, weathering product, their decomposition and related changes that occur in time and space. Infiltration of rainwater through soil over a long period and the passage through the ground do change the physical and chemical composition of water. The changes or alteration in chemical composition may be uniform in a group of groundwater sources on a local or to certain extend regional scale due probable hydraulic connectivity between the sources (Rao, 2006; Suman Sharma, *et al.*, 2003). The use of environmental stable and radioactive isotopes in groundwater is now well established in the field of water resources. Stable isotopes which are the part of water molecule (Deuterium and Oxygen-18) are very valuable in understanding the origin, source and mixing of groundwater, dynamics of pollution and groundwater surface water linkage. (Navada *et al.*, 1986). Various hydrological process lead to variation in the stable isotopic content of water in different parts of hydrologic cycle (Sukhija *et al.*, 2002). The isotopic composition of groundwater is mainly controlled due to recharge process, interaction with rocks or mixing with other waters. Thus, the stable isotopic composition of groundwater will also help to establish the hydraulic continuity in a particular region. In this context, an attempt was made to use the ion chemistry and stable isotopes of water to understand the hydraulic continuity between groundwater sources in sleeted location in Malabar region of Kerala.

STUDY AREA

Kerala state is broadly classified into three physiographic zones, viz. the high land region situated about 75 m above the mean sea level, the mid land region lying between 7.5 m and 75 m above mean sea level and the lowland lying below 7.5 m. The study area covers Malappuram and Kozhikode districts of Malabar region in northern Kerala. The climate in this region is generally hot and humid with the temperature varies between 30°C and 20°C. Homestead open dug wells are the main source of water in this region. The predominant lithology in this region is laterite. The observation wells selected for the present investigation fall within Chaliyar river basin, the fourth largest river in Kerala. It originates from the Ilambalari hills in Gudalur taluk of Nilgiri district in Tamil Nadu, at an elevation of 2,066 m above Mean Sea Level. This inter state river has a total drainage area of 2,923 km^2 of which 2355 km^2 lie in Kerala and rest in Tamil Nadu. The river has a length of about 170 km (PWD, 1974).

Methodology

Water samples from 30 open dug wells were collected during pre monsoon (March-April) and post monsoon (November-December) seasons in 2006. These observation wells are located in Vazhikadavu, Nilambur, Mambad, Edavanna, Areacode and Vazhakkad villages in the basin.

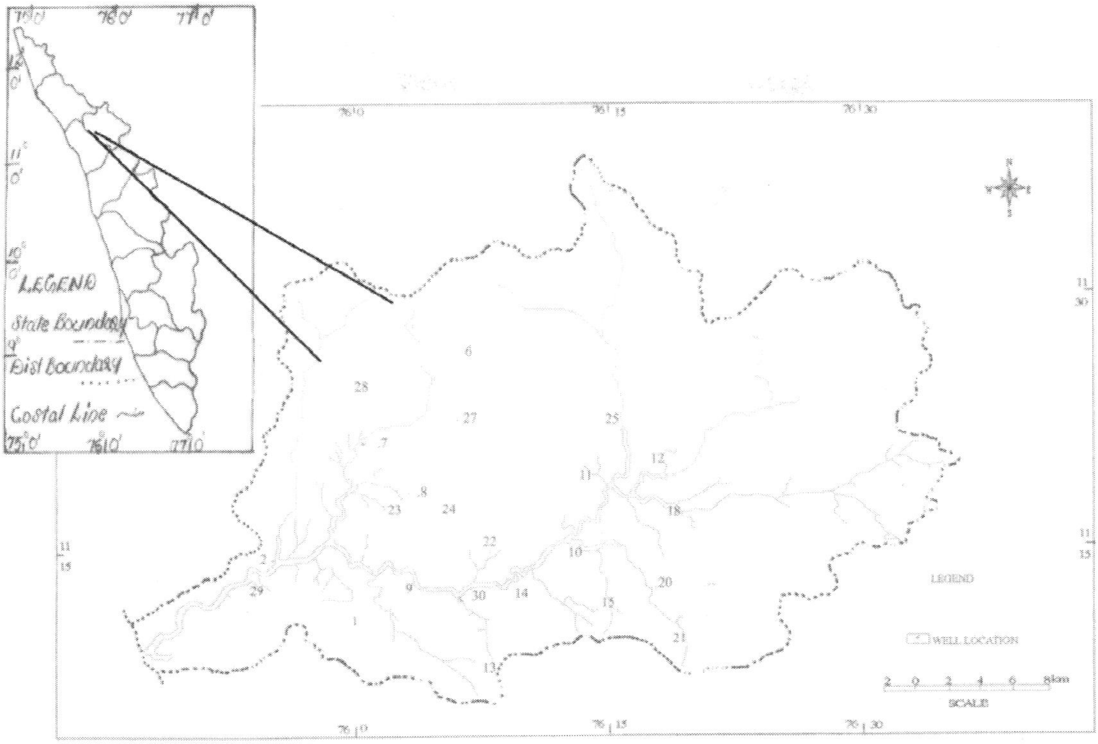

Fig. 1. Location map of the study area.

The location map of the study area with the sampling location is shown in Figure 1. The diameter of the wells varies between 1.3 m and 3.9 m and the total depth from 3.75 m Below Ground Level (BGL) to 12.25 m BGL. The depth to water level in dug wells in the area varies between 1 m to 11.25 m BGL in pre-monsoon and 0.5 m to 10.5 m BGL in post-monsoon with an annual fluctuation of 0.7 m to 5.0 m. Analytical techniques and procedures reported in standard literature were followed for characterization of water samples (APHA, 1998). Anions like chloride, fluoride, nitrate, sulphate and phosphate were analyzed using Ion Chromatograph (Dionex ICS 2500) and cations like Ca and Mg by EDTA complex titration, Na and K by Flame Photometer (Systronics 128) pH and EC were determined using conductivity and pH meters, respectively (Eutech- CON 510 and pH/Ion 510). The major anions and cations were evaluated using Aqua Chem 3.70 software.

Stable isotope measurements were carried out in Isotope Ratio Mass Spectrometer (*FINNIGAN* DELTA ^{PLUS} XP, Thermo Electron Corporation, Germany) at Isotope Hydrology Division of CWRDM using continuous flow measurement techniques. The isotopic concentrations are then expressed as the difference between the measured ratios of the sample and reference over the measured ratio of the reference.

In hydrological investigations, the variations in the stable isotopic compositions with respect to a standard are of interest rather than their absolute concentrations. Universally adopted standard for deuterium and oxygen-18 is SMOW, the Standard Mean Ocean Water.

This refers to hypothetical water whose hydrogen and oxygen isotope ratios are close to the mean isotope ratios of ocean water (Craig, 1961). Isotopic composition of natural waters is expressed as per mil (parts per thousand), the deviation (d) from the standard, SMOW, as shown below:

$$d\ (‰,\ \text{per mil}) = \frac{R_{\text{Sample}} - R_{\text{SMOW}}}{R_{\text{SMOW}}} \times 1000$$

where R denotes the ratio of heavier isotope to lighter isotope.

The global measurements of precipitation have resulted in a general relationship between δD and $\delta^{18}O$ which is expressed as,

$$\delta D = 8\ \delta^{18}O + 10$$

This is known as the Global Meteoric Water Line (GMWL) (Craig, 1961). This is valid for most cases, however, in some closed basins, water vapour is generated with a different isotope signatures, resulting in an alignment of meteoric waters in a line with an intercept different form the equation given above. Depending on the local conditions, a Local Meteoric Water Line (LMWL) is required to be constructed for meaningful interpretation of stable isotope data in precipitation and groundwater sources. The LMWL was prepared using the δD and $\delta^{18}O$ value of WMO data of Calicut station for the year 2005.

Results and Discussion

Hydrochemistry

The chemical analysis of groundwater samples collected during pre-monsoon and post-monsoon periods have been presented in Tables 1 and 2. During pre-monsoon, the dominant anion is HCO_3 followed by Cl and SO_4, but the dominant cations are Na > Ca > Mg and the water type is represented by Na-Ca-Mg-HCO_3. In post-monsoon the dominant anion is Cl followed by HCO_3 and SO_4 and cation dominancy is Ca >Na > Mg and the water type is Ca-Na-Mg-Cl-HCO_3. F and NO_3 are below the detectable level during pre-monsoon while some of the samples in post-monsoon (W-7, W-12, W-17 and W-20) have detectable F and NO_3. During post monsoon, majority of the wells have strong acid anions, especially, chloride, and alkaline earths exceeding alkali metal cations. The change in ion dominancy and water type during post-monsoon indicate the direct influence of monsoon in the study area, groundwater level fluctuations also give evidence for this. The wells having high level of SO_4, NO_3 and having high TDI (total dissolved ions) (W-12, W-17 and W-20) is suspected to have polluted by site specific problem of anthropogenic origin.

Table 1. Chemical composition of groundwater collected during pre-monsoon

Sl.No	Code	pH	EC	HCO_3	Cl	SO_4	Na	K	Ca	Mg	TDI
1.	W-1	7.2	93	0.787	0.112	0.073	0.209	0.015	0.520	0.424	2.787
2.	W-2	7.6	211	1.180	0.770	0.255	0.939	0.038	1.040	0.384	4.607
3.	W-3	6.2	109	0.361	0.138	0.011	0.161	0.003	0.200	0.232	1.105
4.	W-4	7.6	124	0.557	0.403	0.012	0.396	0.010	0.440	0.232	2.050
5.	W-5	6.0	77	0.525	0.345	0.058	0.487	0.026	0.480	0.160	2.080
6.	W-6	5.9	110	0.262	0.114	0.026	0.217	0.008	0.200	0.120	0.947
7.	W-7	7.1	143	0.689	0.638	0.029	0.826	0.077	0.560	0.272	3.090
8.	W-8	6.1	37	0.426	0.250	0.046	0.309	0.008	0.440	0.080	1.559
9.	W-9	7.9	76	0.262	0.390	0.007	0.417	0.038	0.160	0.160	1.435
10.	W-10	7.1	18	0.295	0.154	0.006	0.257	0.010	0.200	0.160	1.082
11.	W-11	6.9	139	0.590	0.669	0.028	0.691	0.005	0.440	0.272	2.695
12.	W-12	7.6	456	0.721	1.213	0.041	1.609	0.153	0.600	0.272	4.609
13.	W-13	6.1	79	0.361	0.339	0.060	0.461	0.036	0.280	0.160	1.696
14.	W-14	6.3	135	0.820	0.528	0.148	0.670	0.079	0.600	0.392	3.237
15.	W-15	5.7	78	0.393	0.405	0.007	0.609	0.015	0.360	0.120	1.910
16.	W-16	6.8	48	0.525	0.075	0.005	0.300	0.015	0.480	0.160	1.561
17.	W-17	6.3	226	0.885	1.183	0.016	1.609	0.051	0.720	0.352	4.816
18.	W-18	7.1	183	1.213	0.751	0.089	0.713	0.031	1.200	0.272	4.269
19.	W-19	6.6	235	1.508	0.479	0.152	1.230	0.023	1.080	0.744	5.217
20.	W-20	6.9	437	2.361	2.323	0.105	1.913	0.051	1.840	1.016	9.608
21.	W-21	6.4	51	0.262	0.182	0.012	0.222	0.005	0.200	0.120	1.003
22.	W-22	7.6	63	0.656	0.176	0.075	0.422	0.033	0.360	0.432	2.153
23.	W-23	7.8	59	0.393	0.113	0.022	0.157	0.010	0.320	0.160	1.174
24.	W-24	7.7	54	0.295	0.071	0.005	0.096	0.010	0.280	0.080	0.837
25.	W-25	7.1	97	0.426	0.234	0.018	0.287	0.015	0.320	0.192	1.492
26.	W-26	7.6	271	1.639	0.758	0.063	0.457	0.010	1.280	0.704	4.911
27.	W-27	7.5	112	0.361	0.507	0.010	0.422	0.046	0.280	0.160	1.785
28.	W-28	6.9	93	0.426	0.243	0.038	0.274	0.061	0.360	0.080	1.483
29.	W-29	7.3	77	0.328	0.117	0.006	0.296	0.010	0.200	0.192	1.149
30.	W-30	7.2	86	0.459	0.147	0.043	0.222	0.018	0.400	0.160	1.448

All values are expressed as meq/l, except pH and EC; pH in pH units, EC: micromhos/cm at 25°C
TDI includes the ionic concentrations of fluoride, nitrate and phosphate which are not shown in the table.

Table 2. Chemical composition of groundwater collected during post-monsoon

Sl.No	Code	pH	EC	HCO_3	Cl	SO_4	Na	K	Ca	Mg	TDI
1.	W-1	5.7	124	0.190	0.099	0.079	0.178	0.018	0.441	0.402	2.121
2.	W-2	6.1	221	0.619	0.541	0.311	0.809	0.046	0.842	0.321	3.489
3.	W-3	6.7	24	0.095	0.118	0.006	0.100	0.013	0.080	0.082	0.484
4.	W-4	5.1	62	0.119	0.223	0.012	0.174	0.013	0.160	0.201	0.896
5.	W-5	5.3	61	0.167	0.223	0.054	0.287	0.031	0.160	0.160	1.081
6.	W-6	6.0	43	0.167	0.090	0.019	0.122	0.013	0.160	0.080	0.641
7.	W-7	4.3	158	0.143	0.715	0.016	0.517	0.248	0.200	0.080	1.921
8.	W-8	5.2	87	0.286	0.327	0.055	0.274	0.013	0.281	0.201	1.433
9.	W-9	5.2	90	0.119	0.594	0.006	0.426	0.061	0.120	0.161	1.488
10.	W-10	5.1	55	0.167	0.144	0.008	0.174	0.015	0.120	0.120	0.748
11.	W-11	5.4	110	0.214	0.403	0.030	0.504	0.041	0.440	0.082	1.715
12.	W-12	5.6	229	0.428	0.932	0.094	0.926	0.294	0.840	0.082	3.598
13.	W-13	5.5	85	0.214	0.330	0.060	0.430	0.049	0.200	0.080	1.364
14.	W-15	6.0	87	0.119	0.377	0.004	0.443	0.018	0.200	0.080	1.242
15.	W-16	5.9	98	0.452	0.062	0.008	0.191	0.015	0.240	0.402	1.371
16.	W-17	5.8	274	0.357	1.110	0.036	0.926	0.072	0.280	0.402	3.182
17.	W-18	5.9	123	0.333	0.485	0.026	0.361	0.026	0.320	0.201	1.751
18.	W-19	6.6	308	1.262	0.580	0.227	0.961	0.072	0.360	0.241	3.702
19.	W-20	6.8	315	0.595	1.730	0.058	1.261	0.041	0.400	0.354	4.438
20.	W-21	6.3	30	0.143	0.127	0.004	0.104	0.013	0.480	0.080	0.940
21.	W-22	6.5	90	0.357	0.206	0.019	0.365	0.015	0.520	0.161	1.643
22.	W-23	5.5	49	0.119	0.146	0.011	0.130	0.015	0.560	0.120	1.103
23.	W-24	4.9	49	0.095	0.082	0.002	0.061	0.013	0.600	0.082	0.927
24.	W-25	5.5	86	0.214	0.377	0.020	0.357	0.028	0.640	0.080	1.717
25.	W-26	7.0	119	0.428	0.251	0.013	0.300	0.013	0.680	0.082	1.767
26.	W-27	6.3	39	0.143	0.101	0.005	0.126	0.013	0.720	0.082	1.186
27.	W-28	6.1	29	0.143	0.099	0.007	0.091	0.013	0.760	0.201	1.306
28.	W-29	6.6	55	0.214	0.124	0.005	0.248	0.013	0.800	0.082	1.481
29.	W-30	6.4	60	0.262	0.124	0.045	0.174	0.018	0.840	0.082	1.544

All values are expressed as meq/l, except pH and EC; pH in pH units, EC: micromhos/cm at 25°C
TDI includes the ionic concentrations of fluoride, nitrate and phosphate which are not shown in the table
W-14 was not collected, as the well was abandoned and reclaimed.

Ion chemistry and hydraulic continuity

Composition, age and temperature similarities may confirm hydraulic interconnections between adjacent springs and wells, whereas significant differences may indicate flow discontinuity (Mazor, 1997). Concentrations of conservative dissolved ions like chloride serve as a potential tool to establish hydraulic connectivity. In a saturated zone the chloride concentration stays constant unless mixing of different water types occur (Mazor, 1997). In the present study, the measured ion concentrations (expressed as epm) are plotted in X-Y diagrams (composition diagrams). The composition diagrams portraying the interrelationships between the ions like Na, K, Ca and Cl with the Total Dissolved Ions (TDI) for pre-monsoon and post monsoon seasons are illustrated in Figs. 2 (a) and 2 (b). Plots show that well water samples vary considerably in their ionic concentration. Significant positive correlation could be observed for selected characteristics like Na, Ca, Mg and Cl versus TDI during pre-monsoon period. However, during post-monsoon season, sodium alone seems to have a significant positive correlation (r^2 = 0.73). The highly distributed points in the post-monsoon (Fig. 2b) are probably due to mixing of many constituents from diffused sources following the monsoon rains.

The data points in the composition diagrams fall as two clusters during pre-monsoon season and post-monsoon periods (A and B). Samples representing cluster A were collected from Edavanna, Areacode and Vazhakkad region and samples representing cluster B were collected from Vazhikadavu and Nilambur region of the study area. Wells representing cluster B is located at a higher elevation compared to those in cluster A in the study area. Cluster A having lower ion concentration and low TDI while cluster B having higher ion concentration and high TDI. These two clusters are well separated in the study area. Most of the samples falling close to each other in each cluster revealing same water type showing possible hydraulic connectivity.

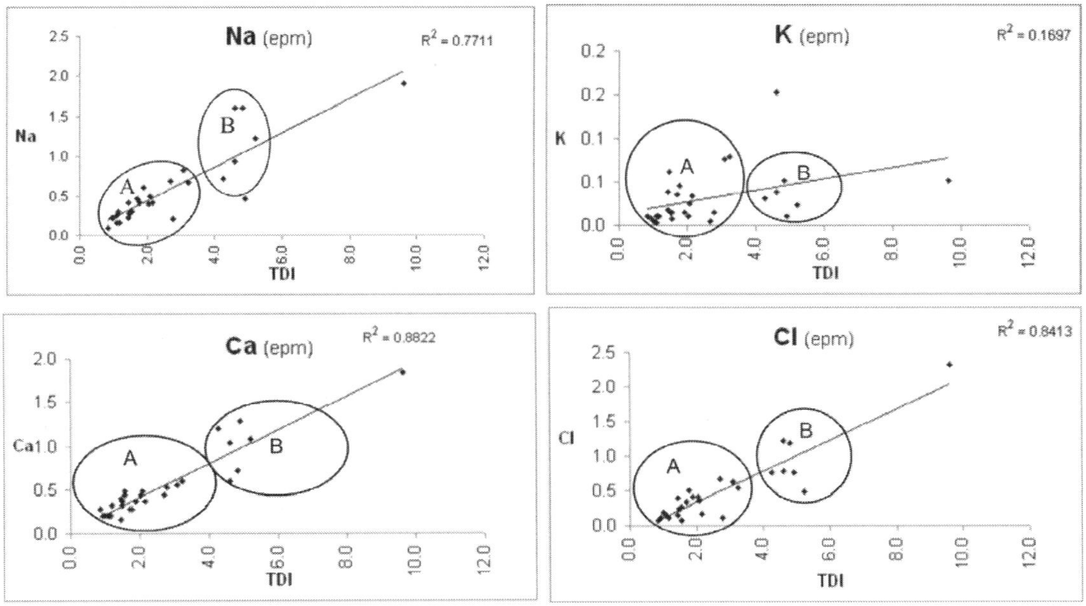

Fig. 2 (a). Composition diagram depicting the relationships between selected ions with TDI (Pre-monsoon).

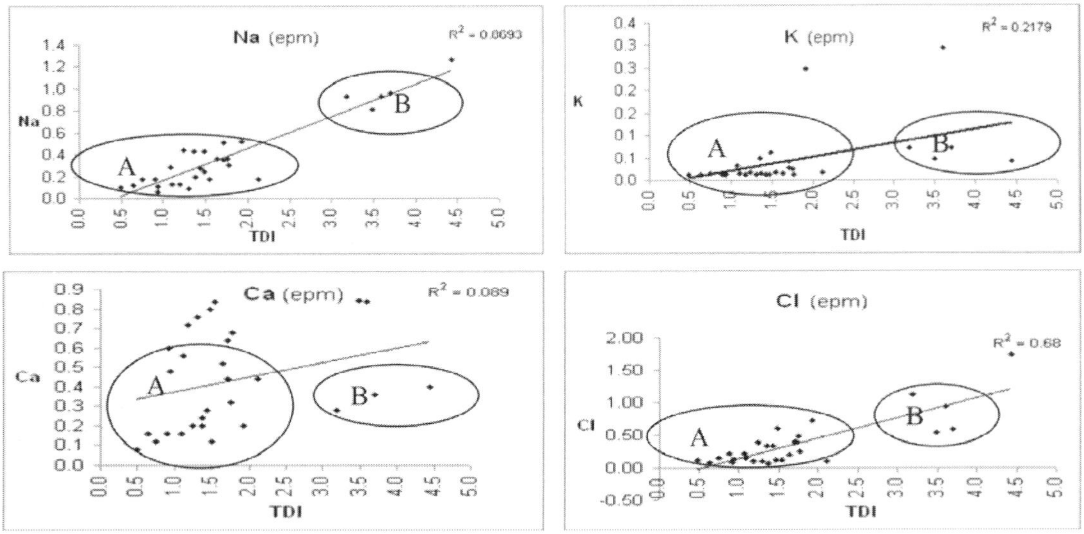

Fig. 2 (b). Composition diagram depicting the relationships between selected ions with TDI (Post-monsoon).

Stable isotopes

The δD and δ^{18}O values of groundwater samples collected during pre monsoon vary from –16.58% to –5.02% and –3.86% to –1.80%, respectively. In post-monsoon period, the values ranges from –26.69% to –10.45% and –4.32% to –2.13%. δD-δ^{18}O plot for groundwater samples is depicted in Fig. 3 (a) and 3 (b). The depleted δD and δ^{18}O values in post-monsoon clearly indicate the effect of monsoon in the study area, the recharge to have occurred from monsoonal

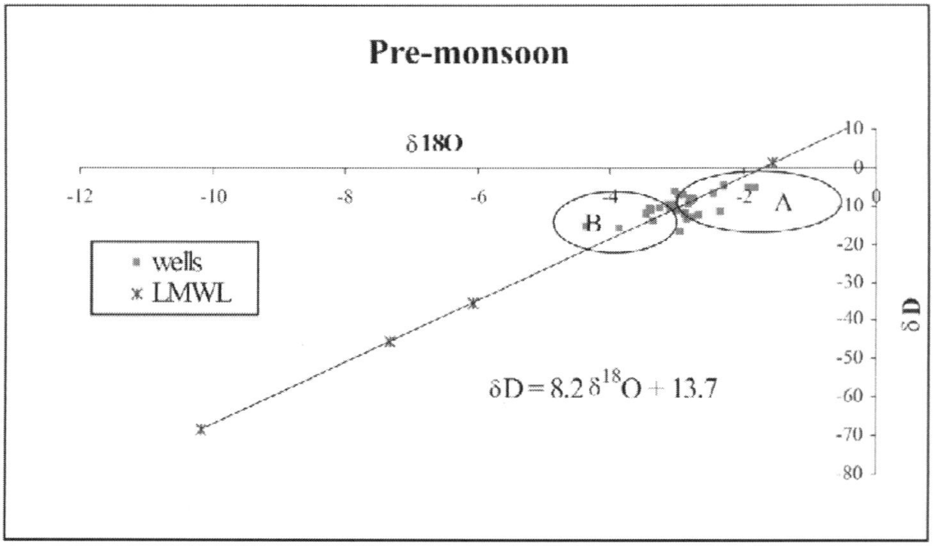

Fig. 3(a). δ^{18}O v/s δD for the groundwater samples—pre-monsoon.

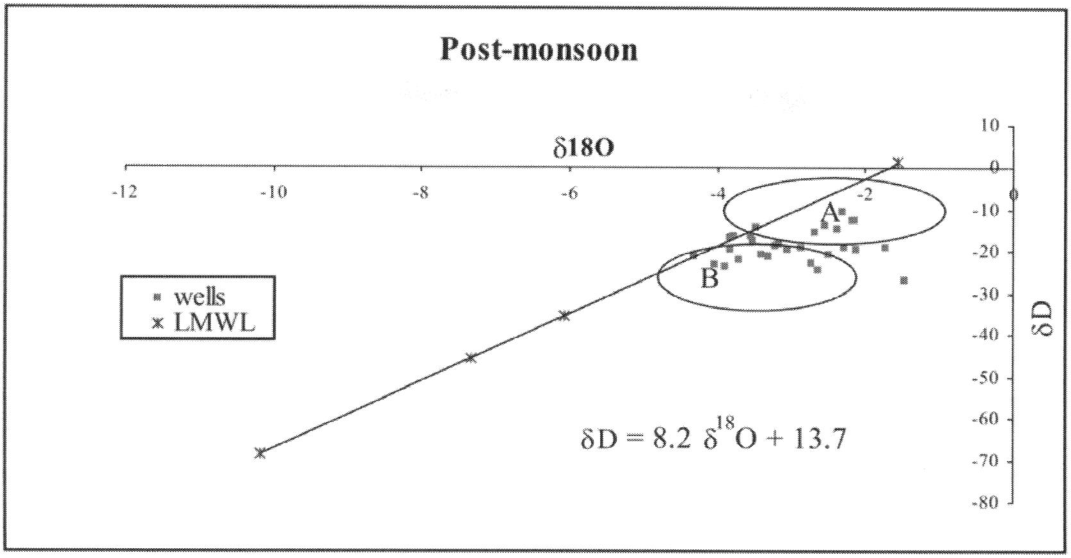

Fig. 3 (b). $\delta^{18}O$ v/s δD for the groundwater samples—post-monsoon.

waters. The isotopic data of groundwater samples representing pre-monsoon falling close to the LMWL, that of post-monsoon indicating an evaporating trend away from the LMWL (Navada *et al.*, 1993). Isotopic pattern clearly demonstrate an altitude effect in the basin whereby the isotopic value (δD and $\delta^{18}O$) depletes towards higher elevation (cluster B). The two clusters (A and B) identified from chemical ion data can also be distinguished from δD-$\delta^{18}O$ plots for both seasons. Wells representing cluster B have depleted stable isotope values compared to cluster A in pre-monsoon and post-monsoon seasons. Stable isotope values of both clusters (except very few wells) exhibit depleted stable isotope content in post-monsoon compared to pre-monsoon and have only slight variation in δD and $\delta^{18}O$ values establishes possible hydraulic connectivity between these water bodies.

CONCLUSION

Water chemistry and stable isotope composition (Deuterium and oxygen-18) were studied in selected dug wells in Malabar region of northern Kerala. Samples were collected in pre-monsoon and post-monsoon seasons. During pre-monsoon, the dominant anion is HCO_3 followed by Cl and SO_4, but the dominant cations are Na > Ca > Mg and the water type is represented by Na–Ca–Mg–HCO_3. In post-monsoon, the dominant anion is Cl followed by HCO_3 and SO_4 and the cation dominancy can be graded as Ca > Na > Mg with the water type represented by Ca–Na–Mg–Cl–HCO_3. The change in ion dominancy and water type during post monsoon indicated the direct influence of monsoon in the study area. The well water samples vary considerably in their ionic concentration. Two distinct water types were identified based on ion chemistry and stable isotope composition. Significant positive correlation could be observed for selected characteristics like Na, Ca, Mg and Cl with Total Dissolved Ions during pre-monsoon period. However, during post-monsoon season, sodium alone seems to have a significant positive correlation. The highly distributed points in the post-monsoon are probably due to mixing of many constituents from diffused sources following the monsoon rains. The composition diagrams with selected ionic concentration against total dissolved ions in different seasons indicate the hydraulic connection between the wells in the study area. Stable isotope composition agrees with the same.

ACKNOWLEDGEMENTS

The authors are thankful to the Executive Director, Centre for Water Resources Development and Management (CWRDM) for permitting to publish this paper. They acknowledge the financial support provided by Department of Science and Technology (DST), Government of India for this study. Authors gratefully acknowledge the help and support rendered by Dr K. Vasu, Former Head, Isotope Hydrology Division, of this Centre.

REFERENCES

APHA, (1998). "Standard methods for the examination of water and wastewater", 19th edition, APHA, Washington D.C.

Craig, H. (1961). Standard for reporting concentrations of deuterium and oxygen-18 in natural waters. *Sci.*, pp. 133 : 1833-34

Mazor, E. (1997). Chemical and isotopic groundwater hydrology – The Applied Approach. (2nd Edn). Marcel Dekkar Inc., New York. pp. 47-143.

Navada, S.V., Nair, A.R., Rao, S.M., Paliwall, B.L. and Doshi, C.S. (1993). Groundwater recharge studies in arid region of Jalore, Rajasthan using isotope technique. *J. Arid Environ.*, **Vol. 24**: pp.125-133.

Navada, S.V., Shivanna, S.K. and Rao S.M. (1986). Application of environmental isotopes in groundwater hydrology. *Indian J. Earth Sci.*, **13:** pp. 223-234.

PWD, (1974). Water Resources of Kerala, Public Works Departments, Government of Kerala, Thiruvananthapuram.

Rao, S.M. (2006). Practical Isotope Hydrology, New India Publishing Agency, New Delhi-110088, p. 201.

Suman Sharma, Kulkarni, K.M., Kulkarni, U.P., Deodhar, A.S., Navada, S.V. and Jain, P.K. (2003). Isotope Hydrochemistry of Groundwater in Purna River Basin, Maharashtra, India. Inter. Symp. On Isotope Hydrology and Integrated Water Resources management Vienna, Australia, 19-23 May 2003. Paper No,-IAEA-CN-104/103.

Sukhija, B.S., Nagabhushnam, P. and Reddy, D.V. (2002). Use of Stable Isotope in Hydrology. *Indian J. Geochemistry*, **Vol. 17:** pp. 89-98.

5

Fracture Detection in Hard Rock Terrains of Villupuram District Using Apparent Resistivity Data — Case Studies

ARULPRAKASAM, V.[*1] **SIVAKUMAR, R.**[2] **LAWRENCE, J.F.**[2] **AND KUMAR, M.S.**[3]

[1] Scientist-B, [3]STA (GP), Central Groundwater Board, SECR, and

[2]Reader in Geology, Department of Geology, Presidency College, Chennai

**E.mail: vaprakasam@yahoo.com*

ABSTRACT

Groundwater resources play in vital role in the economic development of a country like India. As a major part of the country is underlain by crystalline rocks with very limited ground water prospects, delineation of productive fracture zones at depth is necessary for economic cost-effective development of groundwater resources for various uses. Among the available geophysical methods, Electrical Resistivity method is the most widely preferred for groundwater exploration in hard rock areas.

In Electrical Resistivity Surveys, Vertical Electrical Sounding Technique is conventionally adopted for such purpose. Deciphering the actual thickness of the aquifer and delineating the thin or minor high yielding fractures, which form conduits for groundwater flow, is found to be very difficult solely from the interpretation of Vertical Electrical Sounding (VES) data. In many cases, the high yielding zones are found to be well within the part of the sounding curve indicating the bedrock. A non-conventional approach of processing of the Apparent Resistivity data was attempted for the Vertical Electrical Soundings conducted over different hard rock terrain. The interpretation yielded a conductance-depth relation with the yielding aquifer zones hidden in the bedrock. This approach, which had earlier been used successfully in many area of southern and central India, was attempted in Villupuram district of Tamil Nadu. The detailed methodology is discussed in the paper through four case studies.

Keywords: Apparent resistivity, aquifers, hard rock terrains, conductance.

INTRODUCTION

There is an ever-increasing demand for freshwater resources to meet the requirements for industrial, agricultural and domestic sectors. The over–exploitation of groundwater resources and its contamination have put a several stress on the available groundwater resources in the country. Geophysical surveys have found enormous applications in hydrogeological studies. In the present study, an attempt is made to detect the fracture zones in hard rock terrains using "conductance method" in Villupuram district, Tamil Nadu.

BACKGROUND INFORMATION-SALIENT FEATURES OF THE SURVEYED AREA

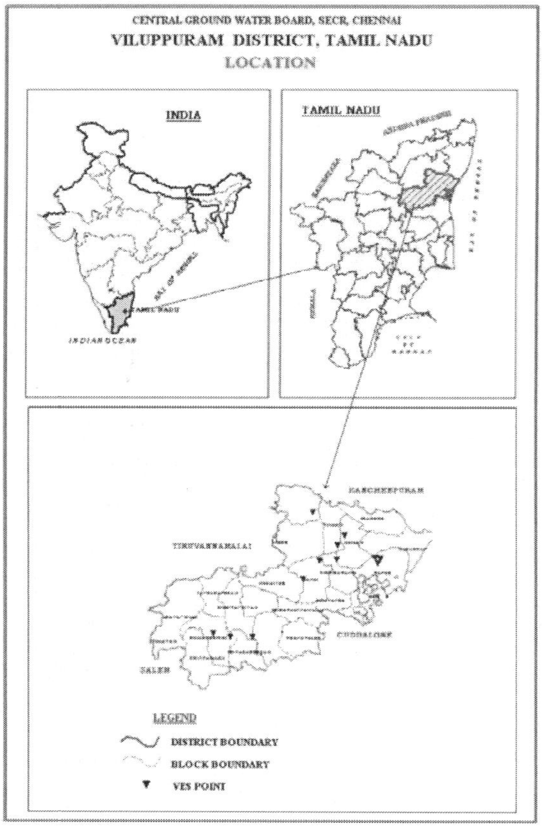

Villupuram district is bounded in the north by Tiruvannamali and Kanchipuram district, south by Cuddalore district and Pondicherry Union Territory, west by Dharmapuri and Salem districts and east by Bay of Bengal. It falls within the co-ordinates of Latitude 11°30' to 12°35' and Longitude 78°37' and 80°00' of Survey of India Topographical map 57P, 58 I, 58J and 58M. The areal extent of the district is 6616 sq.kms. (Fig. 1). The district forms part of Ponnaiyar river basin.

The district enjoys a tropical climate. It receives average annual rainfall of about 1220 mm. About 62 per cent of the annul rainfall is received during the south-west monsoon during June to August, whereas about 34 per cent is received during north-west monsoon during November-December. The area receives rain for about 60 days in a year.

The district is underlain by the crystalline metamorphic complex in the western parts and sedimentary tract in the eastern part of the district. The thickness of sedimentary contact is sheared and faulted.

METHODOLOGY

Under earlier annual action plan, Central Groundwater Board SECR had taken up exploration drilling work in Villupuam district in Tamil Nadu. This was preceded by systematic hydrogeological and geophysical surveys. 20 sites were covered under electrical resistivity surveys for selection of feasible sites through exploratory drilling. Schlumberger configuration had been employed with CRM 500 resistivity meters. Current electrode separation could extend up to a maximum of 600 m. 12 sites have been drilled on the basis of Vertical Electrical Sounding data.

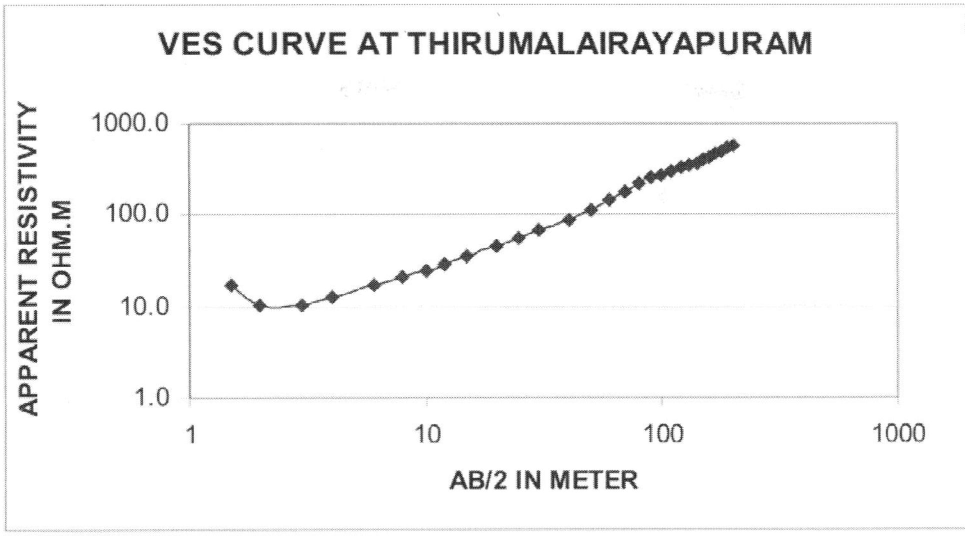

Fig. 1 a. VES curve at Thirumalairayapuram, Villupuram District, Tamil Nadu.

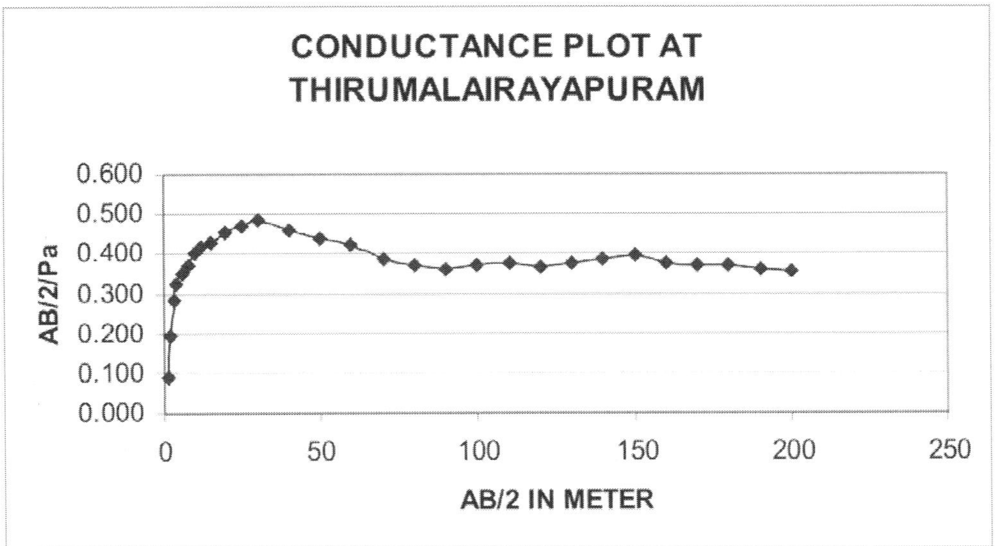

Fig. 1 b. Conductance plot at Thirumalairayapuram, Villupuram Distrct, Tamil Nadu.

The litho logs and depth to aquifer zone along with yield particulars were collected. The comparison of quantitative layer parameters from VES interpretation and borehole particulars revealed that most of the major yielding aquifer zones are in those depths which could not be inferred as aquifer zones from resistivity data.

RESULTS AND DISCUSSION

The resistivity method of geophysical prospecting is based upon the measurement of apparent resistivities in the field and then relating geoelectrical parameters with geological formations at depth. The electrode configurations chosen for resistivity surveys are usually responsible for either the success of the method in solving the problem or otherwise. In symmetrical, four electrode configurations, Schlumberger and Wenner methods are conventionally adopted for groundwater exploration. Mostly Schlumberger configuration is used for vertical electrical sounding. The interpretation of these sounding data leads to a broad picture of the geoelectric nature of subsurface formations. In qualitative interpretation of VES curves, visual inspections of curves, identification of type of curve and demarcation of areas with similar curves are done primarily and quantitative interpretation is done by curve matching techniques. The computerised interpretation is done either by forward modelling or inverse modelling. Forward modelling includes computation of apparent resistivity curves for inferred subsurface resistivity model and matching it with curves of known models. Inversion includes delineating the layer parameters directly from the curve. Nowadays, numbers of computer softwares are available for interpretation of field data utilising various techniques. However, deciphering the actual depth to aquifer zone from their layer parameter is not easy, especially in hard rock areas.

In general, the sounding curves in hard rock areas are of H, A, HA and AA type. Drilling at some sites indicates that the depths inferred from the interpreted results are not matching with actual depths at which major yielding fractures are encountered. In many cases, these fractures are found to be well within bedrock of the sounding curve.

This has prompted to a critical analysis of the basic resistivity data of the VES in a non-conventional way. The approach which comprises examining the raw data of soundings from various hard rock areas and correlating them with yield particulars are detailed in this paper. This approach had already been attempted successfully in many other terrains of India.

CALCULATION OF "CONDUCTANCE" (C) FROM APPARENT RESISTIVITY DATA

Vertical Electrical Soundings helps in delineating since field data thus collected, on processing, gives rise to various geoelectrical layers. The inferences drawn from favourable layer parameters alongwith local hydrogeological conditions give rise to better results in terms of encountering potential zones. It has been observed that in four boreholes drilled at Thirumalirayapuram, Sandisatchi, Thirukkanur, and Thensiruvalur, the major yielding zones fall in the ascending part of the sounding curve. A reexamination of apparent resistivity data has not shown any significant change in trend, which could possibly be isolated and interpreted as a separate geoelectric layer. Consequently, the sounding data for those sites where yield particulars are available were subjected to a calculation in the following manner:

The ratio of a particular current electrode separation (AB/2) to the corresponding apparent resistivity (ρa) is calculated and the ratio is called as "conductance" (C). For example, the conductance (C) for the separation of AB/2=10 m is:

$$C10 = (\rho 10)/10$$

In similar manner, the conductance for all the current electrode separations observed for the sounding is calculated. The plot of conductance verses AB/2 on a linear scale revealed a rising trend at places where high yielding fractures are encountered in the borehole. Table 1 gives details of conductance values along with depth to aquifer zones and their yield, relating to the four sites. The trend remains unchanged even when more number of measurements is observed for a single sounding.

Table 1. Conductance depth relationship, aquifer depth and yield particulars.

Site	Thirumalairayapuram	Sandisatchi	Thirukanur	Thensiruvalur
Geology	Granite Gneiss	Granite Gneiss	Granite Gneiss	Granite Gneiss
Depth drilled in m	200	200	200	200
AB/2 in m	**CONDUCTANCE**			
1.5	0.089	0.238	0.455	0.093
2	0.192	0.345	0.714	0.099
3	0.283	0.448	1.000	0.128
4	0.323	0.615	1.176	0.145
6	0.351	0.923	1.304	0.188
8	0.372	0.825	1.667	0.210
10	**0.402**	0.513	**2.000**	**0.256**
12	**0.420**	0.486	**2.927**	**0.281**
15	**0.427**	0.432	**3.061**	**0.291**
20	**0.454**	0.404	**3.846**	0.259
25	**0.468**	0.377	2.551	0.256
30	**0.485**	0.379	2.222	**0.258**
40	0.457	0.392	2.484	**0.283**
50	0.437	0.400	2.326	**0.297**
60	0.420	0.395	2.273	0.305
70	0.388	0.372	2.310	0.327
80	0.370	0.357	2.369	0.332
90	0.361	0.359	2.143	0.324
100	0.368	**0.367**	2.123	0.312
110	0.374	**0.398**	2.268	0.306
120	0.364	0.392	2.299	0.295
130	**0.377**	0.391	2.355	0.301
140	**0.388**	0.380	2.491	0.297
150	**0.395**	**0.367**	2.517	0.296
160	0.374	**0.381**	**2.516**	0.301
170	0.372	**0.394**	**2.620**	0.308
180	0.368	0.388	**2.710**	0.304
190	0.358	0.383	2.346	0.306
200	0.355	0.375	2.299	0.307
Aquifer depth and yield particulars				
10.5 - 11.00 - 1.5 26.00 - 26.50 - 3.0 140 -141.50 - 5	108 - 108.5 - 3	19.00 - 19.20 - 0.2 178 - 178.50 - 3.5	10.00-10.50 - 1 34 -34.50- 2.2	

CASE STUDIES

THIRUMALAIRAYAPURAM: Thirumalairayapuram site is situated over granite gneiss of Archaean age. Two half Schlumberger soundings were conducted at this site and a borehole was drilled down to 200 m and yielded 5.00 lps of groundwater.

The sounding curve with layer parameters and "conductance" are shown in Fig. 1. In this case also, the aquifer encountered is in the ascending part of the curve. It has indicated rise in conductance values between current electrode separations of AB/2: 10 – 30 and 130 – 150 m. The main yielding zones encountered during drilling are at 10.50 – 11 and 140 – 140.5 m.

SANDISATCHI: Sandisatchi site is situated over granite Gneiss of Archaean era. Two half Schlumberger soundings were taken and a borehole drilled down to 200 mbgl on the basis of recommendations yielded 3.00 lps.

The observed resistivity-sounding curve and interpreted layer parameters along with the plot of "conductance" are shown in Fig. 2. The aquifers are met in the ascending part of curve. The depths at which the major fractures encountered are 108.00 – 108.50 and 153.00 – 153.50 m. The conductance plot has indicated a rising trend between 100 – 110 and 150 – 160 m of AB/2 separation, coinciding with aquifer zones, when duly correlated.

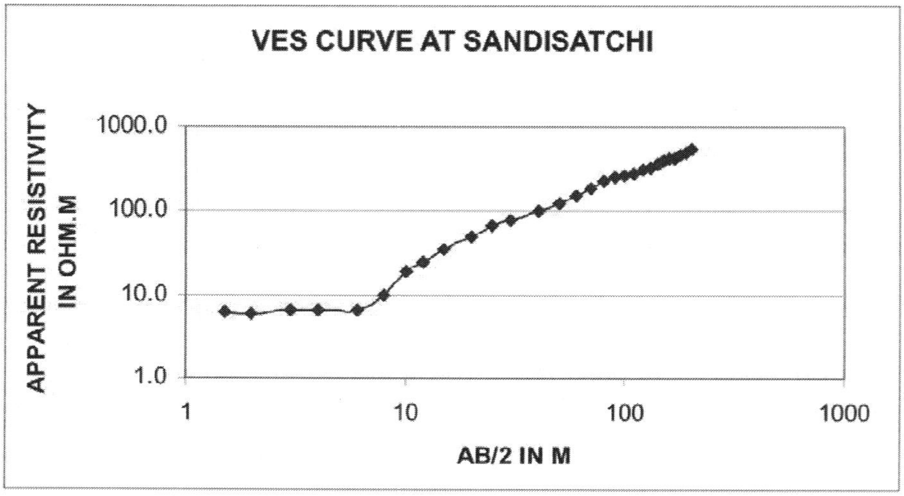

Fig. 2 a. VES curve at Sandisatchi, Villupuram District, Tamil Nadu.

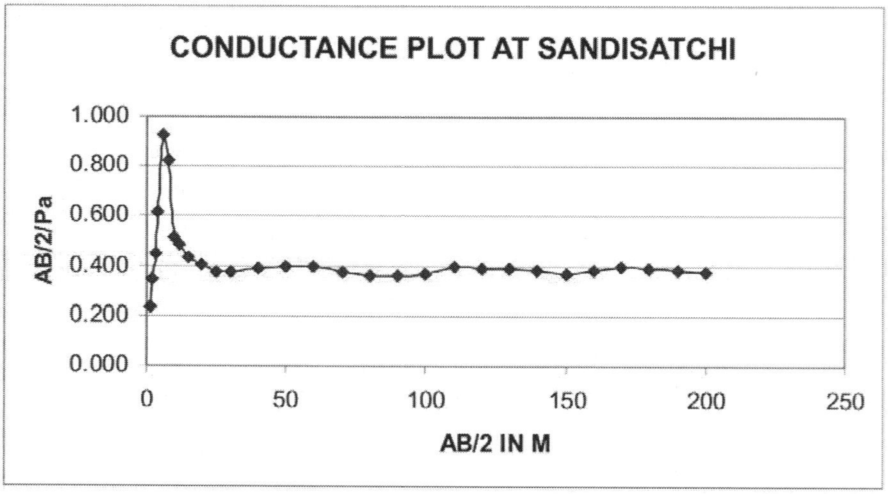

Fig. 2 b. Conductance plot at Sandisatchi, Villupuram District, Tamil Nadu.

THIRUKKANUR: Thirukkanur village is situated over granite gneiss of Archaean age. Four VES were conducted and a borehole drilled down to 200.00 m on the basis of recommendations yielded 3.5 lps. The sounding curve with layer parameters and "conductance" are shown in Fig. 3. In this case also, the aquifer is encountered in the ascending part of the curve. The site was recommended on the basis of conductance calculation, where it has indicated arise in conductance values between current electrode separations of AB/2: 10 – 20 and 170 – 180 m. The main yielding zones encountered during drilling are at 19.0 – 19.20 and 178.00 – 178.50 m. almost coinciding with drilling data.

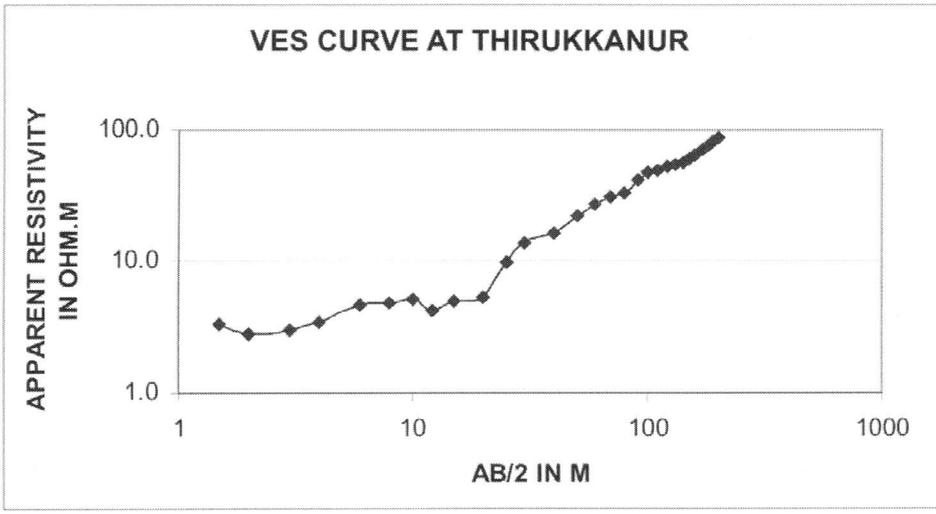

Fig. 3 a. VES curve at Thirukkanur, Villupuram District, Tamil Nadu.

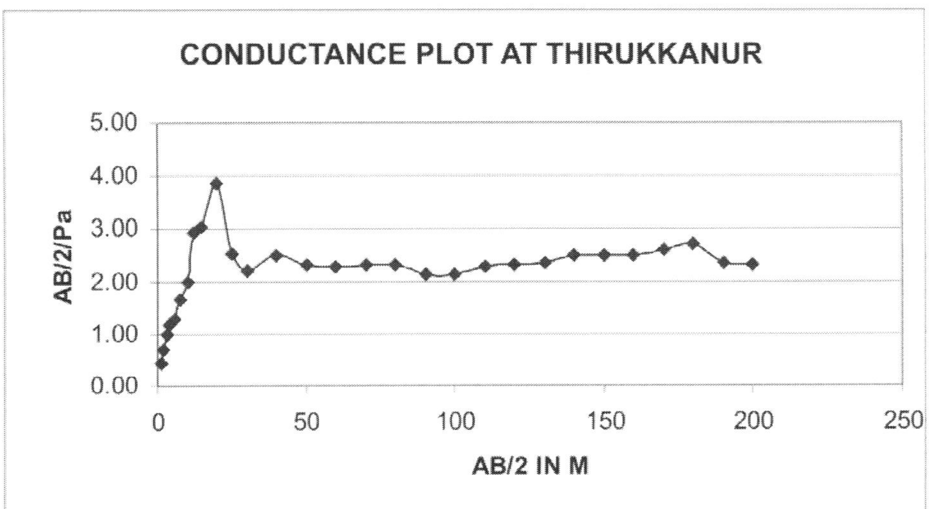

Fig. 3 b. Conductance plot at Thirukkanur, Villupuram District, Tamil Nadu.

THENSIRUVALLUR: Thensiruvallur village is situated over granite gneiss of Archaean age. Three VES were conducted and a borehole drilled down to 200 m on the basis of recommendations, yielded 2.2 lps. The sounding curve with layer parameters and "conductance" are shown in Fig. 4. In this case also, the aquifer encountered in the ascending part of the curve. The site was recommended on the basis of conductance calculation, where it has indicated a rise in conductance values between current electrode separations of AB/2: 10 – 20, and 30 – 40 m. The main yielding zones encountered during drilling are at 10.0 – 10.5 and 34 – 34.5 m almost coinciding with aquifer zones when duly correlated.

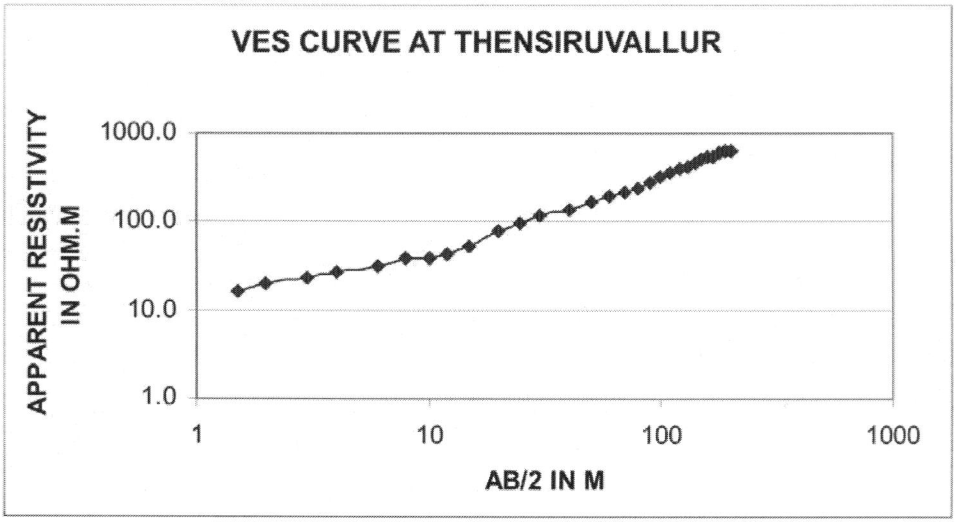

Fig. 4 a. VES curve at Thensiruvallur, Villupuram District, Tamil Nadu.

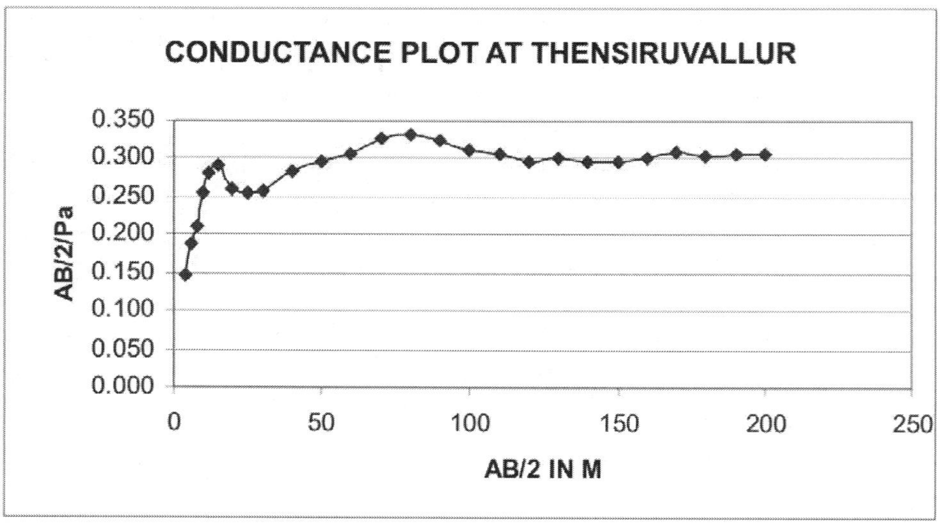

Fig. 4 b. Conductance plot at Thensiruvallur, Villupuram District, Tamil Nadu.

CONCLUSION

The practical utility of the method of calculation of a "conductance" in detecting hidden aquifer zones is demonstrated through its successful application in four case studies in Dharmapuri District, as detailed above.

ACKNOWLEDGEMENTS

The authors are grateful to B.M. Jha, Chairman, and Central Groundwater Board for granting permission to present this paper. They express their sincere thanks to Dr. N.Varadaraj, Regional Director, Shri E. Sampathkumar, Superintendent Hydrogeologist, and Dr. S. Suresh, Scientist-D, CGWB, SECR, Chennai, Dr. G.S. Sukumaran, HOD, Presidency College, Chennai for their able guidance during preparation of this paper. Assistance rendered by the officers and staff of CGWB, SECR, and Chennai in the field surveys as well as in the preparation of this paper is gratefully acknowledged.

REFERENCES

Central Groundwater Board, (1997). Groundwater Resources and Development potential in Villupuram District, Tamil Nadu. Technical report of Central Groundwater Board, SECR, Chennai. Page 92. (Unpublished).

Central Groundwater Board, (2008). Report on Spot Electrical Sounding Survey in Villupuram District, Tamil Nadu. Page 6. (Unpublished).

DRDA, (1998-99). Identification of Recharge area using Remote Sensing and GIS in Tamil Nadu. Technical report of Tamil Nadu Water and Drainage, Chennai, Unpublished).

Orellana, E. and Mooney, H.M. (1966). Master curves of Schlumberger Arrangement. (Published).

6

Identification of Groundwater Flow Direction Using Tritium Analysis of Lower Gundar Basin, Southern Tamil Nadu, India

GOWTHAM, B.* AND LAWRENCE, J.F.

Department of Geology, Presidency College, Chennai

**E.mail: bgowtham23@gmail.com*

ABSTRACT

Environmental isotopes are useful to understand various processes in the hydrogeological system. Environmental isotopes are now used to trace not only groundwater province, but also recharge process, subsurface process, geochemical reaction and reaction rates, groundwater dating and circulation of the groundwater. Among the environmental isotopes tritium is unique and useful in hydrogeological studies as itself is a part of the water molecule. Moreover, it is an unstable isotope with a half-life of sufficient length, that it is widely used for underlying recharge processes. The tritium atom is mixed with precipitation after getting oxidized to water and enters the groundwater through the process of infiltration. Mazor (1997) has reported that the natural production of tritium introduces about 5 TU in precipitation. However, in many sites the tritium values in precipitation is low, for example, in the equatorial region and southern hemisphere the concentration of ^3H in precipitation is about 1-3 TU. In saturated zones, water is isolated from the atmosphere and the tritium concentrations reduce due to radioactive decay. The study area lower Gundar basin falls between 9°00'00" and 9°45'00" North Latitudes and 78°10'00" and 78°45'00" East Longitude. Water samples have been collected from 9 locations, which are spread over in the entire study area. Both poor and good quality waters have been chosen in this study. All water samples have been collected by the standard procedures. The tritium content in groundwater ranges from 0.98 to 4.50 TU. The lowest value falls around Mudukulathur region (Loc. No. 7), which lies in the eastern part of the study area. High tritium (4.50 and 3.99) is observed around Kadaladi and Sayalkudi (Loc. Nos. 8 & 9), which is in south. Moderate to high values occur along the north and central part. From the contour map, it is inferred that the areas having higher tritium concentrations are considered as areas of active recharge and the groundwater flows towards low tritium values. Recharge occurs in areas of high tritium (in water) and will flow towards the area of low tritium. It is also seen that the tritium values decrease with decreasing elevation of the granular zone.

Keywords: Groundwater flow, tritium, Gundar basin.

INTRODUCTION

Environmental isotopes are useful to understand various processes in the hydrogeological system. These are naturally occurring isotopes of elements found in abundance in environment: H, C, N, O and S, which are principal elements of hydrogeological, geological and biological systems. Environmental isotopes are now used to trace not only groundwater province, but also recharge process, subsurface process, geochemical reaction and reaction rates, groundwater dating and circulation of the groundwater. Among the environmental isotopes tritium is unique and useful in hydrogeological studies as itself is a part of the water molecule. Moreover, it is an unstable isotope with a half-life of sufficient length, that it is widely used for underlying recharge processes. Tritium is the heavy isotope of hydrogen. Its symbol is ^3H or T. Tritium occurs in the environment as a result of both natural and man-made activity. It undergoes radioactive decay by beta emission (a low energy beta-radiation of 18.3 kev), resulting transformation in a helium nucleus of mass 3 (^3He). The natural production of tritium results from the interaction of cosmic neutrons with nitrogen in the upper atmosphere producing ^{15}N, which is radioactive and disintegrates into common (^{12}C) carbon and tritium.

Study Area

The Gundar basin lies in the southern part of Tamil Nadu state in Indian subcontinent. The Gundar River originates at an altitude of 500 m near Kothaimalai of Saptur reserve forest belonging to the Varushanadu hills in Tamil Nadu. Entire Gundar river basin covers an area of 5,647 km^2, spreading over in five districts of Tamil Nadu. The area chosen for the present study is the lower Gundar basin that covers part of Ramanathapuram and Virudhunagar districts. The Gundar river coalesces at the Bay of Bengal after covering a distance of nearly 150 km. Geographically, the study area falls between 9°00'00" and 9°45'00" North Latitudes and 78°10'00" and 78°45'00" East Longitudes (Fig. 1). This area is covered in the following Survey of India toposheet numbers 58 K/2, 3, 6, 7, 11 and 12. The study area is bounded by Kariapatti and Aruppukottai blocks of Virudhunagar district in the north and north-west and Mudukulathur and Kadaladi blocks in the east. The Bay of Bengal forms the southern boundary. The total areal extent of the study area is 970 km^2 with a population of 0.8 million. The average annual rainfall of the study area is 700 mm (Gowtham *et al.*, 2003) which is below the state average rainfall of 1020 mm (Ram Mohan, 1984). Agriculture is the main occupation of the people. A large number of rainfed tanks that exist in this region store rainwater, support the agricultural and domestic needs for a short spell after the rainy season.

Tritium in Groundwater Research

An attempt has been made to identify the flow direction of the groundwater in the lower Gundar basin using isotope technique. Among the environmental tracers, tritium is found to be more suitable for the analysis and study of flow direction for the above-mentioned semi-arid region. Chen *et al.* (2003) has traced the flow of groundwater of northern China using tritium content in the groundwater. Moran *et al.* (2000) has studied the groundwater flow of California basin and identified its direction of flow using tritium and helium isotope in groundwater. The concentrations of tritium (^3H) and helium isotopes (^3He and ^4He) were used as tracers of

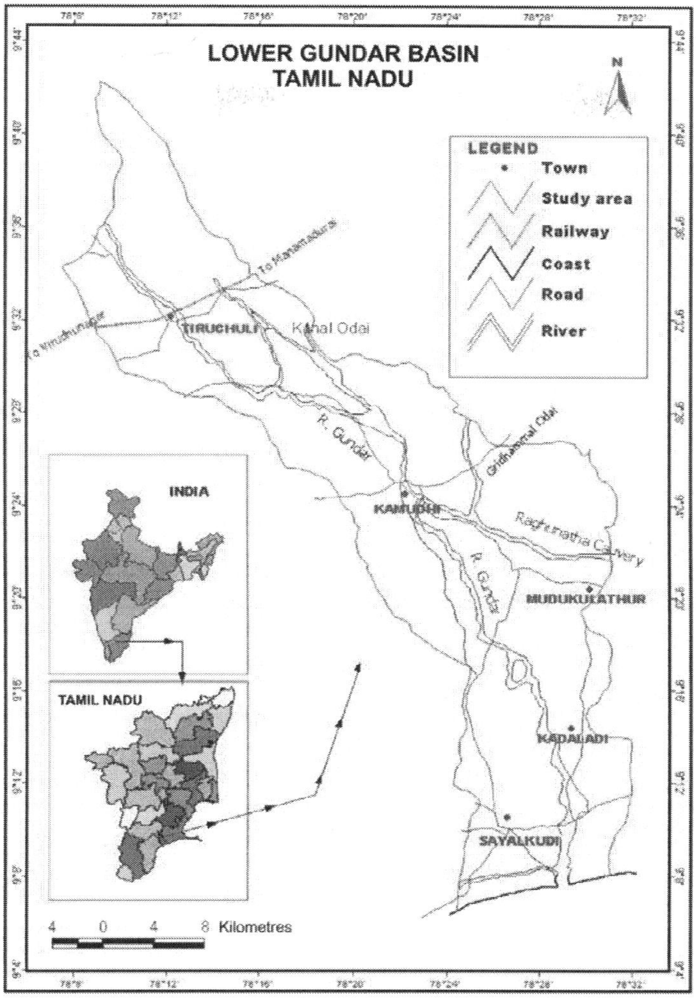

Fig. 1.

groundwater flow in the surficial aquifer system (SAS) beneath Everglades National Park (ENP), south Florida (Price *et al.*, 2003).

Geology

Lower Gundar river basin comprises hornblende biotite gneiss, mica gneiss, feldspathic gneiss, garnetiferous gneiss, pegmatite and quartzite. The regional foliation of these rocks varies from NNE – SSW to WNW – ESE and E – W directions with dip of SE, NE and south directions. These rocks are coarse grained, friable and highly weathered (Varadaraj *et al.,* 1986). Calcareous sand stones (Mio-Pliocene) occur unconformably over the Archaean basement (Singh *et al.,* 2002) and they are exposed in the north and western region of Kamudi as broad elevated high ground. Major part of the study area is covered by alluvium brought by the rivers flowing from the west of this region. Marine and fluvio-marine sediments occur in southern portion. Laterites are seen as patches within these marine sediments. Major part of the study area is covered by the flood plain. Pediments and buried pediments exist in northern highland areas. Major part of the study area is covered by the flood plain. Pediments and buried pediments exist in northern highland areas (CGWB, 1998).

Tritium Analysis of Lower Gundar Basin Groundwater

Water samples have been collected from 9 locations, which are spread over in the entire study area. Both poor and good quality waters have been chosen in this study. All water samples have been collected by the standard procedures prescribed by Mazor (1997). Before sampling, the water is pumped out for at least 30 minutes to avoid the atmospheric influence and to attain stable temperature. Collected samples are kept in dark room to avoid biological influence. The samples have been analysed in National Institute of Hydrology (NIH), Roorkee, for tritium and the results have been listed in Table 1.

From the above results, a tritium concentration map has been prepared (Fig. 2). The tritium content in groundwater ranges from 0.98 to 4.50 TU.

Table 1: Environmental tritium in lower gundar basin groundwater

S. No.	Location	Water Level (Amsl)*	Tritium *(TU)*
1.	Sakkarakottai	75.5	1.90
2.	Chetticulam	72.8	3.19
3.	Periya Kattangkudi	77.0	3.53
4.	Tiruchuli	56.8	2.11
5.	Annaikulam	46.6	2.60
6.	Kil Ponnaikulam	22.7	1.50
7.	Mudukulathur	15.4	0.98
8.	Kadaladi	2.7	3.99
9.	Sayalkudi	0.6	4.50

Amsl-above mean sea level: TU tritium Unite

The lowest value falls around Mudukulathur region, which lies in the eastern part of the study area. High tritium (4.50 and 3.99) is observed around Kadaladi and Sayalkudi, which is in south. Moderate to high values occur along the north and central part. From the contour map, it is inferred that the areas having higher tritium concentrations are considered as areas of active recharge and the groundwater flows towards low tritium values (Fig. 2).

Groundwater Level

The long-term water level data of this area, for the period between 1992 and 2001 have been collected from Tamil Nadu Water Supply and Drainage Board (TWAD) and Groundwater division of PWD and average annual water levels and fluctuation have been computed. From this computation water level, contour maps have been prepared for both pre- and post-monsoon seasons (Figs. 3a and 3b). The depth the water level varies from 3.5 m to 23.7 m (bgl) during pre-monsoon period and 2.9 m to 23.10 m (bgl) in post-monsoon. The water level is deeper in

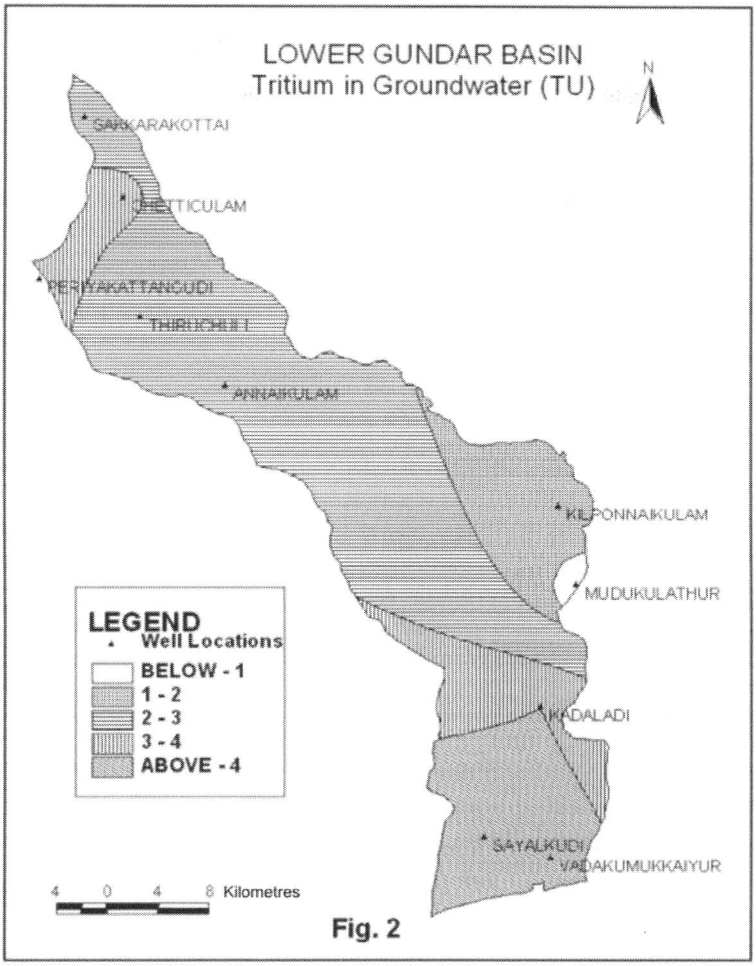

Fig. 2. Spatial variation of TU in the study area.

topographically elevated regions and shallower in sand dunes area. From the water level maps, it is inferred the groundwater flow direction is north-west to south-east.

Results and Conclusion

It is observed that tritium value from Periya Kattangudi to Mudukulathur gradually reduces to minimum (Fig. 2). Generally, tritium content will be higher in areas of recharge and lower around areas of discharge. Recharge occurs in areas of high tritium (in water) and will flow towards the area of low tritium (Yurtsever and Payne, 1979). It is also seen that the tritium values decrease with decreasing elevation of the granular zone. This gives an idea that the water recharges in Periya Kattangudi had reaches Mudukulathur). The distance between these two locations is about 45 km. From the water level contour maps, it is inferred that the flow of the groundwater is from north-west to south-east. The tritium in groundwater reduces from the

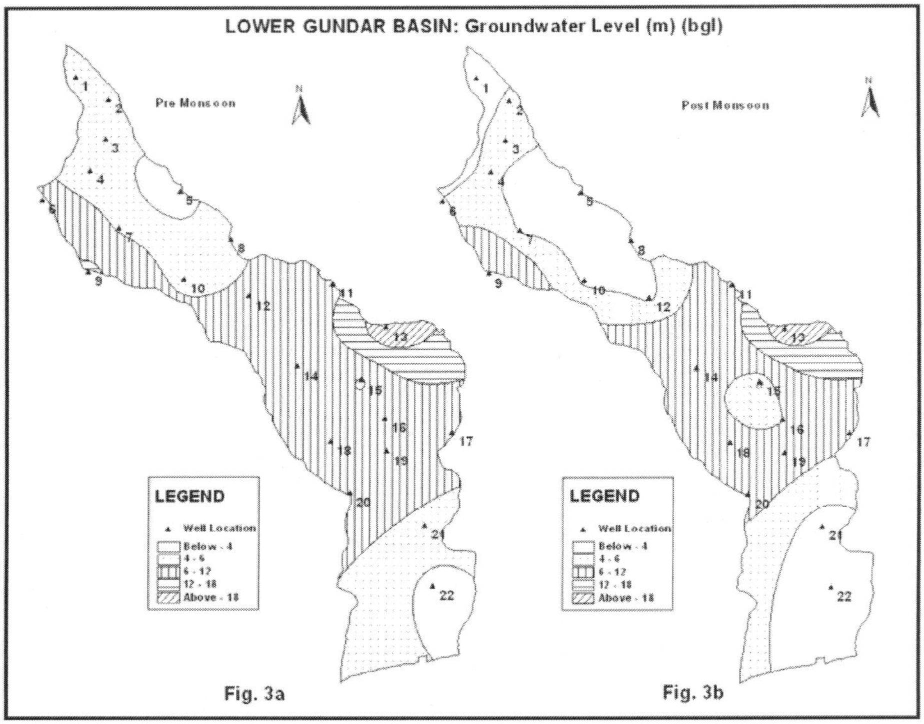

Fig. 3a Fig. 3b

north-west to south-east and the reduction follows the groundwater flow direction. Hence, the flow direction of groundwater is confirmed as from NW – SE. From the above study, it is inferred that tritium content in groundwater may used to trace the flow direction of groundwater and areas of recharge and discharge.

REFERENCES

C.G.W.B. (1998). Departmental Report entitled Groundwater exploration in Tamil Nadu and UT of Pondicherry as on 31.03.1996, Central Groundwater Board (SECR), Chennai, pp. 255.

Environmental Isotope Data, World Survey of Isotope Concentrations in Precipitation, V 1(1969) and Col. 7 (1983).

ESRI. (1988 and 1989). ARC/INFO User Manual Books, (Understanding FIS pp. (ix-xi). Starter Kit Users guide, Chapter 6–7, ARCEDIT, Users Guide Chapter 1, 4-5, ARCPLOT: Users guide, 3-4, SML, Users guide, Chapter 1, 8, User manual books, (understanding FIS), ix–xi.

Geyh, M.A. (2000). Groundwater - Saturated and Unsaturated zone. Environmental Isotope in Hydrological cycle, Technical Documents in Hydrology, **Vol. 4**, **No. 39**, UNESCO.

Gowtham, B. and Lawrence, J.F. (COFIS 2003). "Hydrogeochemical Studies for salt water intrusion in Lower Gundar Basin, Tamilnadu, India" International Conference on Coastal and Fresh water Issues Organized by Institute of Ocean Management, Anna University, Chennai, pp. 27.

Chen, J.Y., Tang, C.Y., Shen, Y.J., Sakura, Y., Kondoh, A. and Shimada, J. (2003). "Use of water balance calculation and tritium to examine the dropdown of groundwater table in the piedmont of the North China Plain (NCP)", *J. Environ. Geo.*, **No. 44:** 564-571.

Moran, Jean E. and Bryant Hudson, G. (2000), "Using Groundwater age and other isotopic signatures to delineate groundwater flow and stratification, Report of the U.S. Department of Energy by the University of California, Lawrence Livermore, National Laboratory under contract No. W-7405-ENG-48.

Laura Toran. (1982). Isotopes in groundwater investigations, *J. Groundwater*, **Vol. 20**, **No. 6:** pp. 740-745.

Mazor, E. (1997). Chemical and Isotopic Groundwater Hydrology. The applied approach, 2nd edition, Mercal Dekker Inc., New York, pp. 413.

Top, R.M.Z., Happell, J.D. and Swart, P.K. (2003). Use of tritium and helium to define groundwater flow conditions in Everglades National Park, *Water Resour. Res.*, **39(9):** 1267.

Ram Mohan, H.S. (1984). A Climatological Assessment of the Water Resources of Tamil Nadu. Ind. *J. Power and River Vallet Dev.*, pp. 58-63.

Singh, S.B., Jimmay Stephan, Srinivas, Y., Singh, U.K. and Singh, K.P. (2002). An integrated geophysical approach for groundwater prospecting. A case study from Tamil Nadu, *J. Geol. Soc. India*, **Vol. 59:** pp. 147-158.

Varadaraj, N. and Parthasarathy, T.N. (1986). Groundwater resources and Development potential of Ramanathapuram District, Tamil Nadu State, C.G.W.B Report, pp. 42.

Yurtsever, Y. and Payne, B.R. (1979). Application of environmental isotopes to groundwater investigation in Qatar. In: Isotope Hydrology 1978, IAEA, Venna, **2:** 465-490.

7

Groundwater Quality Profiling by Borehole Geophysics—Case Studies in Cuddalore District, Tamil Nadu

Badrinarayanan, T.S.*[1] **and Venkatesan, R.**[2]

[1]*Consultant Hydrogeologist, 97, Agraharam Street, Kollidam–609102*
Sirkali Taluk, Nagai District, Tamil Nadu

[2]*Research Scholar, Vinayaga Mission Research Foundation, Deemed University, Salem, Tamil Nadu*

**E.mail: tsbadri56@gmail.com*

ABSTRACT

The Chidambaram town in Cuddalore district of Tamil Nadu is blessed with the "Lord Nataraja temple," but the nature has cursed Chidambaram area with brackish water. The study area, namely, the Chidambaram and Kattumannarkudi talukas of Cuddalore district are coastal talukas and sedimentary terrain with an areal extent of about 250 km^2. The hydrogeological setting is such that the fluvio marine origin sediments constitute almost the entire area ultimately resulting in poor quality of groundwater. The quality of water is poor and saline in most parts of Potnovo, Keerapalayam and Kumaratchi panchayat unions. In some parts, the shallow aquifer water is good, drinkable, but in most parts of Chidambaram town area like Kanagasabainagar, Sivasakthinagar and Mariappanagar even the shallow aquifer water is very poor, saline with rotten egg smell due to the presence of H_2S. Due to poor quality problems, the exploration of freshwater in saline environment becomes challenging to the Geoscientists. The surface method of geophysical techniques may not give the quality scenario of the area. One of the subsurface methods of groundwater exploration, namely, the well logging facilitates continuous recording of physical response versus depth by a sensor when it moves inside the borehole. Among the several methods of well logging, the common methods used for groundwater exploration is electrical well logging. The present paper deals with the exploration of deep confined aquifers by electrical well logging techniques in 14 select locations of part of Chidambaram and Kattumannarkoil taluks. The purpose of investigation is to explore the deep confined freshwater aquifers for drinking water supply and irrigation purposes. Tube wells have been constructed, developed and completed. All the 14 tube wells have discharged good freshwater with copious yield. The resistivities obtained at select depths by electrical logging in all the 14 bore wells have been prepared as a resistivity contour map. An innovative attempt has been made to analyze the quality of ground water by studying the iso resistivity map in depth. The study reveals that the deep fresh water occurs at very deeper depths of 260 to 280 m below ground level in the south eastern Jeyangondapattinam area. The deep confined freshwater aquifer occurs at a depth range of 160 to 180 m bgl in and around Chidambaram town.

Keywords: Confined aquifers, electrical well logging, fluvio marine sediments.

INTRODUCTION

The study area, a part of Chidambaram and Kattumnnarkudi taluks in Cuddalore district, is a coastal tract and the total geographic extent is about 250 sq km. The case study villages fall in Keerapalayam, Kumaratchi and Potnovo unions. The area is drained by the rivers Vellar and Kollidam which is a distributory of the river Cauvery [Figs. 1 and 2]. The topography of the area is almost flat with a gentle slope towards east. Since the quality of groundwater of

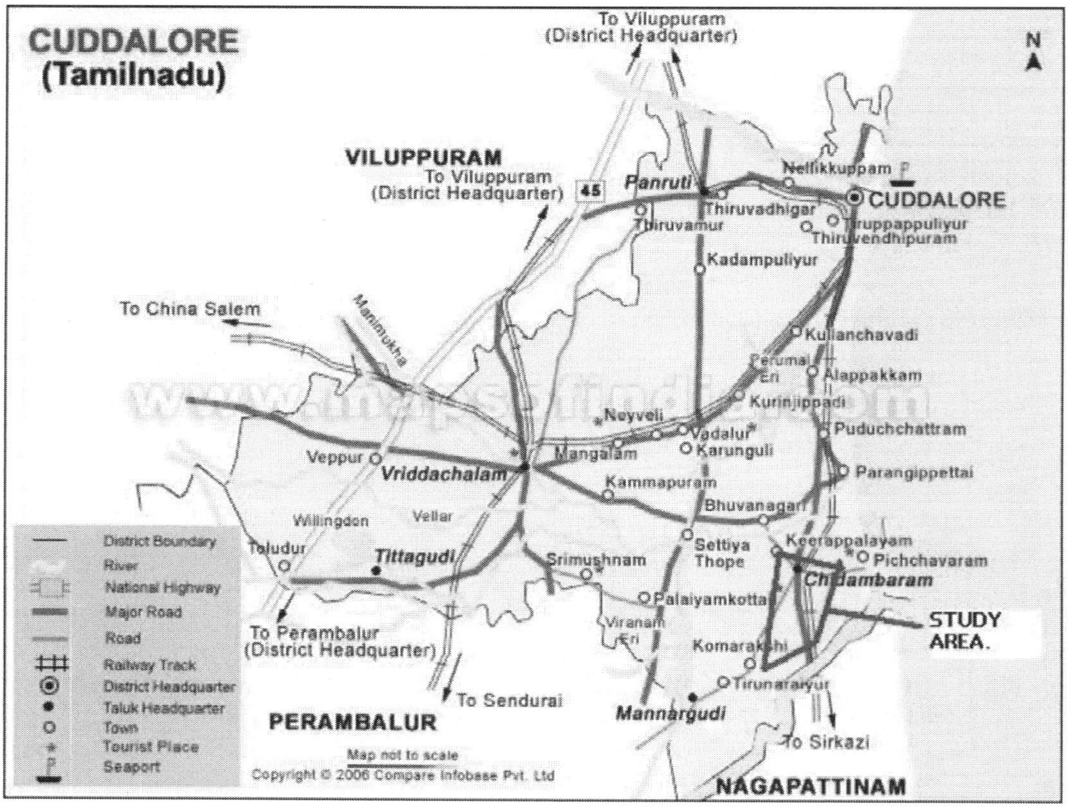

Fig. 1. Cuddalore district map showing study area.

Fig. 2. Google earth image showing the study area.

Chidambaram suburbs is very poor, detailed subsurface explorations have been carried out at 14 locations. An integrated hydrogeological and electrical well logging study has been conducted to explore the deep freshwater confined aquifers.

GEOLOGY AND HYDROGEOLOGY

The entire Chidambaram taluk is covered by Quaternary formations of Recent to Sub recent age. The unconsolidated sedimentary formations of fluvio marine origin which includes colluvium, black clayey soils, fine to coarse grained sands, silts, clays, laterites and lateritic gravels. The fluvial sediments occupy the flood plains of Vellar and Kolidam rivers, consisting mainly sand and sandy loams [Fig. 3]. Along the coast of Potnovo wind blown sands in the form of sand dunes are observed. The gravimetric and seismic surveys carried out by the Oil and Natural Gas Corporation [ONGC] and the Geological Survey of India [GSI] in the coastal sedimentary tract revealed a major fault running at a distance of 1.5 km west of Potnovo, merely suggesting the existence of a grabben like structural trough.

The unconsolidated Quaternary sediments consisting of laterites, fluvial and coastal alluvium and sand form the principal and potential aquifers. Groundwater occurs under water table as well as under confined conditions. The aquifers are extensive and interconnected. The aquifers of porous formations are of two types. One is pheratic aquifer and the other is confined aquifer. The shallow pheratic aquifers occurring within 10 to 12 metres below ground level is generally fresh in nature and below 12 m the quality is poor. The aquifers occurring 100 m below ground

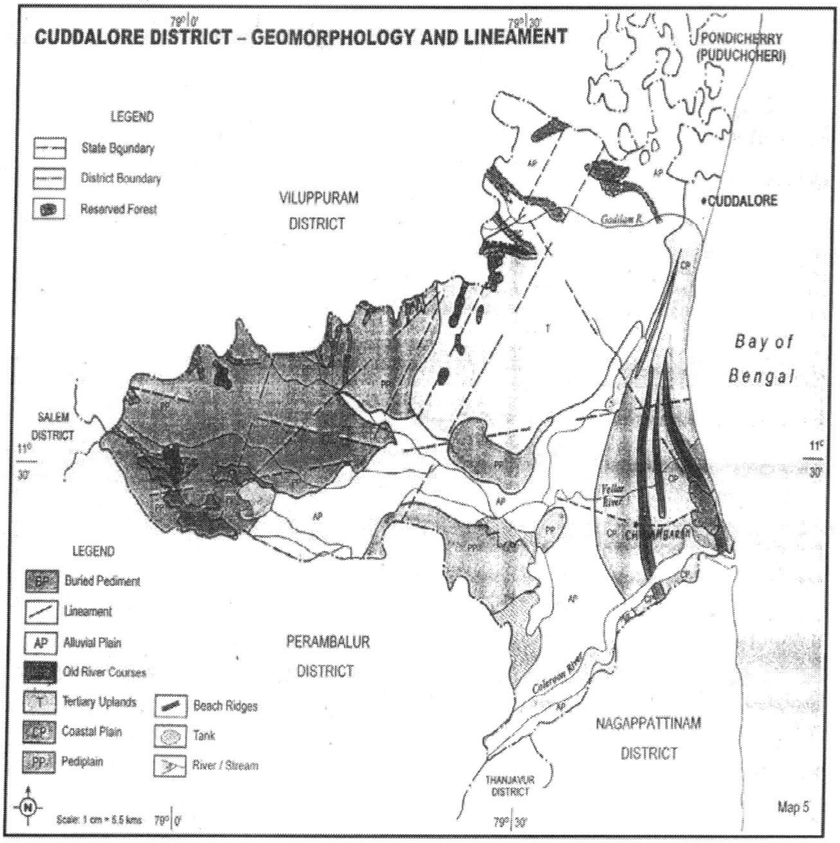

Fig. 3. Geomorphology map of cuddalore district.

level are grouped as deep confined aquifers. Even the aquifers encountered at these depths in and around Chidambaram town area are poor. The pre-monsoon depth to water table ranges from 1 to 2 m below ground level. The main source of groundwater recharge is the precipitation. The rivers Vellar and Kollidam also contribute recharge to the aquifer systems. The area experiences maximum rainfall during north-east monsoon [October to December]. The annual normal rainfall of Cuddalore district is 1212 mm.

GROUNDWATER EXPLORATION

The groundwater exploration may be broadly classified as aerial, surface, subsurface and esoteric methods. Groundwater exploration comprises a number of stages. The first stage is identification of suitable well sites by integrated hydrogeological and geophysical method while the next stage is drilling and the last stage is development. Since it is a sedimentary terrain with quality problem, the saline water zones have to be sealed by constructing the well through well assembly consisting of slotted and plain pipes. For this purpose, identification of precise disposition of various lithological layers, various aquifer zones, and their interface and water quality changes are essential.

ELECTRICAL WELL LOGGING

There are several methods of well logging. The common methods employed in groundwater exploration is the spontaneous potential log [SP] and resistivity logs – lateral log, short normal and long normal logs. In this study, electrical well logging of 14 exploratory or pilot bore holes have been carried out using portable spot logger with SP and resistivity modes. Under resistivity log modes both short normal [SN-16''] and long normal [LN-64''] have been conducted [Figs. 4 and 5]. SP and resistivity measurements have been recorded for every 2 metres. For

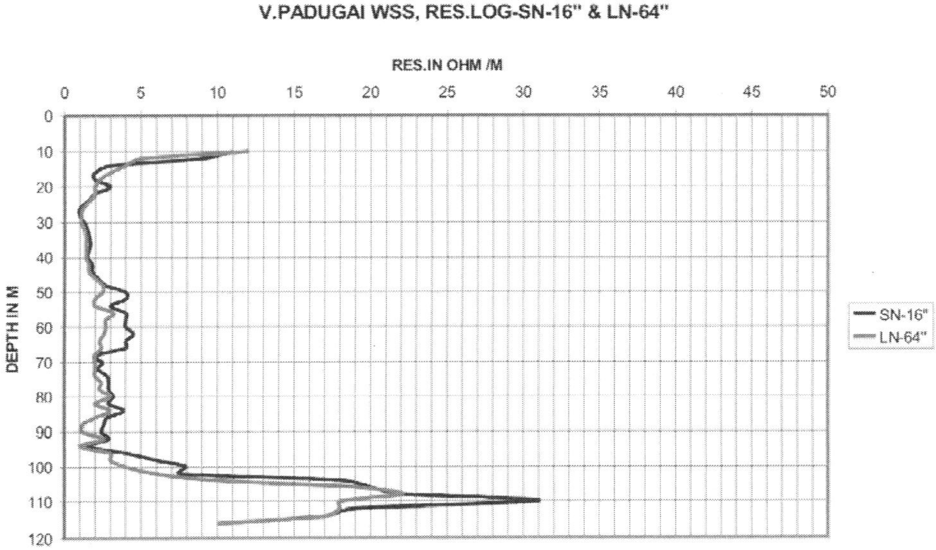

Fig. 4. Vallampadugai WSS RES LOG SN-16" & LN-64".

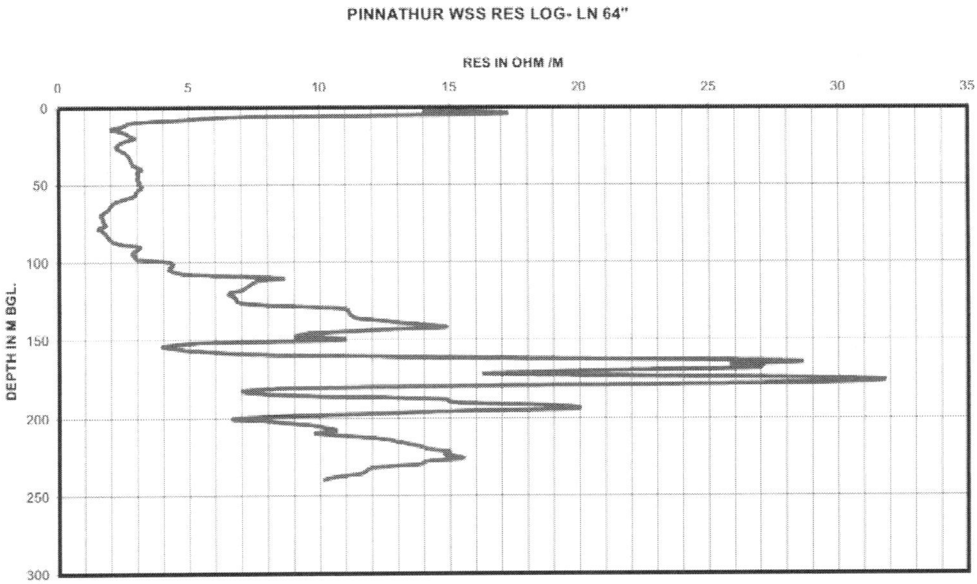

Fig. 5. Pinnathur WSS, RES LOG- LN- 64".

conducting this procedure both DC digital and microprocessor based signal stacking digital resistivity meters– model SSR-MP-AT S of IGIS make, Hyderabad have been used. SP measurements are highly useful in deciphering the saline and clay predominant zones. Single point resistance logs are useful for obtaining information about lithology. Saline water saturated zone produces sharp negative SP and low resistance. Normal resistivity logs are 4 electrode system widely used in groundwater hydrology which measures the apparent resistivity. There are two types of normal resistivity logs. Short normal [SN-16”] and Long normal [LN-64”]. The normal resistivity logs clearly distinguish the clay, saline water and freshwater zones. The interpretation of electric logs has been done in correlation with the litholog of boreholes.

WELL DEVELOPMENT AND COMPLETION

Based on the electrical well logging findings, the tube wells have been constructed. Wells were developed by compressor and the yields of wells have been ascertained by 90° V-notch methods. Finally, the wells were completed. Tube wells yield contour map has been prepared [Fig. 10].

ISO RESISTIVITY MAPS

Commonly and usually the ISO resistivity maps are prepared from the apparent resistivity measurements made on the surface [surface investigation]. An innovative attempt has been made to prepare the resistivity contour maps from the apparent resistivity measurements taken by the electrical well logging – normal resistivity logs of all the 14 boreholes [subsurface investigation]. ISO resistivity maps for the depths of 10, 100, 150, 160, 180 and 200 m have been prepared by the surfer package. The low and high resistivity anomalies have been identified and demarcated [Figs. 6 and 7].

GROUNDWATER QUALITY PROFILE

Resistivity is the inverse of conductivity and as saline water being highly conducting which has low resistivity of 0.1-4 Ωm, the very low resistivity zones may indicate saline water aquifers. Based on the electrical logging resistivity data and ISO resistivity maps, two tentative groundwater quality profiles have been prepared. The I profile is along south-west to north-east direction from Siragilandhanallur in the upstream, Vallathurai, Vallampadugai and Jeyangondappattinam in the downstream [Fig. 8]. The II profile is along south-west to north-east direction from Sakthinagar, Paradesinagar, Mariyappanagar, Muthaiahnagar, Srichakra Avenue and Pinnathur in the downstream [Fig. 9]. In the profile I it can be observed that the very deep confined freshwater aquifer occurs at a depth of 290 m below ground level, i.e., at Jeyangondappattinam. The depth decreases gradually in the upstream directions, i.e. at Siragilandhanallur. In the profile II, the same trend is seen, but the deep freshwater confined aquifers are not encountered at very deeper depths as at Jeyangondappattinam. But the depth of occurrence of deep freshwater confined aquifers increases in the downstream directions.

RESULTS AND DISCUSSIONS

The low and high resistivity anomalies have been correlated with the lithology of boreholes and the shallow fresh water aquifers, the deep saline water zones and the very deep freshwater

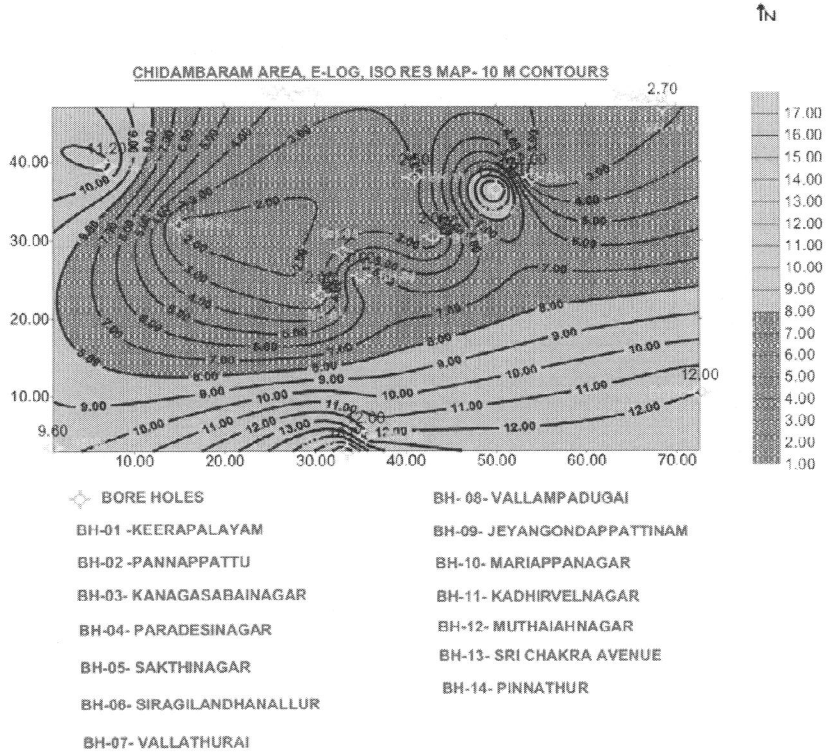

Fig. 6. ISO resistivity map – 10 M contours.

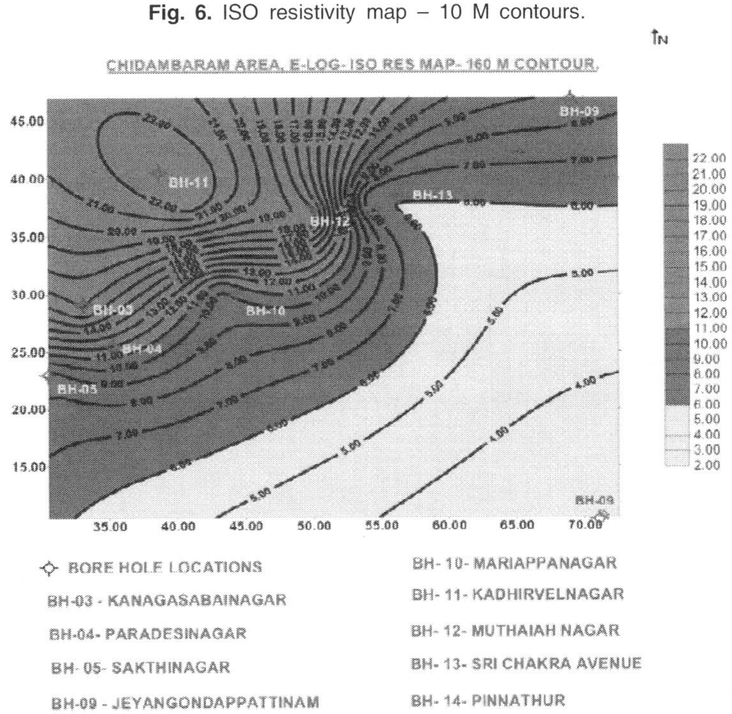

Fig. 7. ISO resistivity map – 160 M contours.

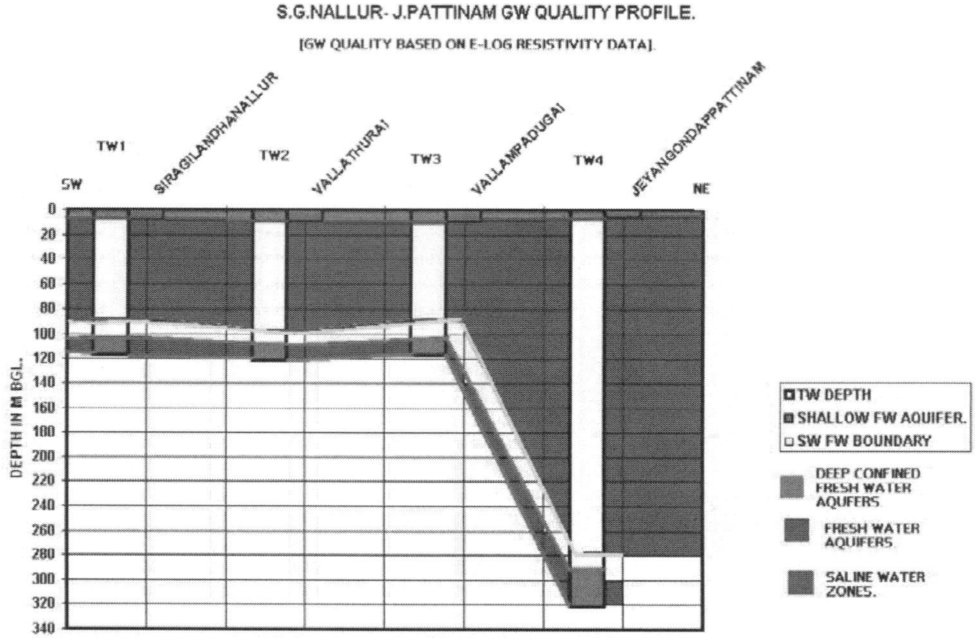

Fig. 8. S.G. Nallur – J. Pattinam GW quality profile.

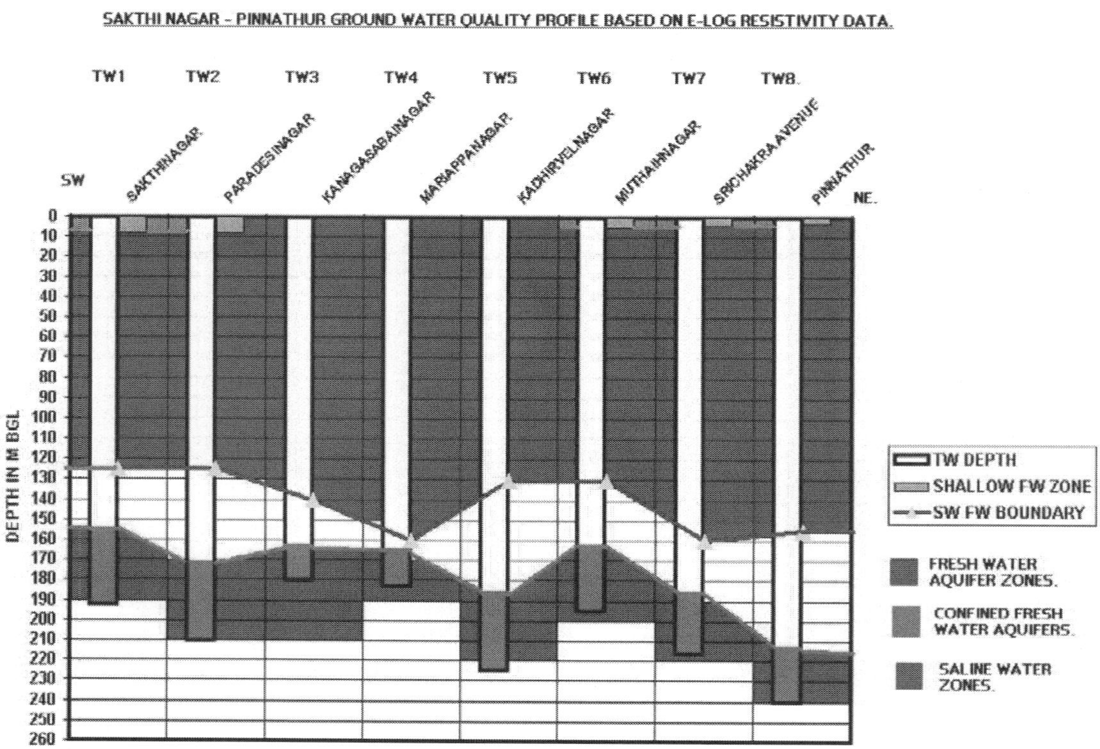

Fig. 9. Sakthinagar – Pinnathur GW quality profile.

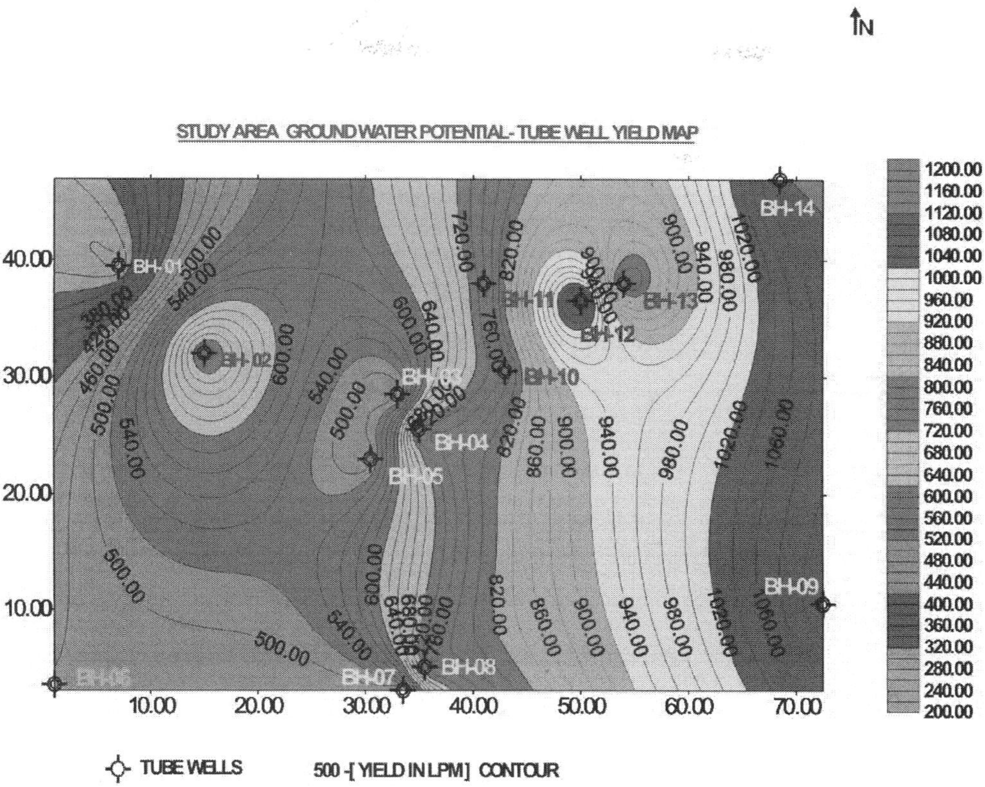

BH-01-KEERAPALAYAM	BH-08-VALLAMPADUGAI
BH-02-PANNAPPATTU	BH-09-JEYANGONDAPPATTINAM
BH-03-KANAGASABAINAGAR	BH-10-MARIAPPANAGAR
BH-04-PARADESINAGAR	BH-11-KADHIRVELNAGAR
BH-05-SAKTHINAGAR	BH-12-MUTHAIAHNAGAR
BH-06-SIRAGILANDHANALLUR	BH-13-SRI CHAKARA AVENUE
BH-07-VALLATHURAI	BH-14-PINNATHUR

Fig. 10. Study area groundwater potential.

aquifer zones have been identified and demarcated. In 6 m contour map three low and four high resistivity anomalies observed. The low resistivity anomalies indicate saline water zone and the high resistivity anomaly zones indicate freshwater zone. The low resistivity anomaly at TW-6 Siragilandhanallur area is due to clayey formations. The south-eastern and north-western zones water is fresh in nature. Up to 10 m contours almost same trend continues. 10 m below ground level the water is generally saline in the entire study area except TW-01 Keerapalayam zones. The 110 m contour map of the study area reveals that except south and south-western zones all are saline water zones. In the 120 m contour map the southern, south-western and north-western zones display high resistivity indicating freshwater zones. The low resistivity at TW-6 is due to clay formations. In the 160 and 180 m contour maps except TW-9 zone, i.e., south-eastern area all the zones are freshwater zones. In the 200 m contour map except TW-9 zone all the other areas are freshwater zones. The low resistivity observed at TW-14 is due to clay. At TW-9

Jeyangondappattinam area the freshwater zone has been encountered at a depth of about 260 m below ground level. Except Keerapalayam area which is situated on south side bank of the Vellar river, where there is no salinity up to a depth of 31 m, all the other 13 tube well water is saline in nature at depth of 30 m. The top shallow aquifers [depth of 10 to 12 m bgl] are fresh in S.G. Nallur, Vallampadugai, Vallathurai, J.G. Pattinam which are situated on the north bank of the Kollidam River and Paradesinagar areas. In all other zones even the top shallow aquifers are saline. The depth of occurrence of deep freshwater confined aquifers varies from 100 to 110 m in the south and south-western zones. In and around Chidambaram town areas the depth of occurrence of deep freshwater confined aquifers range from 150 to 160 m. Groundwater potential of the study area is good. The minimum and the maximum discharge of the tube wells are 276 and 1118 lpm, respectively.

Table 1: ISO resistivity map details.

Sl.No.	Resistivity Contours– Depth in M	Apparent Resistivity in OHM		Resistivity Anomalies		Inference
		Minimum	Maximum	Low	High	
1.	6	1	20	3	4	A high res anomaly sandwitched in a low res anomaly-TW-1,4,6,7,8,9 & 12 freshwater zone- TW-6 low res because of clay.
2.	8	1	16	2	4	TW-1,4,6,7,8,9 & 12 freshwater zone
3.	10	1	17	2	1	A high res anomaly sandwitched in a low res anomaly-TW-1,6,7,8,9 & 12 freshwater.
4.	100	1	22	3	2	TW-6,7,8,10,11 & 12 freshwater zone
5.	110	1	18	3	1	TW-6,7,8 freshwater zones
6.	120	2	16	4	2	TW-2,6,7 & 8 freshwater zone
7.	160	2	22	-	1	TW-3,4,5,10,11,12,13 & 14 freshwater zones
8.	180	3	21	1	2	TW 3,4,5,10,11,12,13 & 14 freshwater zones
9.	200	4	18	1	1	TW-4,11,13 & 14 freshwater zones

TW-9 saline up to 260 metres – Freshwater zone below 260 m.

CONCLUSION

The study area, a part of Chidambaram and Kattumannarkoil taluks of Cuddalore district, in Tamilnadu is a coastal sedimentary terrain with a geographical extent of 250 km^2. The entire area is covered by Quaternary sedimentary formations of fluvial and fluvio marine origin of Recent to Sub recent age. The alluvium and sand form the principal and potential aquifers. The quality of water in most parts of Potnovo, Keerapalayam and Kumaratchi unions is brackish. In some pockets the shallow aquifer water is good and fresh, but in most parts of Chidambaram town areas even the shallow aquifer is very poor and saline. An integrated hydrogeological and subsurface geophysical investigations have been carried out in some parts of the area at 14 locations to explore the deep freshwater confined aquifers for drinking and irrigation purposes. Electrical well logging have been conducted in 14 exploratory bore holes, the deep confined and shallow pheratic freshwater aquifers were explored, tube wells have been constructed, developed and completed. The wells have discharged copious yield in the range of 276 to 1118 lpm and the quality of water is very good, fit for drinking and irrigation purposes. An innovative attempt has been made to prepare the ISO resistivity maps by incorporating the resistivity data obtained from the electrical well logging of 14 boreholes. The fresh and saline water zones have been identified and demarcated based on the resistivity contours. Two groundwater quality profiles have been prepared based on the electro log resistivity data. From the profiles it can be revealed that the depth of freshwater confined aquifers increases considerably towards east that is in the downstream directions. From the ISO resistivity maps, we can get the groundwater quality scenario of the entire area. The thickness of the deep confined freshwater and shallow pheratic aquifers can be delineated. The study reveals that the thickness of shallow freshwater pheratic aquifers range from 10 to 12 m and the very deep freshwater confined aquifers range from 100 to 110, [Vallampadugai zones] 150 to 160 [Chidambaram town areas] and 260 to 290 [J.G. Pattinam zone]. Thus, borehole geophysics plays a vital role in the exploration of freshwater aquifers.

REFERENCES

1. Badrinarayanan, T.S. and Venkatesan, R. (2007). Exploring the shallow and deep fresh water aquifers along the Kollidam river bank in parts of Kumaratchi union, Cuddalore district, Tamilnadu-Hydrocare-Chidambaram [Page 160-173].
2. Badrinarayanan, T.S. and Venkatesan, R. (2007). Exploring the deep fresh water aquifer in a sedimentary saline water environment–A case study from Pannapattu of Chidambaram taluk, Cuddalore district, Tamilnadu--WMEC Hyderabad. [Page 38-45].
3. Badrinarayanan, T.S. and Senthilkumar, G.R. (2006). A systematic groundwater exploration to provide protected safe drinking water supply in a sedimentary terrain – Some case studies in Kammapuram union of Cuddalore district-ICSTEP-Coimbatore.
4. Carg, S.P. Groundwater and tube wells, third edition.
5. CGWB report, (1997). Groundwater resources and development prospects in South Arcot Vallalar district, Tamilnadu.
6. Setty, G.Y., Narayana Raja, V., Arulprakash, V., Subburaj, A. (2007). CGWB, SECR, Chennai Integrated groundwater exploration using drilling and bore hole logging techniques-A case study, WMEC, Hyderabad [Page 61-71].
7. Banerjee, Saon., Trivedi, S.M. and Chandrasekharan, H. (2002). Geoelectrical investigations for the assessment of groundwater salinity in an agricultural farm in Haryana-Hydrology J., **25,** [Page n 41-48].

8

A Study on Trace Metal Pollution in the Gadilam River Basin, Tamil Nadu, India

Prasanna, M.V.,*[1] Chidambaram, S.,[2] Nagarajan, R.,[1] Elayaraja, A.,[3] and Rajalingam, S.[1]

[1]*School of Engineering and Science, Curtin University of Technology, Miri, Sarawak, Malaysia–98009*

[2]*Department of Earth Sciences, Annamali University, Annamali Nagar–608002*

[3]*School of Business, Curtin University of Technology, Miri, Sarawak, Malaysia–98009*

**Email: prasanna@curtin.edu.my*

ABSTRACT

To understand the variation of trace metals such as Iron (Fe), Lead (Pb), Nickel (Ni), Bromide (Br), Iodide (I) and Aluminium (Al) in the study area after the monsoon period, samples were collected and analyzed. The concentration levels of different parameters have been compared with the WHO (1993) standard. Interpretation of data shows that some trace metals such as Al, Ni and Pb exceed the acceptable limit of WHO (1993) standard.

Keywords: Trace metals, pollution, permissible limit, Gadilam river.

INTRODUCTION

Trace element studies of groundwater have attracted researches for a variety of reasons, its significance related to public health problem, their need in relation with plant growth and the mechanism of metal transport in aqueous environment. Trace element in groundwater are defined as chemical elements dissolved in water in minute quantities, always or almost always in concentration less than 1 mg of trace element in one litre of water (USGS, 1993). In igneous and sedimentary rock several trace elements can be accommodated in the crystal structure of different minerals. Fe, Mn, Cu, etc. are expected to occur as major trace elements in the rocks and sediments. The distribution and occurrence in the groundwater depends upon the degree of weathering and mobility of these elements during weathering (Handa, 1986). The study area Gadilam river basin is located in Cuddalore and Villupuram district, Tamil Nadu, India. The area is bounded between latitudes 11°30'N – 11°55'N and longitudes 79°0'E – 79°47'E. It

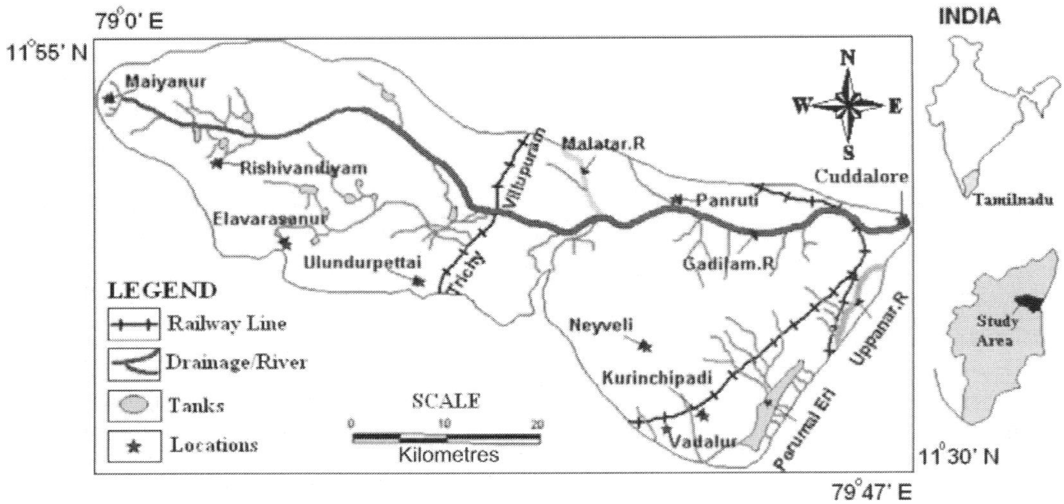

Fig. 1. Location map of the study area.

covers a total area of about 1,394 sq.km (Fig. 1). The location of Neyveli mining region and Veeranam Lake from which groundwater extraction is at its maximum falls in this environment. There are group of industries sparsely distributed which generate multifacet chemicals and raw materials is located along the downstream of the river Gadilam. The Gadilam river basin covers different stratigraphic units, viz. Archaean, Cretaceous, Tertiary to Recent alluvium (Fig. 2). Archaean complex consists of garnetic gneiss, charnockite, leptinites and schists. Limestone constitutes the cretaceous formation and cuddalore sandstone of the tertiary age are the litho

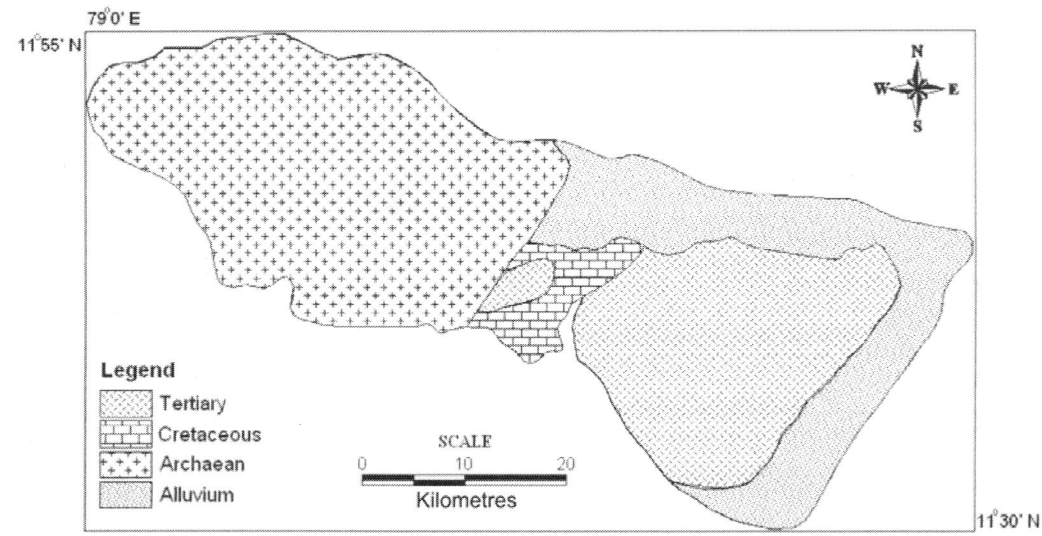

Fig. 2. Geology map of the study area.

units of the area. The western part of the study area is composed of hard rock and the eastern part by sedimentary formation. It is inferred as a faulted contact between the hard rock and the sedimentary formation (Aravindan *et al.*, 2004a). The basin lies between the Ponnaiyar in the north and Vellar in the south.

MATERIALS AND METHODS

Twenty-nine groundwater and seven surface water (river water) samples were collected and analysed during post-monsoon period (Fig. 3). Before estimation of dissolved metals the samples were filtered through 0.45 μm membrane filter paper using vacuum pumps. The samples were then spiked with the standard solution of the metals chosen for analysis. This standard addition method was adopted to achieve greater accuracy in the analysis of trace metals such as Fe, Pb, Ni, Br, I and Al. Fe was analysed in double beam Atomic Adsorption Spectrophotometer (AAS), (Model-SL 176, Elico Limited, Hyderabad, India) with spectral range 185-930 nm and ± 0.5 nm accuracy in Department of Earth Sciences, Annamalai University. Pb, Ni and Al was analysed in AAS (Model-Shimadzu-AA-6800) and Br and I was analysed in ion electrodes (Model-CONSORT) at School of Environmental Sciences, Jawaharlal Nehru University, New Delhi.

RESULT AND DISCUSSION

To understand the variation of trace metals such as Fe, Pb, Ni, Br, I and Al in the study area after the monsoon period, samples were collected and analysed (Fig. 4).

IRON

Iron is an essential element in the metabolism of animals and plants. If present in water in excessive amounts, however, it forms red oxyhydroxide precipitates that stain laundry and

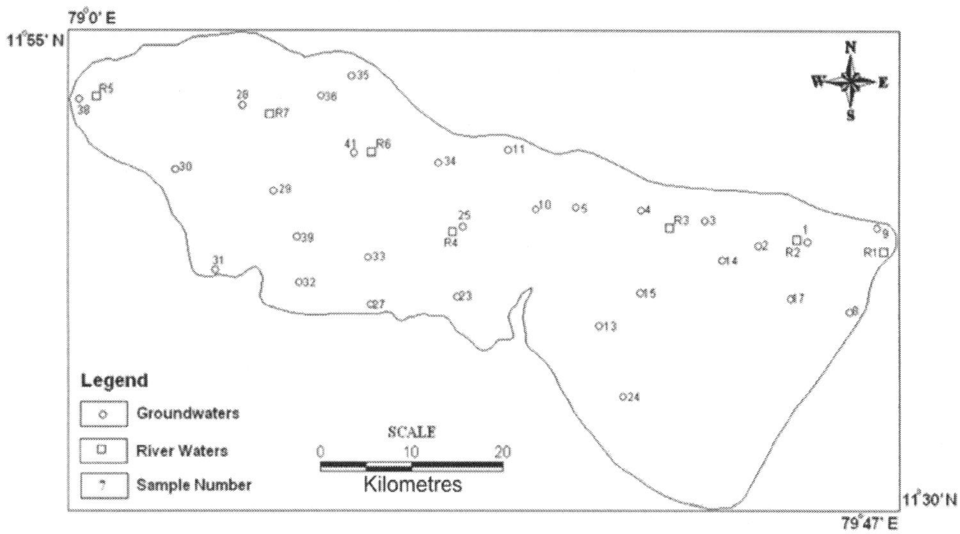

Fig. 3. Sample Location map

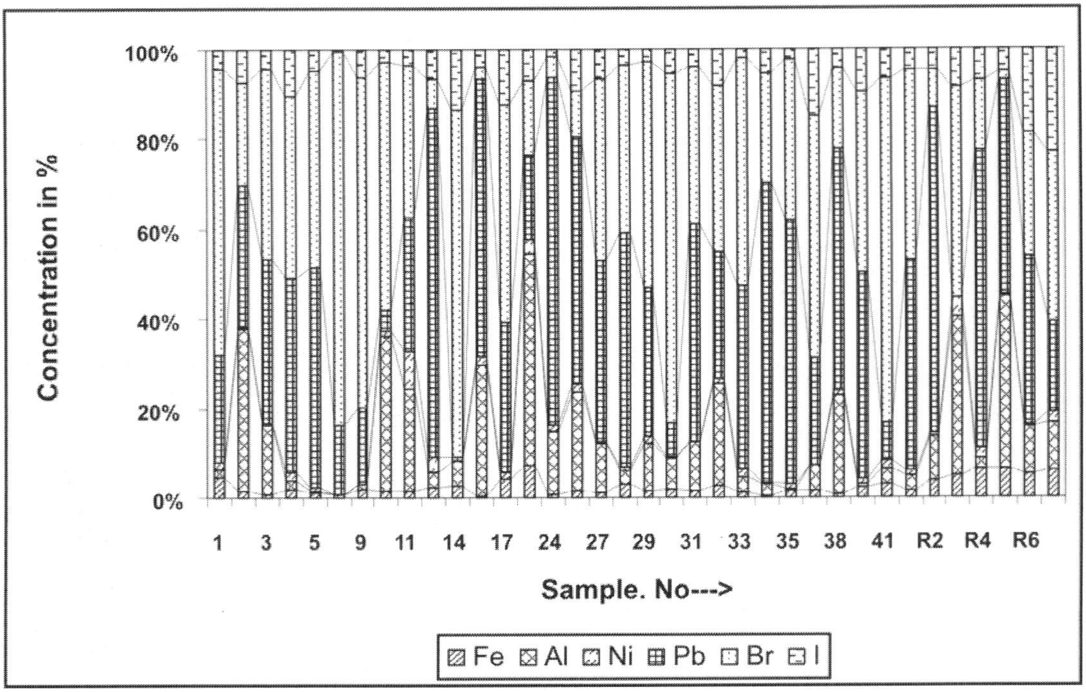

Fig 4. Trace metal concentrations in the study area.

plumbing fixtures and, therefore, it is an objectionable impurity in domestic and industrial water supplies. For this reason, iron determinations are commonly included in chemical analyses of water. A recommended upper limit for iron in public water supplies is 0.3 mg/L (NAS-NAE, 1972).

Fe play a biogeochemical role in the life cycle of plants and animals, their presence in groundwater with high concentration is undesirable to use (Ballukaraya and Ravi, 1999). In Alluvium formation, Fe was ranges from 0.09 mg/L to 0.41 mg/L with an average of 0.17 mg/L (Table 1). Higher concentration of Fe in groundwater derived from nearby laterites capping in Cuddalore and Panruti regions. In Tertiary formation, Fe was ranges from 0.02 mg/L to 0.29 mg/L with an average of 0.10 mg/L. Higher concentration of Fe in groundwater may be attributed to the weathering of marcasite and iron oxide coasting in sediments which might have contributed this ion (Anandhan, 2005). In Archaean formation, Fe ranges from 0.03 mg/L to 0.19 mg/L with an average of 0.11 mg/L. The Fe in all of the surface water bodies (river water) exceeds the average levels in the world's rivers (Taylor and McLennan 1985) with ranging from 0.11 mg/L to 0.41 mg/L with an average of 0.21 mg/L. Generally, higher concentration was observed in Alluvium followed by Tertiary and Archaean formation.

ALUMINIUM

Aluminium is the most abundant metallic element and constitutes about 8% of the earth's crust. Aluminium salts are widely used in water treatment as coagulants to reduce organic matter, color, turbidity and microorganism levels. Aluminum occurs in substantial amounts in many silicate igneous rock minerals such as the feldspars, the feldspathoids, the micas and many amphiboles.

In Alluvium formation, Al ranges from 0.011 mg/L to 2.800 with an average of 1.054 mg/L, whereas in Tertiary formation, it ranges from 0.004 mg/L to 2.300 mg/L with an average of 0.917 (Table 1). In Archaean formation, Al ranging from 0.032 mg/L to 2.740 mg/L with an average 0.741 mg/L. In surface water, it ranges from 0.097 mg/L to 2.380 mg/L with an average of 0.723. Generally, higher concentration was observed in Archaean followed by Alluvium and Tertiary formation.

Table 1: Comparison of trace element with WHO (1993) standard.

Parameters	WHO(1993) Standard		Alluvium (in mg/l)	Tertiary (in mg/l)	Archaean (in mg/l)	Surface water (in mg/l)
	Maximum Acceptable	Minimum Acceptable				
Fe	0.3	1	0.090-0.410	0.020-0.290	0.030-0.190	0.110-0.410
Al	0.3	—	0.011-2.800	0.004-2.300	0.032-2.740	0.097-2.380
Ni	0.01	—	0.001-0.742	0.015-0.134	0.005-0.163	0.005-0.134
Pb	0.05	0.1	0.370-8.519	0.001-5.556	0.370-6.667	0.001-4.444
Br	—	—	1.630-45.600	0.204-2.020	0.693-4.580	0.118-4.000
I	—	—	0.230-0.693	0.111-0.353	0.180-0.644	0.267-0.552

NICKEL

The concentration of nickel in drinking water is normally less than 0.02 mg/L, although nickel released from taps and fittings may contribute up to 1 mg/L. In special cases of release from natural or industrial nickel deposits to the ground, the nickel concentrations in drinking water may be higher (WHO, 2005). In Alluvium formation, Ni ranges from 0.001 mg/L to 0.742 mg/L with an average of 0.130 mg/L. In Tertiary formation, Ni ranging from 0.015 mg/L to 0.134 mg/L with an average of 0.089 mg/L, whereas in Archaean formation, it ranging from 0.005 mg/L to 0.163 mg/L with an average of 0.072 mg/L. The Ni in all of the surface water bodies (river water) exceeds the average levels in the world's rivers (Taylor and McLennan, 1985). It ranges from 0.005 mg/L to 0.134 mg/L with an average of 0.066 mg/L. Generally, higher concentration was observed in Alluvium followed by Archaean and Tertiary formation.

LEAD

A study by Schock (1980) of the factors that might affect the corrosiveness of water toward lead pipe indicated that maintaining a tap-water lead concentration below 50 μg/L might be possible by careful control of pH and alkalinity in water supplies. The guideline value for Pb is 0.01 mg/L (WHO, 2003). In Alluvium formation, Pb ranges from 0.370 mg/L to 8.519 mg/L with an average of 3.128 mg/L, whereas in Tertiary formation, it ranges from 0.001 mg/L to 5.556 mg/L with an average of 2.531 mg/L. In Archaean formation, Pb ranges from 0.370 mg/L to 6.667 mg/L with an average of 3.229 mg/L. In surface water it ranges from 0.001 mg/L to 4.444 mg/L with an average of 2.328 mg/L. Generally, higher concentration was observed in Alluvium followed by Archaean and Tertiary formation.

BROMIDE

Bromine is similar in chemical behaviour to chlorine but is much less abundant. Bromine in natural water is always present as the bromide ion Br^-. The concentration present in seawater is 65 mg/L, an amount sufficient to make bromide an important constituent. Bromide concentrations in rainwater and snow range from about 5 to more than 150 µg/L, and a wider range is given for river water by Fuge (1973). Bromine is extracted commercially from seawater and from brines. The presence of small amounts of the element in freshwater is not known to have any ecologic significance. Bromide is generally most conservative constituents in waters, i.e., they are little affected by chemical reactions such as mineral precipitation or adsorption.

In Alluvium formation, Br ranges from 1.630 mg/L to 45.600 mg/L with an average of 8.606 mg/L. Samples shows higher concentration of Br, which is fall along the coast, may be the possibility of seawater intrusion. In Tertiary formation it ranges from 0.204 mg/L to 2.020 mg/L with an average of 0.761. In Archaean formation, it ranges from 0.693 mg/L to 4.580 mg/L with an average of 2.856 mg/L. In surface water it ranges from 0.118 mg/L to 4.000 mg/L with an average of 1.184. Generally, higher concentration was observed in Alluvium followed by Archaean and Tertiary formation.

IODIDE

Although iodine is not a particularly abundant element, it is widely distributed. It is essential in the nutrition of animals, including humans, and concentrations in natural water have received considerable attention. In Alluvium formation, I value ranges from 0.23 mg/L to 0.69 mg/L with an average of 0.42 mg/L. In Tertiary formation, iodide value ranges from 0.11 mg/L to 0.35 mg/L with an average of 0.26 mg/L, whereas in Archaean formation it ranges from 0.18 mg/L to 0.64 mg/L with an average of 0.38. In surface water, I range, from 0.26 mg/L to 0.55 mg/L with an average of 0.36 mg/L. Generally, higher concentration was observed in Alluvium followed by Archaean and Tertiary.

CONCLUSION

The present study evaluated the trace metal content and toxicity in groundwater collected in Gadilam river basin. Metal contamination may lead to a potential danger for the health of human and animal populations in the vicnity of these area. The results of the study show that the groundwater in these area is significantly contaminated by Al, Ni and Pb which may give rise to various health problems and diseases. The detected levels of total metal contamination in some of the samples were found to exceed international threshold values.

ACKNOWLEDGEMENTS

The authors wish to express thank to DST-SERC for provided the necessary financial support to carry out this study. We are very much thankful to the CWRDM, Kozhikode for their cooperation. We are also thankful to the SES, JNU, New Delhi Dept. of Earth Sciences, Annamalai University for sample analysis.

REFERENCES

Anandhan, P. (2005). Hydrogeochemical studies in and around Neyveli mining region, Tamilnadu, India. Ph.D. Thesis, Department of Earth Sciences, Annamalai University, pp.189.

Aravindan, S., Manivel, M. and Chandrasekar, S.V.N. (2004a). Groundwater quality in the hard rock area of the Gadilam river basin, Tamilnadu. *J. Geol. Soc. India.* **63**: 625-635.

Ballukraya, P.N. and Ravi, R. (1999). Characterization of groundwater in the unconfined aquifers of Chennai City, India. Part I Hydro geochemistry. *J. Geol. Soc. India.* **54:** 1-11.

Fuge, R. (1973). Bromine-I, in Wedepohl, K.H., ed.; Berlin, Handbook of geochemistry. II-3 Springer-Verlag, pp. 351-1-351-4.

Handa, B.K. (1986). Water quality and water pollution problems on the Indian sub pollution and monitoring. Csio Chandigarh, pp. 473-492.

National Academy of Sciences-National Academy of Engineering. (1972). Water quality criteria 1972: Washington, D.C National Academy of Sciences. pp. 594.

Schock, M.R. (1980). Response of lead solubility to dissolved carbonate in drinking water. *J. Am. Water Works Association.* **72:** 695-644.

Taylor, S.R. and McLennan, S.M. (1985). The Continental Crust: its Composition and Evolution. *Blackwell*, Oxford, UK. pp. 312.

USGS. (1993). Nation water summary 1990–1991; stream water quality. *U.S. Geol. Suru.* Water supply paper no. 2400. pp. 590.

WHO. (1993). Guidelines for drinking water quality. World Health Organization, *Geneva.* pp.171.

WHO. (2003). Lead in drinking-water. Background document for preparation of WHO Guidelines for drinking-water quality. Geneva, World Health Organization (WHO/SDE/WSH/03.04/9).

WHO. (2005). Nickel in drinking-water. Background document for development of WHO Guidelines for drinking-water quality. Geneva, World Health Organization (WHO/SDE/WSH/05.08/55).

9

Groundwater Quality Assessment of Andimadam Block, Perambalur District, Tamil Nadu

PRABHAKARAN, R.,* SUBRAMANIAN, P., CHIDAMBARAM, S., KUMAR R.S., SRINIVASAMOORTHY, K. AND SELVARAJ, B.

Department of Earth Sciences, Annamalai University, Annamalai Nagar–608002

**Email: geoprabha@yahoo.com*

ABSTRACT

A study on geochemical characterization of groundwater and its suitability for drinking and irrigation purposes was carried out in Andimadam block of Tamil Nadu. Representative of 34 groundwater samples were collected from open wells and bore wells. The samples were analysed for various water quality parameters such as pH, electrical conductivity, total dissolved solids, calcium, magnesium, sodium, potassium, bicarbonate, sulphate, chloride and phosphate. Sodium, Calcium and Bicarbonate are the dominant ions in the groundwater chemistry. The chemical composition of groundwater is controlled by rock water interaction. The chemical quality was evaluated for drinking use following the guidelines of WHO and ISI. Groundwater is generally potable except in few locations. As per USSL classification of water quality, mostly it falls in C_1S_1 and C_2S_1 category.

Keywords: Groundwater, quality, assessment, Andimadam, Perambalur.

INTRODUCTION

The water is the foremost requirement of all human beings. Groundwater is one of the most important and widely distributed natural resources which is considered as supplemental resource to surface water to meet the domestic, agricultural and industrial requirements. Rapid growth of population, expansion of irrigation and increasing trend of industrialization have contributed towards rising demand for groundwater in many areas. The quality of water is getting importance due to different types of pollution in many ways by natural, agricultural and anthropogenic activities. Geochemical processes in groundwater involve the interaction of rocks with water, leading to the higher concentration of chemical elements in water. The principles governing the chemical characteristics of groundwater were well documented in many parts of the world (Saha *et al.*, 2008, Subba Rao, 1983). The present work is aimed in assessing the groundwater quality with respect to drinking and irrigation purposes.

STUDY AREA

The Andimadam block which falls in Vellar sub-basin, located in the central part of Cauvery basin. It lies in between longitudes 79°15' to 79°30'E and latitudes 11°22' to 11°30'N. Physiogrpahically the area is flat with gentle slope towards the eastern direction. This area shows dentritic, trellis and combination of dentritic and parallel drainage patterns. Andimadam block is underlain by geological formations ranging in age from cretaceous to recent, i.e., Kallamedu, Niniyur and Cuddalore formations. They are composed of sandstone, mottled sandstone, ferrunginous sandstone, sandstone with clay intercalation, limestone, marl, shell, laterites, alluvium and red soils (Fig. 1).

METHODOLOGY

During the pre-monsoon season, June 2008, thirty-four groundwater samples were collected. The samples were collected in 1000 mL pre-washed bottles with airtight lids. The samples were analysed for pH, EC, TDS, major cations and anions by adopting standard procedures (APHA, 1995). To assess the groundwater suitability for drinking purposes, it is interpreted with WHO (1996) and ISI (1995) guidelines. For its suitability for irrigation purposes, SAR, RSC and Na% were calculated and plotted in the USSL diagram by using WATCLAST (Chidambaram et al., 2003).

RESULTS AND DISCUSSION

Hydrogeochemistry of Groundwater

pH of the water is an indicator of its quality and geochemical equilibrium for solubility calculation (Hem, 1985). The pH analysed samples varies from 5.69 to 7.89 with an average of 6.52. The electrical conductivity depends upon temperature, concentration and types of ions present in the water (Hem, 1985). The EC values varies from 77.77 to 827.23 ms/cm with an average of 334.79 ms/cm. The difference between the values may reflect the wide variation in the activities and chemical processes prevailing in the region. The total dissolved solids ranges from 49.72 mg/L to 529.43 mg/L with an average of 214.27 mg/L. The water samples have been classified based on the concentration of TDS (ICMR, 1975). According to ICMR classification most of the study area samples fall in desirable for drinking. Sodium is the dominant cation present in groundwater followed by Ca, K and Mg. Na is found to be the dominant cation it may be due to the weathering of alkali feldspar in rocks. Na concentration varies from 3.3 to 222.6 mg/L. with an average of 16.34 mg/L. The results of chemical composition of groundwater are given in Table 1. The Ca concentration is varying from 5 mg/L to 87 mg/L with an average of 31.03 mg/L. The concentration of Ca is due to interaction of minerals like feldspar, amphiboles, pyroxenes and weathering process. Potassium in groundwater is generally less due to its higher solubility

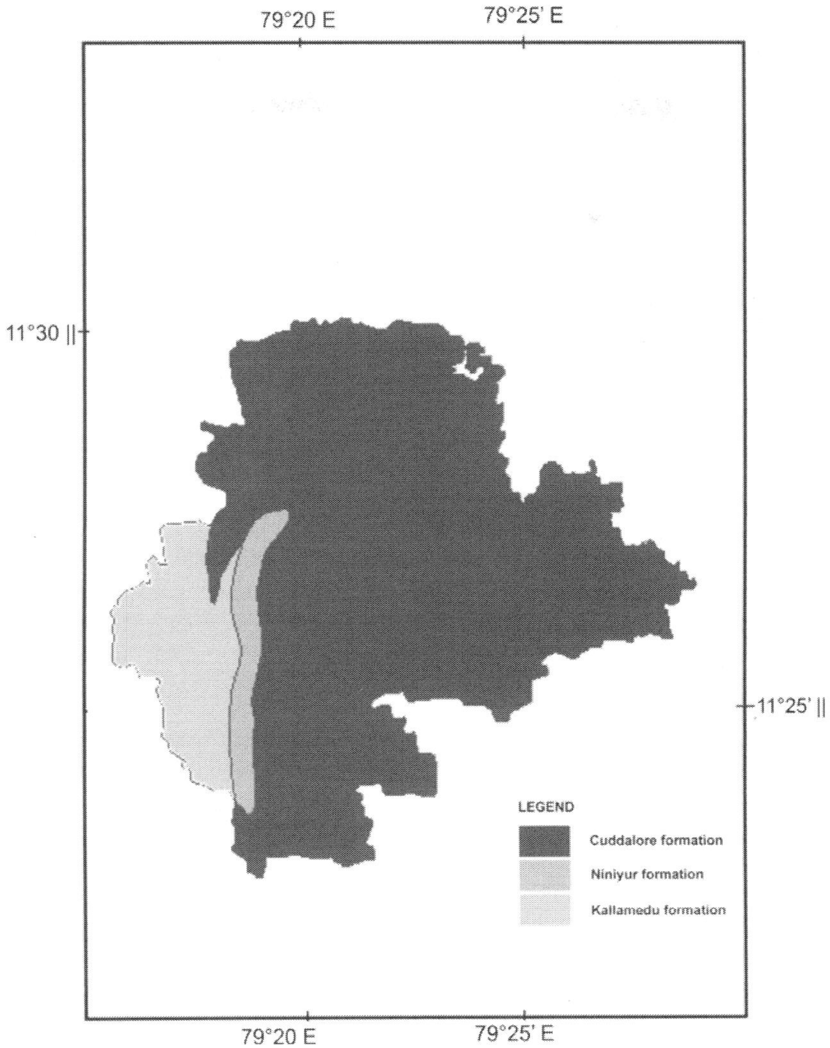

Fig. 1. Geology map of the Andimadam block.

(Bouwer, 1978). K concentration is varying from 0.98 to 26.7 mg/L with an average of 3.85 mg/L. Magnesium in groundwater is mainly due to the leaching of magnesium bearing minerals. Mg concentration is found to varies from 1 to 25.6 mg/L with an average of 7.91 mg/L.

Bicarbonate is the dominant anion followed by chloride, phosphate and sulphate. The bicarbonate concentration in groundwater is varying from 11.23 to 398.2 mg/L with an average of 89.64 mg/L. The Cl concentration is varying from 22.3 to 267.8 mg/L with an average of 67.16 mg/L. Cl concentration in groundwater is due to the influence of domestic sewage and industrial effluents (Karanth, 1987, Craig and Anderson 1979). The phosphate concentration is varying from 0.011 to 0.67 mg/L with an average of 0.19 mg/L. Sulphate concentration is varying from 0.02 to 0.89 mg/L with an average of 0.11 mg/L. Apart from the natural rock sources,

Table 1: Chemical composition of groundwater (All values in mg/L except EC (µs/cm) and pH)

S.No	Name of the Location	EC	pH	Ca	Mg	Cl	K	HCO_3	SO_4	Na	PO_4	H_4SiO_4	TDS
1.	Varadharajanpettai	239.5156	6.17	16	22	68.2	2.5	36.6	0.05	8.9	0.34	1.2	153.29
2.	Andimadam	303.9219	6.02	12	5.6	56.99	2.6	112	0.09	7.4	0.33	0.1	194.51
3.	Vilandai (E)	578.7969	6.4	56	11.2	77.9	2.3	214	0.89	7.9	0.34	2.2	370.43
4.	Periyakrishnapuram	209.5781	6.43	33	6.7	45.7	2.4	36.6	0.06	8.9	0.67	2.5	134.13
5.	Thirukalapur	827.2344	6.34	14	11.1	88.09	2.2	398.2	0.89	14.2	0.35	2.6	529.43
6.	Karukkai	338.4672	6.11	22	5.2	36.7	2.2	144.2	0.089	5.6	0.23	2.6	216.619
7.	Nagampanthal	379.3328	6.45	16	5.2	44.7	4.3	168.9	0.043	6.6	0.13	1.2	242.773
8.	Viluthudiyan	384.2625	6.1	64	11.5	48	6.7	113.6	0.078	6.7	0.45	1.6	245.928
9.	Olaiyur	319.9844	6.34	56	8.9	59	6.9	73.2	0.06	5.7	0.23	1.7	204.79
10.	Alagapuram	412.7656	6.89	35	9.9	30.7	5.7	178	0.03	7.8	0.63	2.11	264.17
11.	Sriraman	181.75	6.08	5	1	59.9	2.3	36.6	0.04	11.4	0.23	2.15	116.32
12.	Rangiyam	487.1563	7.1	35	13.5	99.09	5.6	144.6	0.05	16.7	0.22	2.62	311.78
13.	Kodukkur	479.4688	7.89	67	5.8	35.7	13.5	189	0.03	7.7	0.53	1.1	306.86
14.	Vallm	464.6875	6.2	87	6.9	68.9	3.5	124.9	0.08	7.8	0.22	1.6	297.4
15.	Kuvagam	217.9234	5.69	34	1.2	59	3.4	36.6	0.031	6.6	0.34	1.7	139.471
16.	Ayyur	199.1875	6.84	34	13.5	44.8	2.2	24.4	0.194	7.8	0.53	2.256	127.48
17.	Kattathur	566.9938	6.03	45	11.6	267.8	26.7	24.4	0.048	12	0.22	1.808	362.876
18.	Kavarapalayam	77.76875	6.54	5	3.3	22.3	2.2	11.23	0.02	6.6	0.022	1.3	49.772
19.	Andimadam (W)	214.3609	6.66	11	2.4	56.7	2.1	61	0.047	4.6	0.044	1.4	137.191
20.	Vilandai	422.8	6.75	15	11.4	68.9	2.2	167.9	0.06	5.7	0.032	1.6	270.592

Table 1: Continued.

S.No	Name of the Location	EC	pH	Ca	Mg	Cl	K	HCO_3	SO_4	Na	PO_4	H_4SiO_4	TDS
21.	Kovilvalkai	335.2656	6.8	13	12.4	99	2.1	61	0.049	26.7	0.021	2.4	214.57
22.	Agaram	173.55	5.94	15	5.7	67.9	2.4	12.2	0.051	7.7	0.021	2.5	111.072
23.	Silampur	154.3781	7.32	22	5.2	24.6	2.1	35.7	0.03	8.7	0.022	2.55	98.802
24.	Idayakuruchi	234.1719	6.54	45	1.1	45.7	2.1	48.8	0.048	7.7	0.022	1.5	149.87
25.	Marudur	393.7375	6.47	56	6.8	68.9	5.2	112.9	0.07	5.7	0.022	1.6	251.992
26.	Variyankaval	224.7703	5.9	22	2.9	60.4	2.4	36.6	0.041	20.3	0.012	1.6	143.853
27.	Elaiyur	240.1906	6.45	11	1.1	45.7	1.1	89.8	0.07	4.4	0.052	1.6	153.722
28.	North pudukkudi	210.1344	6.59	21	2.5	78.6	5.2	24.4	0.124	6.3	0.022	1.54	134.486
29.	Melur	194.8781	6.9	11	15.2	45.6	1.1	45.89	0.08	4.5	0.012	2.44	124.722
30.	Devanur	198.3188	6.3	34	12.6	56.7	1.1	12.2	0.053	8.8	0.011	2.56	126.924
31.	Kulathur	771.4094	7.01	56	25.6	67.8	1.1	119	0.03	222.6	0.012	2.66	493.702
32.	Ayyur (W)	216.7859	7.02	22	3.3	37.7	0.98	67.93	0.08	5.6	0.023	2.11	138.743
33.	Kuvathur	190.1609	6.55	19	1.2	46.7	1.2	48.8	0.061	3.3	0.022	2.62	121.703
34.	Anikudhichan	539.2875	6.79	46	5.6	198.9	1.2	36.6	0.092	56.7	0.012	1.24	345.144
	Max	**827.234**	**7.89**	**87**	**25.6**	**267.8**	**26.7**	**398.2**	**0.89**	**222.6**	**0.67**	**2.66**	**529.43**
	Min	**77.7688**	**5.69**	**5**	**1**	**22.3**	**0.98**	**11.23**	**0.02**	**3.3**	**0.011**	**0.1**	**49.772**
AVERAGE		**334.794**	**6.51794**	**31.0294**	**7.91471**	**67.155**	**3.84647**	**89.6397**	**0.11056**	**16.3412**	**0.18747**	**1.89012**	**214.268**

sulphates could be introduced through the application of sulphatic soil conditioners (Karanth, 1987).

Suitability for Drinking Purpose

The data obtained from the chemical analyses were evaluated in terms of its suitability for drinking and irrigation purposes. Table 2 shows the range of ionic concentration in the groundwater samples of the study area and prescribed specification of WHO (1996) and ISI (1995).

Table 2: Comparison of chemical composition of water with WHO (1996) and ISI (1995) in mg/L (except EC and pH)

Parameters	Maximum permissible limit (WHO)	Maximum permissible limit (ISI)	Range of observed values	Suitability of drinking
EC	1400	–	77.77 - 827.23	Suitable=all
pH	6.5 – 8.5	6.5 - 9.2	5.69 - 7.89	Suitable=all
TDS	1000	1500	49.77 - 529.43	Suitable=all
Ca	500	200	5 - 87	Suitable=all
Mg	–	100	1 - 25.6	Suitable=all
Na	200	–	3.3 - 222.6	Not Suitable=31
K	–	–	0.98 - 26.7	
HCO_3	–	600	11.23 - 398.2	Suitable=all
SO_4	400	400	0.02 - 0.89	Suitable=all
Cl	250	1000	22.3 - 267.8	Not Suitable=17

According to WHO (1996) standard Na, Cl exceed the maximum limit in one location of the study area. According to ISI standard, all the parameters are within the maximum permissible limit. Generally, the groundwater samples of the study area are suitable for domestic and drinking purposes except few locations.

Suitability for Irrigation

The suitability of groundwater for irrigation purposes are mainly based upon the estimation of the parameters like SAR, Na% and RSC. Total Na concentration and EC are important in classifying the water for irrigation purposes (Wilcox, 1955). Sodium percentage is calculated against major cations and expressed in terms of SAR. Na is an important cations which in excess deteriorates the soil structure and reduces crop yield (Srinivasamoorthy, 2005). SAR values (Richards 1954) of the groundwater falls in excellent (0-10) category. Excess of magnesium and calcium ions tend to precipitate as carbonate. This is designated as residual sodium carbonate (RSC). According to Richards (1954), majority of the study area groundwater samples falls in good category (<1.25 $meql^{-1}$) indicating fit for irrigation purpose. Few representations were also noted in medium to bad water (1.25 – 2.5 and >2.5). The irrigation water with high RSC is considered to be deleterious to the physical properties of the soil (Gurugnanam *et al.*, 2009). According to chloride classification by Stuyfzand (1989), majority of samples in groundwater fall in freshwater (0.84-4.23) category.

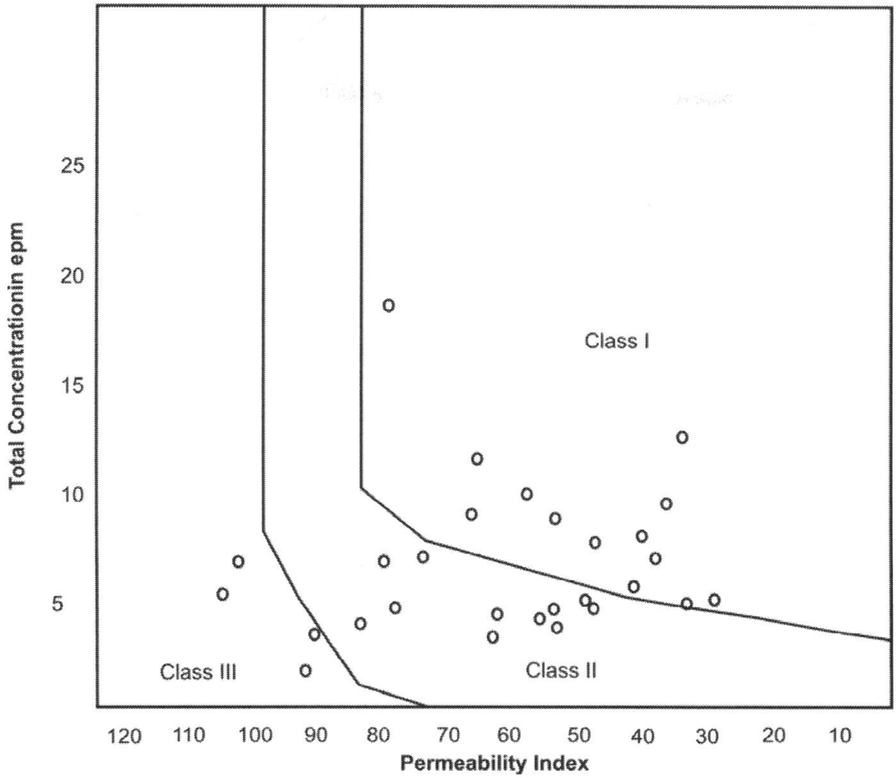

Fig. 2. Doneen's permeability index for groundwater (after Doneen, 1948).

Permeability index is an important factor which influences quality of irrigation water in relation to soil for development in agriculture. Based on permeability index, Doneen (1948) classified the groundwater as class I, class II and class III to find out suitability of groundwater for irrigation purpose. In groundwater of the study area (Fig. 2) most of the samples falls in class I and class II indicating waters are good to moderate for irrigation purpose. Minor representations are also fall in class III indicating poor quality for irrigation.

Salinity Hazard

The salinity of groundwater and SAR also determines its utility for agricultural purpose (USSL, 1954). The study area samples fall between C_1S_1 and C_2S_2 fields. Few water samples fall in C_3S_1 (Fig. 3) indicating high salinity and low alkali hazard satisfactory for plants having moderate salt tolerance on soils.

CONCLUSION

The results reveal that the chemical composition of groundwater of Andimadam block is near acidic to alkaline nature. All the samples in the study area, sodium is the dominant cation and

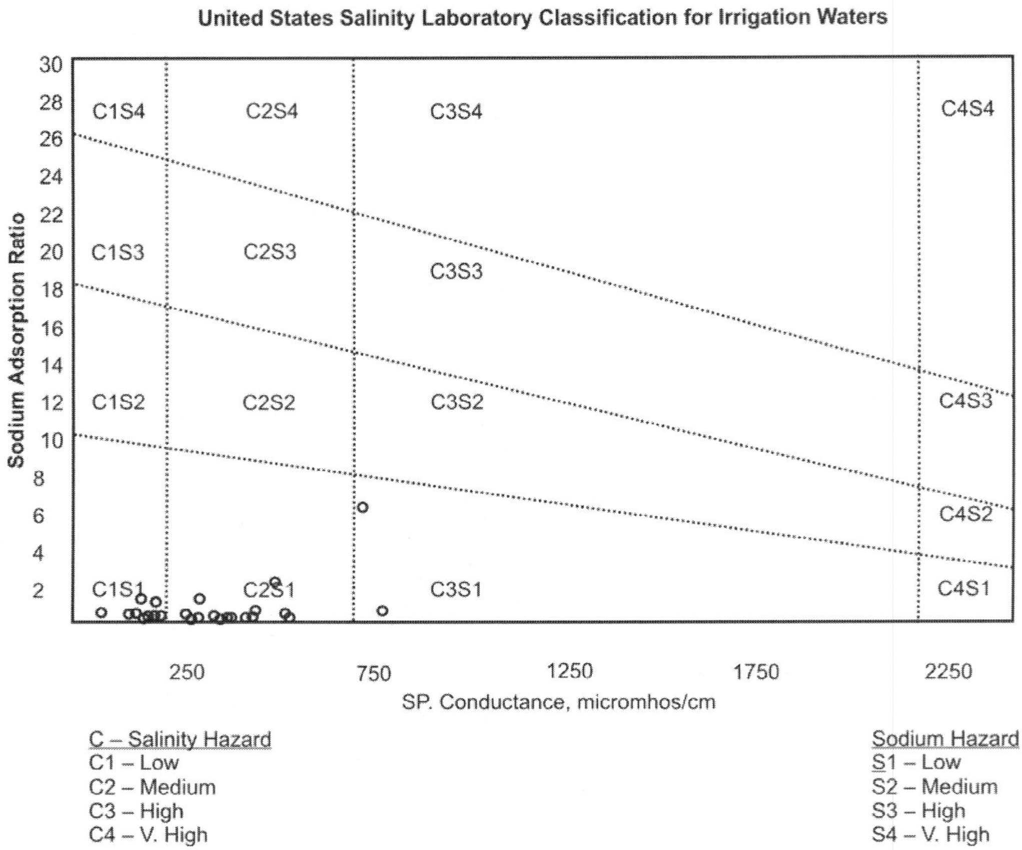

Fig. 3. USSL classification for the groundwater (after USSL, 1954).

bicarbonate is the dominant anion in groundwater. In general, groundwater is suitable for domestic and irrigation purposes. Concentrations of chemical constituents in groundwater were compared with drinking water standard of WHO (1996) and ISI (1995). Parameters like pH, EC, TDS, Ca, Mg, HCO_3 and SO_4 in groundwater at most places are within maximum permissible limit.

ACKNOWLEDGEMENTS

The second author thank to Dr. S. Chidambaram, Reader, Department of Earth Sciences, Annamalai University for permitting analytical work to his laboratory.

REFERENCES

Apha, (1995). Standard methods for the examination of water and wastewater, 19th edition. APHA, Washington DC, USASS.

Chidambaram et al., (2003).

Craig, E., and Anderson, M.P. (1979). The effects of urbanization of groundwater quality. A case study of groundwater **17:** 456-562.

Doneen, L.D. (1948). The quality of irrigation water. *California Agriculture Dept.* **4-11:** 6-14.

Gurugnanam, B., Suvedha, M., Vasudevan, S., Murugaih, M. and Karthikeyan, C. (2009). Groundwater quality investigation in Veeranam catchment area, Tamil Nadu, *Indian J. Appl. Geoche.*, **Vol. 11, No.1:** pp.76-83.

Hem, J.D. (1985). Study and interpretation of chemical characteristic of natural water. U.S. *Geol. Sur. Water Supply Paper*, **2254:** pp. 264.

Herman Bouwer, (1978). Groundwater Hydrology International Student Edition.

ICMR, (1975). Indian Council of Medical Research: Manual of standards of quality for drinking water supplies.

ISI (1995). Indian standard specification for drinking water. ISI New Delhi, IS 10500-1995.

Karanth, K.R. (1989). Groundwater assessment development and management. Tata McGraw-Hill Publ., New Delhi, pp. 720.

Ramappa, R. and Suresh T.S. (2000). Quality of groundwater in relation to agriculture practices in Lakapavani river basin, Karnataka India. In: Proceedings of International seminar on Applied Hydrogeochemistry, Annamalai University, pp.136-142.

Richards, L.A. (1954). Diagnosis and improvement of saline and alkali soils, USDA handbook **60:** pp. 160.

Saha, D., Dhar, Y.R. and Sikder, P.K. (2008). Geochemistry of groundwater of the Pleistocene aquifer of mid Ganga Basin, A case study for south Ganga plain, Bihar, Journal of Geological Society of India. **Vol. 71, No.4:** pp. 473-484.

Srinivasamoorthy, K., Chidambaram, S., Ananthan, P. and Vasudevan, S. (2005). Application of Statistical analysis of the hydrogeological study of groundwater in hard rock terrain, Salem District, Tamil Nadu, *J. Geoche.*, **Vol. 20:** pp.181-190.

Subba Rao, N. (1983). Hydrogeology and Hydrogeochemistry of Visakhapatnam basin, India, *J. Water, Air and Soil Pollution*, **Vol. 16,** pp. 317-329.

U.S. Salinity Laboratory, (1954). Diagnosis and Improvement of Saline and Alkali Soils, U.S. Dept. Agriculture Handbook, pp. 60.

WHO, (1996), International standard for drinking water, 3rd edition, Geneva, pp. 70.

Wilcox, L.V. (1995). Classification and use of irrigation water, U.S. *Geol. Dept. Agri. Arc.* **969:** pp. 19.

10

Physico-chemical Analysis of Groundwater Around Sathya Dye Industry, Tiruppur

Bakkialakshmi, S.,*[1] Shanthi, B.,[2] Menaka, T.[1] and Prema, V.[1]

[1]*Department of Physics, Annamalai University, Annamalai Nagar–608002*

[2]*Centralized Instrumentation and Services Laboratory (CISL), Annamalai University, Annamalai Nagar–608002, Tamil Nadu, India*

**Email: bakkialakshmi66@yahoo.com*

ABSTRACT

Environmental degradation has become a serious problem all over the world today. The growth of population and resultant rise in poverty is the most important causes for the change of environment. The economic development, urbanization, industrialization, availability of natural resources and lifestyle of people also affect the environment. For the present paper the waste water samples were collected from the outlet of Periandippalayam Sathya dying factory (Near to Coimbatore) situated at southern district. Well water samples were collected at eight different locations. The water samples collected for our study are used for drinking and washing purposes by the people living in that area. Some of the physical parameters determined by the analysis of groundwater samples are appearance, odour, turbidity and some of the chemical parameters are pH. Conductivity, total dissolved solids, total hardness, etc. These parameters are tabulated and represented by graph. It can be concluded that the waste water discharged from the dye industry needs efficient treatment in reducing or neutralizing the toxic elements present in it, thus making it suitable for irrigational and drinking purpose.

Keywords: Physico-chemical analysis, conductivity of water, pH, total hardness, dye industry and water pollution.

INTRODUCTION

Subramaniyan *et al.* (2007) have studied the biodiversity of caynobacteria in industrial effluents. Four different effluents, such as dye, papermill, pharmaceutical and sugar were selected. The physico-chemical characteristics of all the effluents studied were more or less similar. 26 species of cyanobacteria were recorded in common to all effluents analyzed. Jameel and Sirajudden (2006) have studied the risk assessment of physico-chemical contaminates in groundwater of Pettavaithalai area, Tiruchirappali. Tamil Nadu. For the sugar mill effluents, the physico-chemical parameters have been analyzed. The methods employed for the determination of physico-chemical characteristics of groundwater were followed as given in the standard book of "American Public Health Association" [APHA] (1975).

Location of Tiruppur area has a strategic importance in the production of weaving products. Periandippalayam, is placed at a distance of 10 km from Tiruppur in Coimbatore District. Uthukkuli is located at the western side. Palladam is located at the northen side and Avinasi is located at the sourthern side of our experimental area. The nearby town of Periandippalayam is Tiruppur. The intensity of pollution well water samples were collected at eight important locations around the Sathya Dying Factory, Periandippalayam, Coimbatore District. The location map of Sathya Dye Industry is given is Fig. A.

Experimental

Sample analysis

All the water samples were collected in pre-cleaned acid-washed and dried water bottles and taken to the laboratory for analysis.

The water analysis was done by the Tamil Nadu Water Supply and Drainage (TWAD) Board, District water testing laboratory, Nagoor, Nagapattinam.

Results and Discussion

The results of various physico-chemical analysis of groundwater sample around dye industry are provided. Various physical examinations such as appearance, odour, turbidity and electrical conductivity are done and the results are given in Table 1.

All of our samples are clear and unobjectionable except S1 and S2. S1 and S2 are the 1st and 2nd dye water samples. These are just like the effluents. The turbidity values are below the desirable limit except S1 and S2. TDS values for the water samples at S1 and S2 are 11060 and 5145. All other samples S3, S4, S5, S6, S7 and S8 exceeds the desirable limit and also exceeds the limit of, cause for rejection when exceeds. It is as shown in Fig. 1. The electrical conductivity values ranges from 15800 to 2240 mho/cm for our location. According to the classification made by the united state salinity laboratory, EC greater than 2250 category of water is considered poor. This water is not suitable for irrigation under ordinary condition but may be used occasionally under very special circumstances. (1995). In the EC values at all the light sports are higher than the permissible limit, as shown in Fig. 2.

The observed chemical parameters are given in Table 2. It can be observed that the pH of the well water varies from 7.6 to 11.5 in our location. The lowest value 7.6 occurs at the sample station S7. The highest value 8.5 (within the permissible limit) occur at S2. The sampling

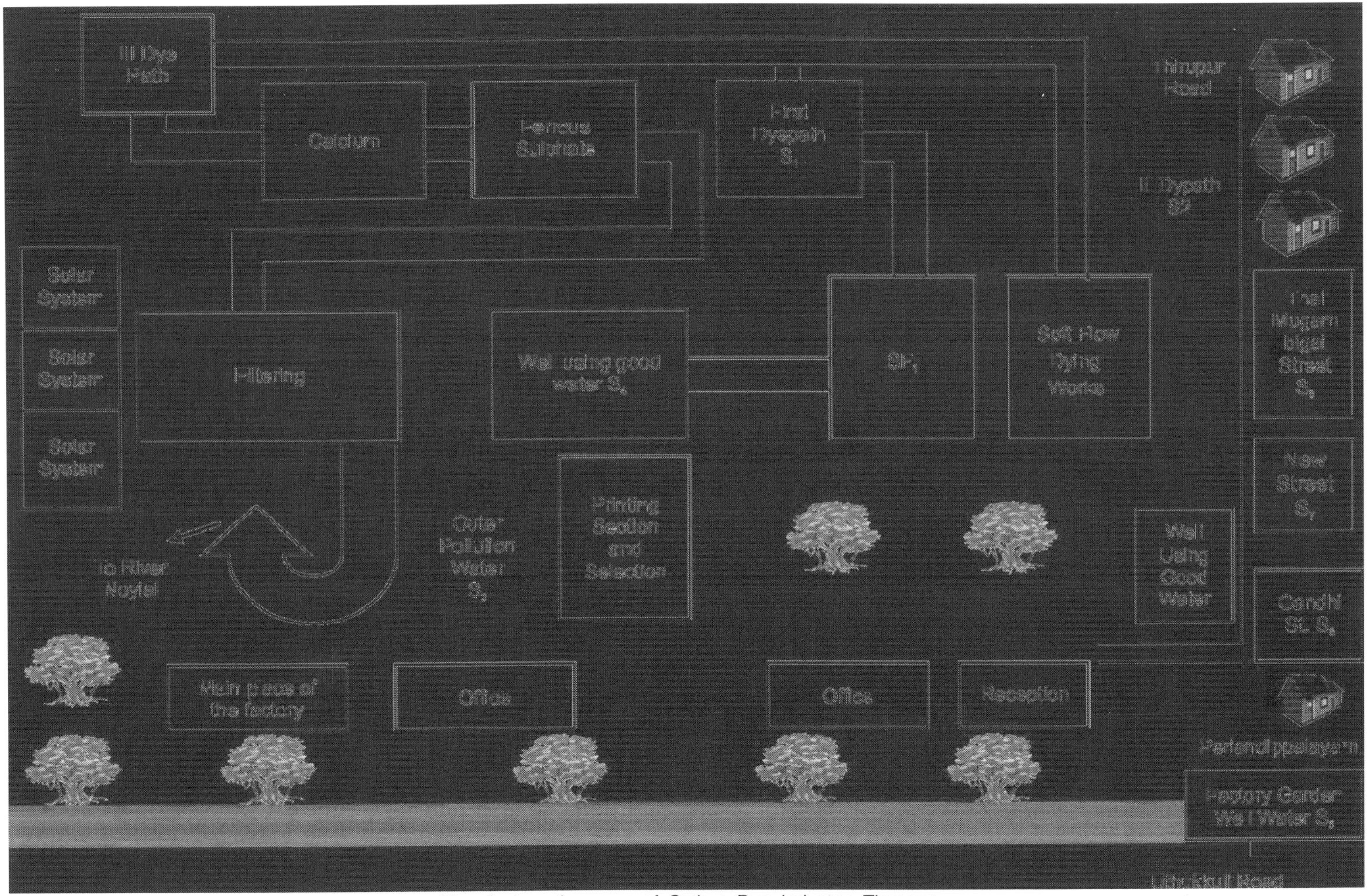

Fig. A. Location map of Sathya Dye Industry, Tiruppur

Table 1: Physical parameters

Sample No.	S_1	S_2	S_3	S_4	S_5	S_6	S_7	S_8	Maximum Permissible Limit	
Distance (km)	0.0	0.25	0.50	0.75	1.0	1.5	2.0	2.5	**A**	**B**
Appearance	Clear	Clear	Clear	Clear	Clear	Clear	Clear	Clear	Colourless	Colourless
Odour	Objectional	Objectional	None	None	None	None	None	None	Unobjectionable	Unobjectionable
Turbidity NTU	20	37	0	0	0	0	0	0	1	10
Total Dissolved Solids (mg/L)	11060	5145	5488	2940	2569	2800	3122	1568	500	2000
Electrical Conductivity (mho/cm)	15800	7350	7840	4200	3670	4000	4460	2240	-	-

A – CPHEEO Standards Desirable limit
B – CPHEEO Standards cause for rejection when exceeds

Table 2: Chemical parameters

Sample No.	S_1	S_2	S_3	S_4	S_5	S_6	S_7	S_8	Maximum Permissible Limit	
Distance (km)	0.0	0.25	0.50	0.75	1.0	1.5	2.0	2.5	**A**	**B**
pH	9.6	8.5	11.5	8.3	8.0	8.0	7.6	8.3	7.0	6.5
Alkalinity pH ($CaCO_3$)	60	24	160	16	0	0	0	10	-	-
Total alkalinity ($CaCO_3$)	500	504	960	600	520	540	580	400	200	600
Total hardness ($CaCO_3$)	3750	1200	1500	600	720	440	820	352	200	600

A – CPHEEO Standards Desirable limit
B – CPHEEO Standards cause for rejection when exceeds
Results are expressed in mg/L

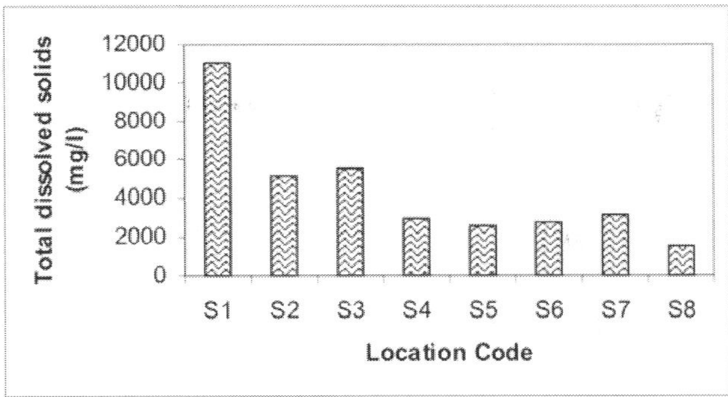

Fig. 1. Variation of TDS with location code.

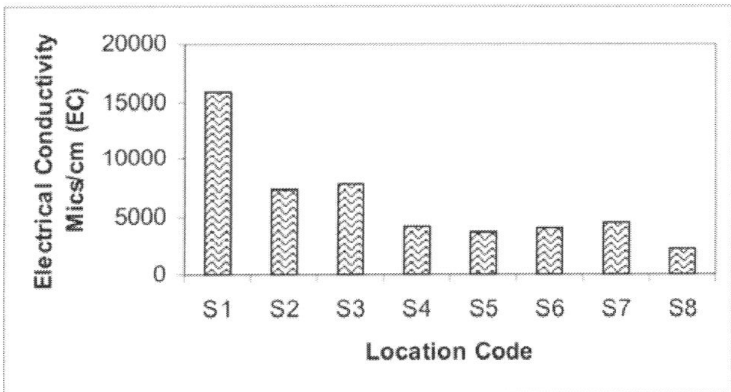

Fig. 2. Variation of electrical conductivity with location code.

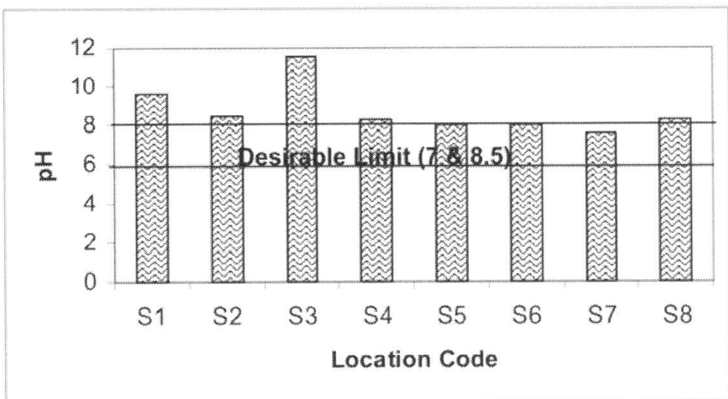

Fig. 3. Variation of pH with location code.

station S1 (9.6) and S3 (11.5) have the pH value greater than the permissible limit. The variation of pH with location code as shown in Fig. 3.

The hardness in water is mainly due to its calcium and magnesium contents (1986). The hardness of water in our experimental location is tabulated in Table 2. The ranges of hardness value is 352 to 3750. Salt water with hardness less than 100 mg/L may have lower buffer capacity and more corrosive in nature. Whereas hardness above 800 mg/L cause scale deposition

Table 3: Concentration (ppm) of heavy metals in water collected at various locations

Sample No.	S_1	S_2	S_3	S_4	S_5	S_6	S_7	S_8	Maximum Permissible Limit	
Distance (km)	0.0	0.25	0.50	0.75	1.0	1.5	2.0	2.5	**A**	**B**
Calcium (Ca)	1200	320	400	160	240	120	240	88	75	200
Magnesium (Mg)	180	96	120	48	29	34	53	32	30	150
Iron (Fe)	2.16	0.00	0.00	0.00	0.00	0.00	0.00	0.00	0.1	1.0
Manganese (Mn)	0.00	0.00	0.00	0.00	0.00	0.00	0.00	0.00	0.05	0.5
Free Ammonia (NH_3)	8.00	4.00	4.00	8.00	0.00	0.00	0.00	0.00	-	-
Nitrite (NO_2)	0.00	0.00	0.00	0.00	1.50	1.50	1.00	1.00	-	-
Nitrate (NO_3)	69	46	57	46	46	92	92	92	45	100
Chloride (Cl_2)	4630	2012	1800	772	752	748	1040	320	200	1000
Fluoride (F_2)	1.0	1.2	1.1	1.0	1.2	1.1	1.3	1.0	1.0	1.5
Sulphate (SO_4)	170	128	85	170	33	124	85	170	200	400

A – CPHEEO Standards desirable limit
B – CPHEEO Standards cause for rejection when exceeds
Results are expressed in mg/L

in the distribution system and results in excessive soap consumption and subsequent sun formation. It may also cause, cardiovascular diseases. Srinivasa Rao and Venkateswaralu, have reported that the hardness values 186-920 mg/l. According to them few samples have very high hardness values, which are due to the presence of high salt content (2000, 2001). The concentration of heavy metals in water collected at various sampling stations as shown in the Table 3. The variation of calcium, magnesuim & fluoride are given Figs, 4, 5 and 6 respectively. The

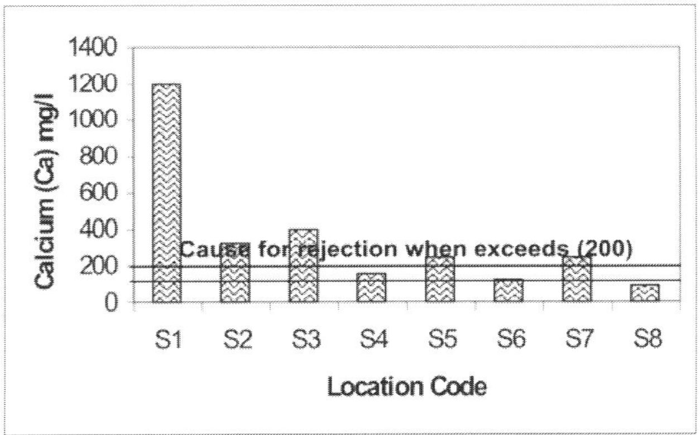

Fig. 4. Variation of calcium with location code.

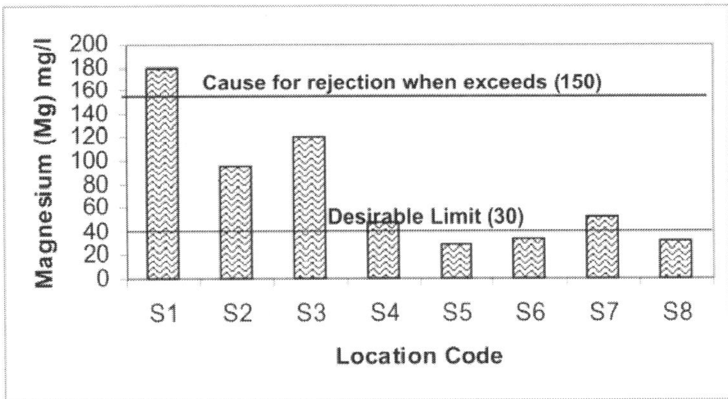

Fig. 5. Variation of magnesium with location code.

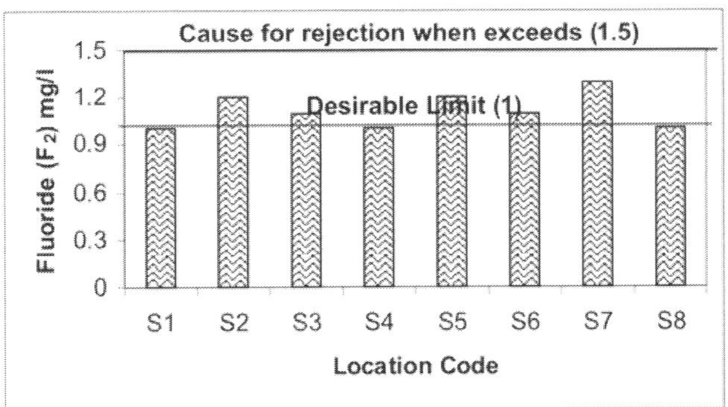

Fig. 6. Variation of fluoride (F_2) with location code.

concentration of some heavy metals like Ca, Cl2 in wastewater are higher than the prescribed limits given by FAO, ISI and WHO. The higher concentration of these heavy metals present in the wastewater samples may be toxic to plants. As such the wastewater is unsuitable for irrigation purpose unless it is treated properly. If this is untreated, effluents are allowed to mix with river water, then, the habitants using this water for drinking purpose would be much suffered of heavy metals dose. Hence, industrial waste should be properly treated before it mixes in water bodies and reservoirs.

The presence of fluoride at about 1 mg/L in drinking water is known to protect against dental caries but high levels of fluoride cause mottling of the dental enamel (1). The water from sampling stations S2, S3, S5, S6 and S7 have above 1 mg/L of fluoride. The water from these stations are not suitable for drinking purpose. Hardness of water appears to have a beneficial affect against cardiovascular disease. The Sathya dye industry does not discharge the wastewater directly. The wastewater is filtered and only one part is made to go out of the industry. The remaining part of the wastewater is evaporated by using the solar system, which is kept inside the industry. This is appreciable. The solar system used by the dye industry is shown in Figs. 7 and 8.

Fig. 7. Solar treatment before the process.

Fig. 8. Solar treatment after the process.

CONCLUSION

The pollution crisis and the energy requirement are mere different sides of the same coin. Setting up or expansion of the dye industries has a profound influence on society and environment, both in their positive and negative aspects. Keeping in view of all these results, it can be concluded that the wastewater discharged from the dye industry needs efficient treatment in reducing or neutralizing the toxic elements presents in it, thus making it suitable for irrigational and drinking purpose. Hence, it is recommended that water must be thoroughly analysed and studied before being used for domestic applications.

REFERENCES

APHA, (1975) Standard method for the examination of water and waste. **American Public Health Association** (Washington) 1975.

Edgier, G.P. (1973). Atomic absorption, Newsletter, 12. pp. 115.

Hussain. J., *et al.*, (1995). *Asian J. Chemistry*, **13(3):** pp. 1114-1122.

Jameel, A. and Sirajudden, J. (2006). Risk assessment of physico-chemical contaminants in groundwater of Pettaviathali area. Tiruchirappalli, Tamil Nadu, *En v monitoring assess*, **123(1-3):** 299-312.

Raman N and Murugesan, A. (2001). *Asian J. Chemi.* **13(3):** pp. 995-1000.

Srinivasa Rao and Venkateswaralu, (2000). *Indian J. Environ Protection*, **20(3):** pp. 161-164.

Subraminayan, Vijayakumar, Nooruddin Thyaddin and Chokkaiya Manoharan. (2007). Biodiversity of cyanobacteria in industrial effluents. **38:** 1-8.

Trivedy and Goel, K.K. (1986). Water pollution 2nd edition.

11

Hydrochemical Assessment of Groundwater in Thirumanimuttar Sub-basin – Implication of Water Quality Degradation by Anthropogenic Activities

VASANTHAVIGAR, M., SRINIVASAMOORTHY, K.,* VIJAYARAGHAVAN, K., RAJIVAGANTHI, R., CHIDAMBARAM, S., ANANDHAN, P., MANIVANNAN, R. AND VASUDEVAN, S.

Department of Earth Sciences, Annamalai University, Annamalai Nagar–608002

**Email: moorthy_ks@yahoo.com*

ABSTRACT

A hydrochemical investigation was conducted in the Thirumanimuttar sub-basin an important Tributary of river Cauvery in Tamil Nadu to identify the factors influencing the chemical quality of groundwater in a hard rock terrain. Groundwater samples were collected from two different seasons post- and pre-monsoon, respectively. The major ion chemistry of groundwater is dominated by cations like Na, Mg, Ca^{2+} and K^{2+} in anions Cl, HCO_3^- and SO_4^2 were found to be dominated. An increase in salinity is observed along the groundwater flow direction. The occurrence of saline groundwater, as witnessed by the high electrical conductivity (EC), is attributed to the long residence time of water and the dissolution of minerals, as well as evaporation of rainfall and irrigation return flow. Saturation index of minerals indicates oversaturation and equilibrium state of silicate, carbonate and fluoride minerals during both seasons. Sulphate minerals revealed undersaturated during both seasons. Water Quality Index (WQI) worked out decipher poor of groundwater during Post-Monsoon Season (POM) season when compared with Pre-Monsoon (PRM) due to impact of anthropogenic activities.

Keywords: Groundwater geochemistry, anthropogenic, water quality index, sodium percentage.

INTRODUCTION

Water quality analysis is one of the most important aspects in groundwater studies. It helps to understand the change in due to rock water interaction or any type of anthropogenic influence to demarcate groundwater suitability for drinking purposes. The chemical parameters of groundwater play a significant role in classifying and assessing water quality which mainly depends on the quality of recharged water, atmospheric precipitation, inland surface water and subsurface geochemical processes. Temporal changes in the origin and constitution of the recharged water, hydrologic and human factors, may cause periodic changes in groundwater quality. Water pollution not only affects water quality but also threats human health, economic development and social prosperity (Milovanovic, 2007). Assessing risk involves identifying the hazard associated with a particular occurrence, action or circumstance and determining the probability of that hazard occurring (Smith 2001). The protection of groundwater resources is a recent emerging issue in semi-arid and arid environments.

The study area being an hard rock terrain gains importance due to its presence in rain shadow region receives major part of rainfall from north-east monsoon. The surface water sources are generally precarious to get their supply during monsoon seasons and in non-monsoonal periods people have to largely depend on groundwater resources for their domestic, agricultural and industrial activities. The area has been marked as drought prone area by United National Development Project. 70% of study area is dominated by agricultural activities and the rest by industries manufacturing chemicals, paints, dying, bleaching and bus body building are sparingly distributed all over the study area which disposes industrial and hazardous wastes nearby agricultural lands which craft a major threat to the adjoining groundwater environments. Saturation index (SI) approach, is used to predict the reactive mineralogy of the subsurface from groundwater data without collecting the samples of the solid phase and analyzing the mineralogy (Deutsch 1997).

The groundwater investigation is to be given priority, to characterize the hydrogeochemical processes activated in the study area, with reference to natural and man-made activities and to classify water on the basis of Water Quality Index (WQI). Water quality index is useful in determining the spatial and temporal changes in water quality and in quantifying the effectiveness of water quality regulations and protection programmes.

STUDY AREA

Thirumanimuttar sub-basin lies between North Latitudes 10°58′ and 11°48′and East Longitudes 77°53′ and 78°28′ with a total drainage of about 2030 km^2. It originates at Manjavadi pass in Salem district and configures river Cauvery at Paramathi in Namakkal district (Fig. 1). The mean temperature varies from 33°C to 45°C. The normal annual rainfall over the study area is about 1590 mm. The geology of the area is underlined by rocks representing metasediments of Achaean age with Charnockites, granitic gneiss, calc granulites, syenites and ultra basics as major exposures. The following geomorphic units were identified, viz. Alluvial Plain, Bajadas, Colluvial Plain, Deep Pediment, Hill, Pediplain and Structural hill. The drainage type is dentritic and trellis thickness of weathered zone varies from 5 to 15 m along upstream and 3 to 30 m due downstream in locations like Namagiripetai and Singlandapuram. The occurrence and movement of groundwater in the hard rocks are controlled by the secondary porosity and groundwater occurs under water table to semi confined condition.

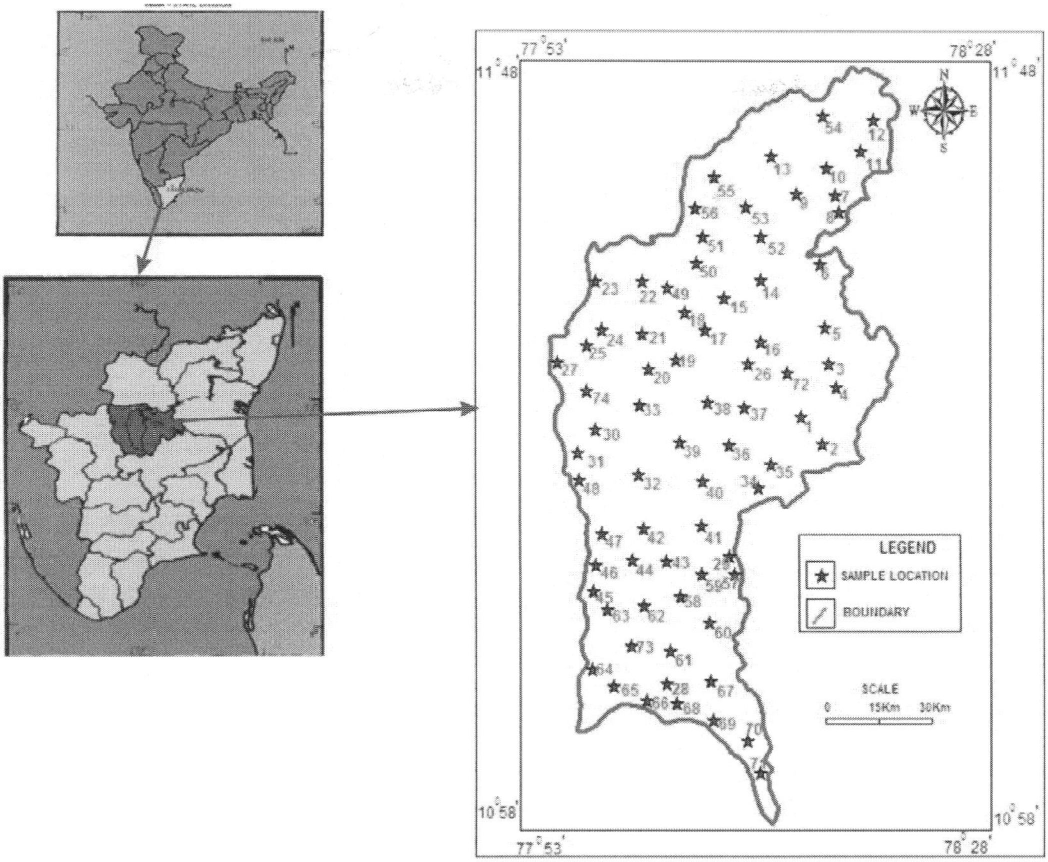

Fig. 1. Location and groundwater sample points in the study area

MATERIALS AND METHODS

To get a brief insight about the quality of groundwater and to determine the spatial and temporal variation of water quality parameters a total of 148 groundwater samples were collected for two different seasons, viz. Pre-monsoon (PRM) during May 2008 and Post-monsoon season (POM) during January 2008. The samples were analyzed for Ca, Mg, Na, K, HCO_3, Cl, SO_4, PO_4, H_4SiO_4, F, pH, EC, TDS by using standard procedures (APHA 1995). pH, TDS and EC were measured by Systronics Water Quality Analyzer 371. F was analyzed using portable consort electrochemical analyzer model C933. SO_4, PO_4, H_4SiO_4 was determined by digital spectrophotometer model GSS 700A (Electronic Corporation of India). Na and K were determined by using flame photometer (Systronics mk-1/mk-III). Ca, Mg, HCO_3 and Cl were analyzed by titrimetric method. The charge balance calculated was generally < 10% and ratio of TDS/EC are within acceptable limits (0.8), confirming the reliability of the analytical results.

Table: 1: Statistics of groundwater chemistry during post- and pre-monsoon season (all values in mg/L except pH and EC in µs/cm)

Parameters	pH	Ca^{2+}	Mg^{2+}	Na^{+}	K^{+}	HCO_3	PO_4	H_4SiO_4	Cl^-	SO_4	NO_3	F^-	TDS	EC
Post-monsoon Season														
Max	8	464	377	598	322	1068	3	37	1730	758	136	1	3729	5826
Min	7	8	5	6	1	92	0	7	24	25	1	0	408	638
Average	8	57	113	234	50	376	1	19	429	218	38	1	1535	2399
Std. Dev.	0	58	90	165	61	224	1	6	357	180	40	0	764	1193
Pre-monsoon Season														
Max	9.22	179	158.4	1497.0	277.0	893.8	1.5	33.0	1602.3	126.8	133.0	4.2	3490	5453
Min	6.9	16	11.91	0.8	0.8	73.22	0.475	2	86.84	1	4	0.03	303	473
Average	7.29	70.8	48.27	312.27	26.39	413.7	0.649	17.068	475.91	28.69	38.03	1.14	1433	2239
Std. Dev.	0.35	34	28.89	330.71	55.53	164.1	0.162	5.7703	356.64	27.9	29.08	0.78	754	1179

RESULTS AND DISCUSSION

Groundwater, in general, is alkaline during post-monsoon season with pH ranging from 7 to 8 and acidic to alkaline during pre-monsoon season with pH ranging from 6.9 to 9.22 during post-monsoon season. Electrical conductivity values ranges from 638 to 5826 and 473 to 5453 during post- and pre-monsoon season, respectively. Higher EC was recorded during post-monsoon season along groundwater flow direction. Bicarbonate was noted high during post-monsoon season (1068 ppm); derived from the action of CO_2 upon the basic material of soil and granitic rock. Chloride was higher in post-monsoon season (1730 mg/L) due to leaching from upper soil layers due to industrial and domestic activities and dry climates (Srinivasamoorthy, 2008).

Sulphate was higher in post-monsoon season (758 mg/L) due to the action of anthropogenic activities (Saxena, 2004). PO_4 was within the permissible limit during both seasons. Fluoride was higher during pre-monsoon (4.2 mg/L) due to leaching from fluoride rich rocks found in the source rocks of the study area. Nitrate was higher during post-monsoon season (136 mg/L) in agricultural areas due to leaching from plant nutrient and nitrate fertilizers (Freeze and Cherry, 1979). Sodium was higher during pre-monsoon season (1497 mg/L) indicating weathering from plagioclase bearing rocks of the study area (Hem, 1985). Potassium is higher in post-monsoon season (322 mg/L) due to weathering of feldspar and clay from aquifer matrix. Calcium was higher during post-monsoon season (464 mg/L) due to dissolution of precipitates of $CaCO_3$ and $CaMg(CO_3)_2$ during precipitation and recharge (Datta and Tyagi, 1996, Lakshmanan *et al.*, 2003). Magnesium concentration of the study area was higher during post-monsoon season (377 mg/L) derived from dissolution of magnesium calcite, gypsum and/or dolomite from source rock (Garrels, 1976).

GOUNDWATER QUALITY PARAMETERS

Water Quality for Drinking Purposes

Groundwater chemistry has been utilized as a tool to prospect water quality for drinking and irrigation purposes (Edmunds *et al.*, 2002). WQI is an important parameter for demarcating groundwater quality and its suitability for drinking purposes (Tiwari and Mishra 1985; Subba Rao, 1997). WQI is defined as a technique of rating that provides the composite influence of individual water quality parameters on the overall quality of water for human consumption. The standards for drinking purposes as recommended by WHO (1993) and BIS 10500 (1991) have been considered for the calculation of WQI. For computing WQI, three steps are followed. In the first step, each of the 12 parameters (TDS, HCO_3, Cl, SO_4, PO_4, NO_3, F, Ca, Mg, Na, K, Si) has been assigned a weight (w_i) according to its relative importance in the overall quality of water for drinking purposes (Table 2).

The maximum weight of 5 has been assigned to the parameters like nitrate, Total dissolved solid, Chloride, Fluoride and Sulphate due to their major importance in water quality assessment. Bicarbonate and Phosphate is given the minimum weight of 1 as it plays an insignificant role in the water quality assessment. Other parameters have a weight between 1 and 5 depending on their importance in water quality determination. In the second step, the relative weight (W_i) is computed from the following equation:

Table 2: Relative weight of chemical parameters

Chemical parameters	Indian Standard (BIS 10500, 1991)	Weight (wi)	Relative weight $W_i = \frac{w_i}{\sum_{i=1}^{n} w_i}$,
Total dissolved solids	500	5	0.116
Bicarbonate (mg/L)	-	1	0.023
Chloride (mg/L)	250	5	0.116
Sulphate (mg/L)	200	5	0.116
Phosphate (mg/L)	-	1	0.023
Nitrate (mg/L)	45	5	0.116
Fluoride (mg/L)	1	5	0.116
Calcium (mg/L)	75	3	0.070
Magnesium (mg/L)	30	3	0.070
Sodium (mg/L)	-	4	0.093
Potassium (mg/L)	-	2	0.047
Silicate (mg/L)	-	2	0.047
		$\Sigma w_i = 41$	**$\Sigma w_i = 0.953$**

$$W_i = \frac{w_i}{\sum_{i=1}^{n} w_i},$$

where,

- W_i is the relative weight,
- w_i is the weight of each parameter and
- n is the number of parameters.

Calculated relative weight (W_i) values of each parameter are given in (Table 2). In the third step, a quality rating scale (q_i) for each parameter is assigned by dividing its concentration in each water sample by its respective standard according to the guidelines laid down in the BIS 10500 (1991) and the result multiplied by 100:

$$q_i = (C/S_i) \times 100$$

where,

- q_i is the quality rating,
- C_i is the concentration of each chemical parameter in each water sample in mg/L, and
- S_i is the Indian drinking water standard for each chemical parameter in mg/L according to the guidelines of the BIS 10500, 1991.

For computing the WQI, the SI is first determined for each chemical parameter, which is then used to determine the WQI as per the following equation

$$SI_i = W_i \cdot q_i$$

$$WQI = \sum SI_i$$

where,

- SI_i is the sub-index of ith parameter,
- q_i is the rating based on concentration of ith parameter and
- n is the number of parameters.

Water quality types were determined on the basis of WQI. The computed WQI values ranges from 37.94 to 298.96 and 41.35 to 291.94 for POM and PRM, respectively. The WQI range and type of water can be classified as follows:

Range	Type of water
< 50	excellent water
50-100	good water
100-200	poor water
200-300	very poor water
>300	water unsuitable for drinking

Table 3: Comparison of groundwater quality with drinking water standards (India and WHO)

Parameters	Indian Standard	Percent Compliance POM	PRM	WHO limits	Percent Compliance POM	PRM
pH	6.5-8.5	100	98.6	7.0-8.0	97.30	85.14
Sp. Conductivity (mS/cm at 25°C)	-	-	-	-	-	-
Total dissolved solids	500	97.29		1,000	28.38	32.43
Total hardness as $CaCO_3$ (mg/L)	300	59.46	36.50	-	-	-
Bicarbonate (mg/L)	-	-	-	-	-	-
Chloride (mg/L)	250	40.54	29.73	250	40.54	29.73
Sulphate (mg/L)	200	48.65	100	250	66.22	100
Phosphate (mg/L)	-	-	-	-	-	-
Nitrate (mg/L)	45	67.57	66.21	50	32.43	68.92
Fluoride (mg/L)	1	60.23	45.95	-	-	-
Calcium (mg/L)	75	85.14	59.46	75	85.14	59.46
Magnesium (mg/L)	30	9.46	25.68	30	9.46	25.68
Sodium (mg/L)	-	-	-	200	48.65	51.35
Potassium (mg/L)	-	-	-	-	-	-
Silicate (mg/L)	-	-	-	-	-	-

The chemical analysis of groundwater and the per cent compliance with ISI and WHO is represented in the (Table 3) which indicates that majority of the samples exceed the permissible limit prescribed by ISI and WHO. In POM 4% of samples represent excellent water, 41% of samples indicate "good water", 35% of samples shows "poor water" and 5% of samples shows "very poor water". During PRM only one sample represent "excellent water", 36% of samples shows "good water", 48.6 % of samples shows "poor water" and remaining 13.5 % of samples shows "very poor water". In pre-monsoon season samples exhibit poor quality in greater percentage when compared with post-monsoon. This may be due to effective leaching of ions, over exploitation of groundwater, direct discharge of effluents and agricultural impact (Paulami Sahu P.K. Sikdar, 2008).

The WQI was tested with chlorides and EC parameters selected as pollution indicators. The observed high values of chloride and EC correspond to the same WQI, indicating the quality of groundwater in the study area. During PRM season bad quality water is observed along NS direction where the contamination of aquifers due to anthropogenic impact derived from agricultural activities is noted. During POM season very poor water quality is confined along the NW and central part of the study area in locations like 11, 19, 33, 36, 37, 44, 48, 49, 50, 56. This may be due to the presence of industrial contamination and landfill activities observed in the study area along the above said sample locations. Poor water is noted along major part of the study area along the groundwater flow direction.

DISEQUILIBRIUM INDICES

Mineral equilibrium calculations for groundwater are useful in predicting the presence of reactive minerals in the groundwater system and estimating mineral reactivity. By using the saturation index (SI) approach, it is possible to predict the reactive mineralogy of the subsurface from groundwater data without collecting the samples of the solid phase and analyzing the mineralogy (Deutsch, 1997).

In the present study, to determine the chemical equilibrium between minerals and water, SI of calcite and dolomite were calculated. If the groundwater is saturated with respect to a mineral, it is prone to deposit (precipitation) some of the solute loads. On the other hand, if it is unsaturated (SI < 0) it will take more minerals into the solution (dissolution). Hence, the SI of a mineral is calculated based on the following equation (Lloyd and Heathcode, 1985):

$$SI = \log\left(\frac{IAP}{K_s}\right)$$

where,

- IAP is the ion activity product and
- Ks is the solubility product of the mineral.

Disequilibrium indices log (IAP/KT) was calculated by using PHREEQC geochemical model software for those minerals and other solids stored in the model data book for which the dissolved constituents are reported in groundwater analysis. Disequilibrium indices log (IAP/KT) was calculated to determine whether water is in thermodynamic equilibrium log (IAP/KT = 0), oversaturated log (IAP/KT > 0), or undersaturated (IAP/KT < 0), with respect to certain solid phases (Trusdell and Jones, 1973).

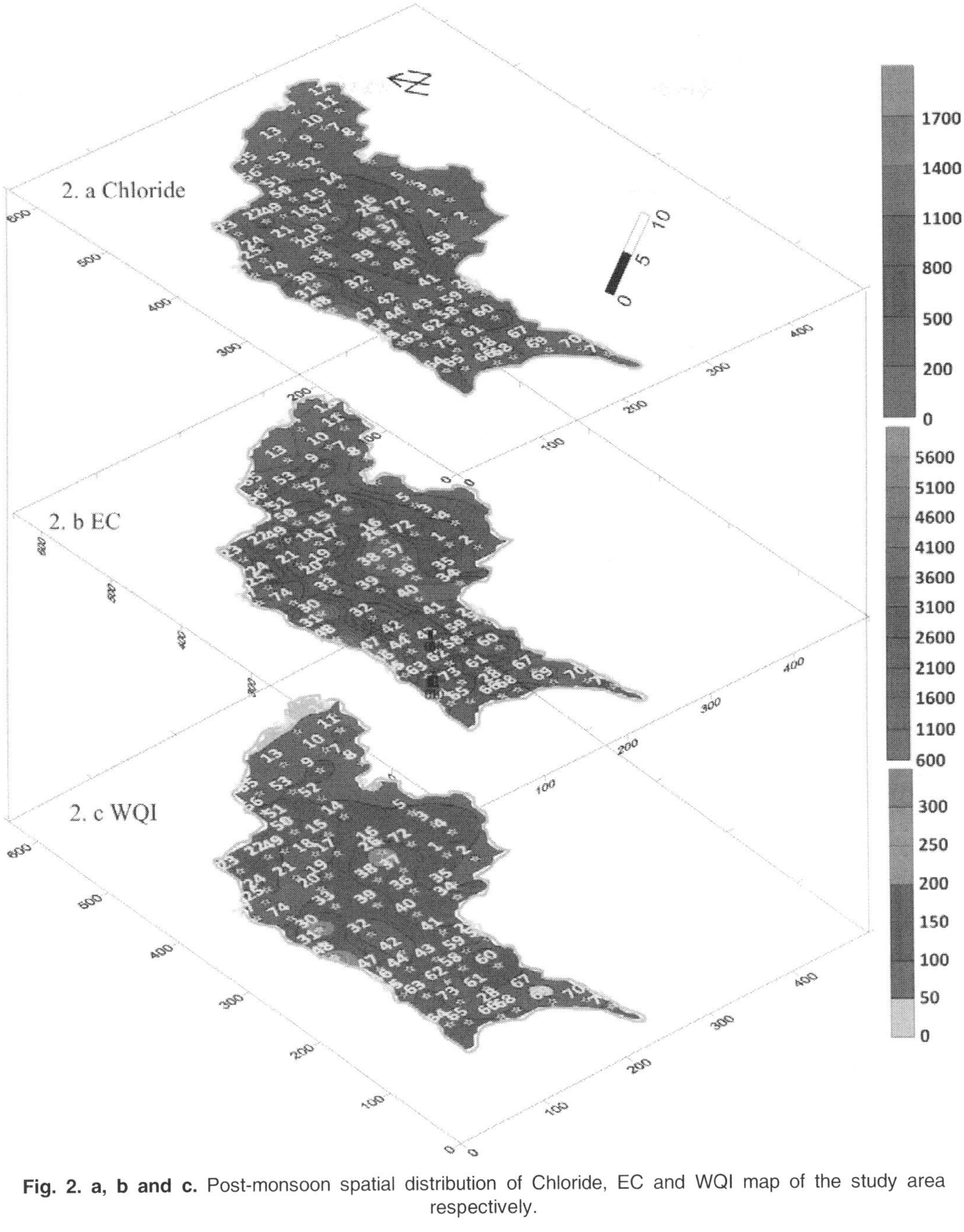

Fig. 2. a, b and c. Post-monsoon spatial distribution of Chloride, EC and WQI map of the study area respectively.

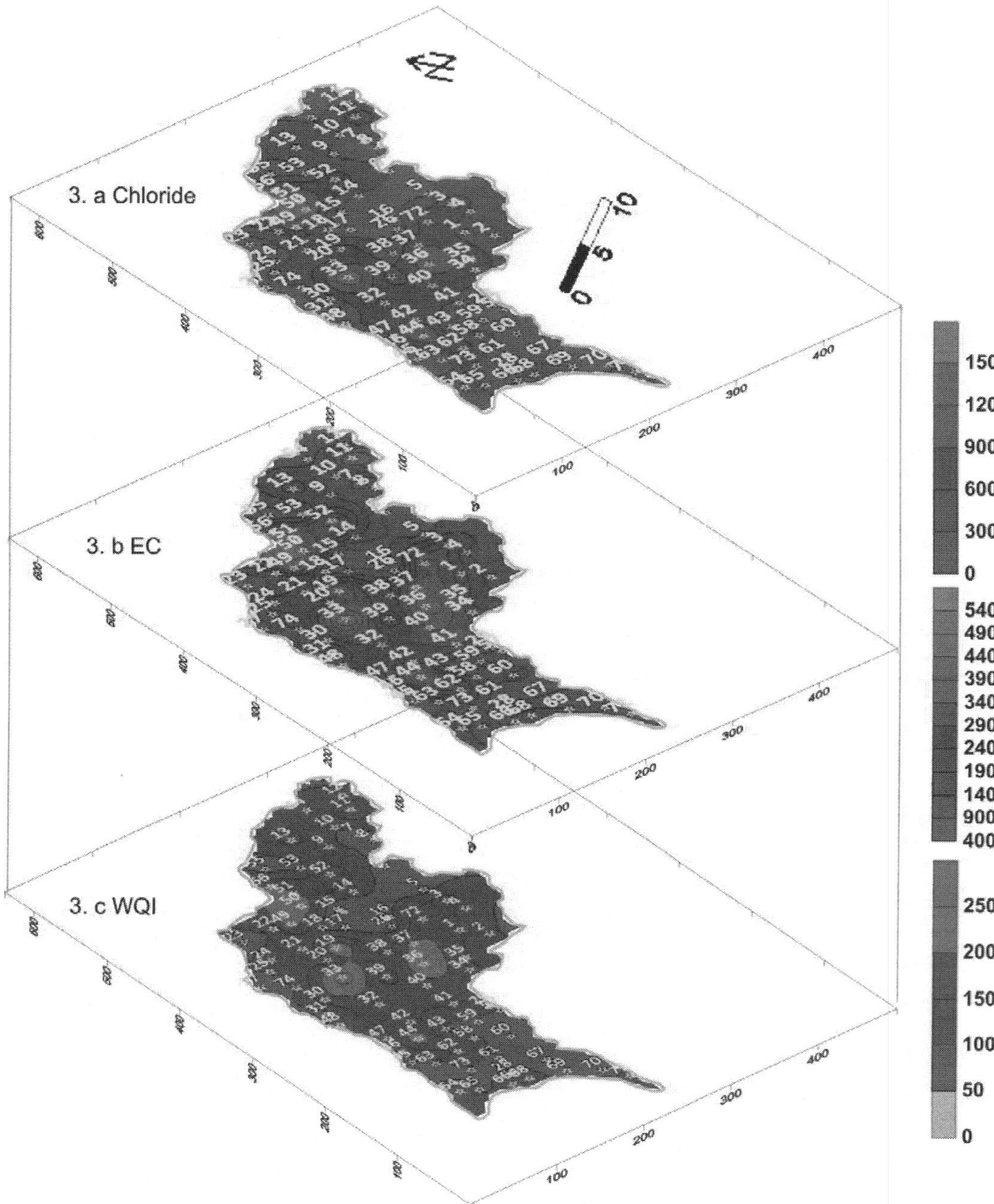

Fig. 3. a, b and c. Pre-monsoon spatial distribution of Chloride, EC and WQI map of the study area respectively.

The disequilibrium indices of carbonate minerals figure (Fig. 4a and b) Aragonite, calcite, dolomite and magnesite were represented from the data bank of PHREEQC. The samples in pre-monsoon season are oversaturated to equilibrium with aragonite, calcite, dolomite and magnetite with minor indications in undersaturation state. The samples in post-monsoon season follow the same trend as in pre-monsoon. The general carbonate results shows that calcite, dolomite and magnetite are saturated to oversaturated, which is indicated by excess input of Ca and Mg ions mainly derived from silicate weathering process (Srinivasamoorthy 2008).

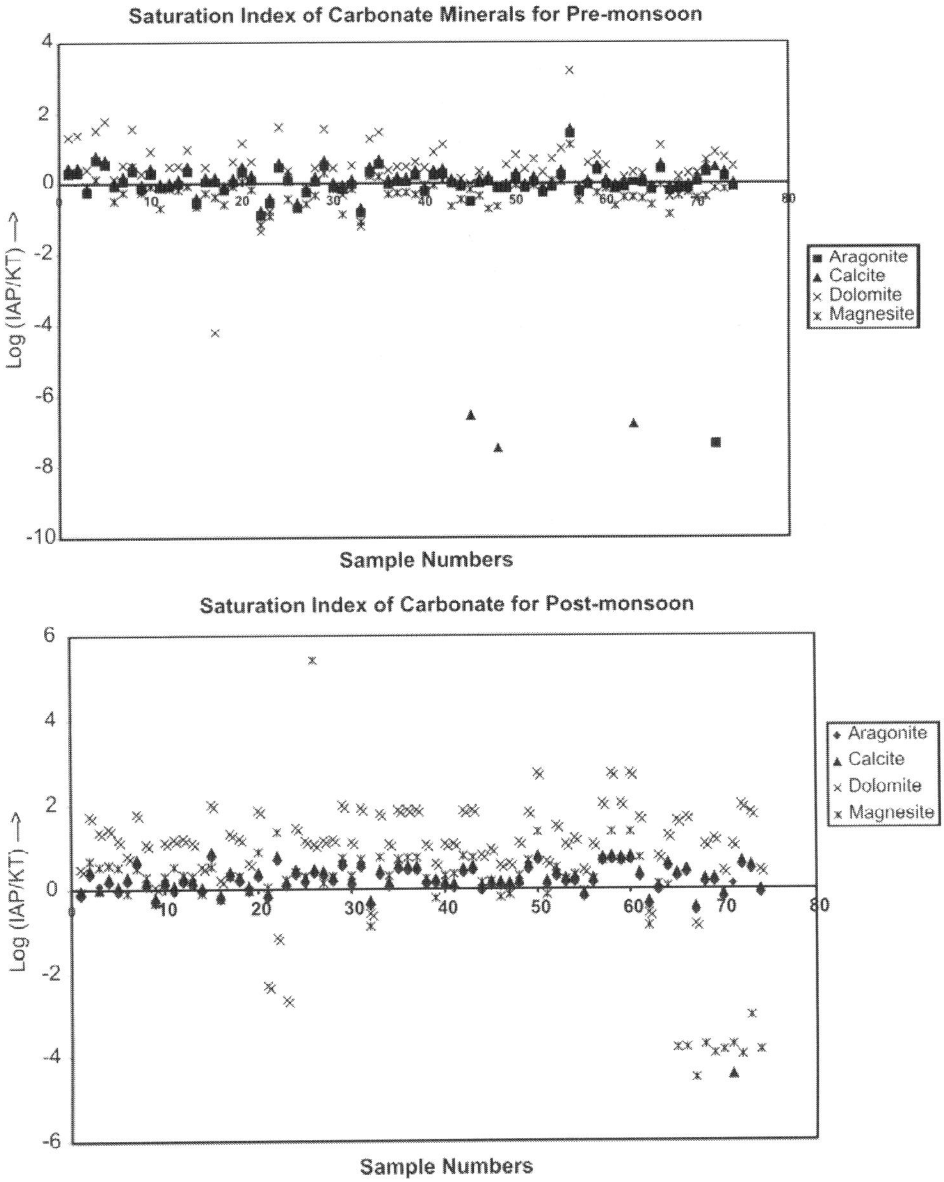

Fig. 4. a and b. Saturation index of carbonate minerals for pre- and post-monsoon season.

The disequilibrium indices of Fluoride (Fig. 5a and b) Fluorapatite, FCO_3 apatite and Hydroxyapatite were represented. The status of groundwater in pre-monsoon indicates that all the minerals are in the state of near equilibrium to oversaturation during both seasons which indicate greater dissolution of fluoride rich minerals from aquifer matrix.

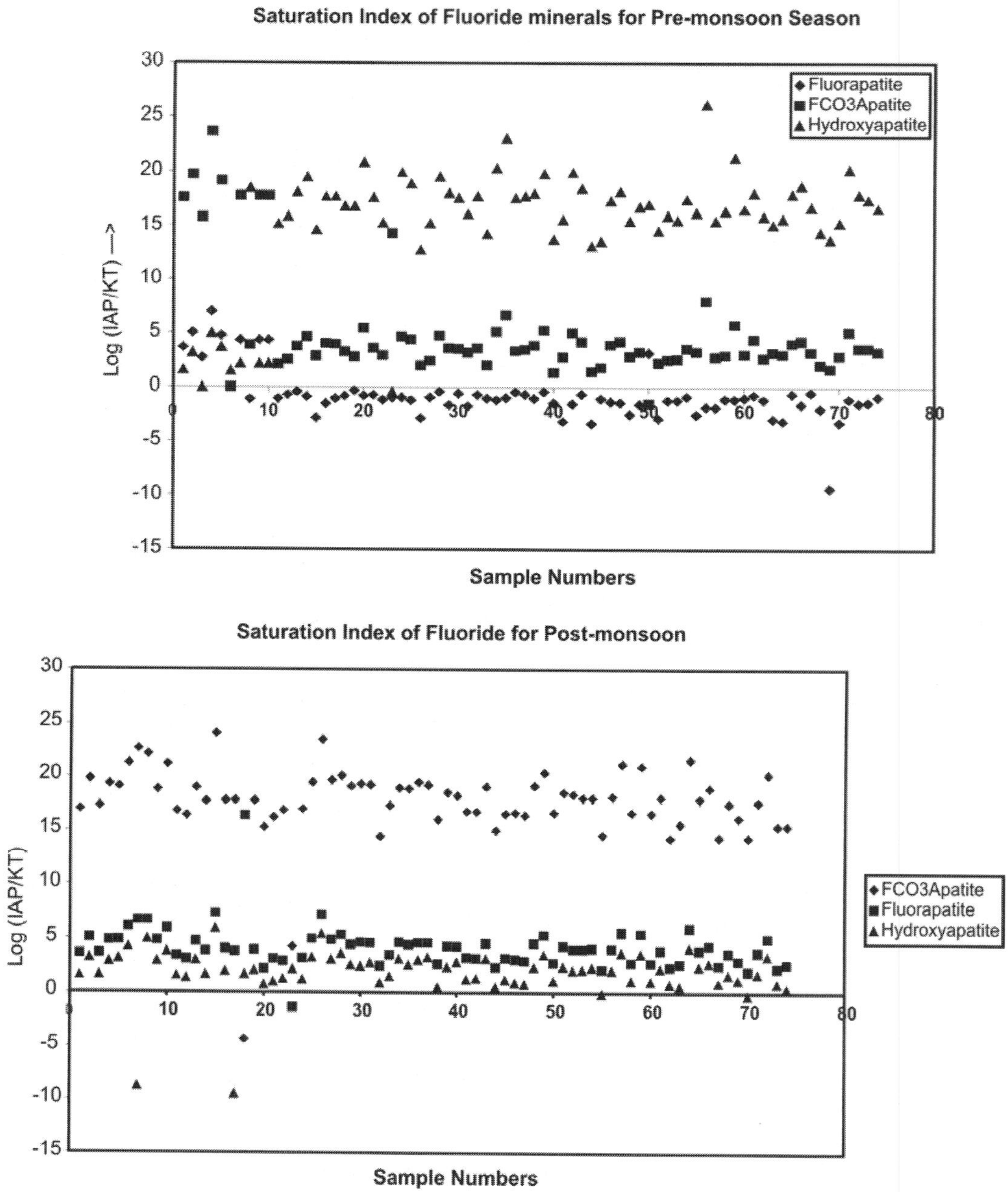

Fig. 5. a and b. Saturation index of fluoride minerals for pre- and post-monsoon season.

The disequilibrium indices of Silicate minerals (Fig. 6a and b) Chalcedony, Quartz, SiO_2(a) and Cristobalite were represented for dissociation factors in the study area. During both seasons, silicate minerals represent the state of undersaturation to oversaturation due to the Ca and Mg released from silicate weathering process.

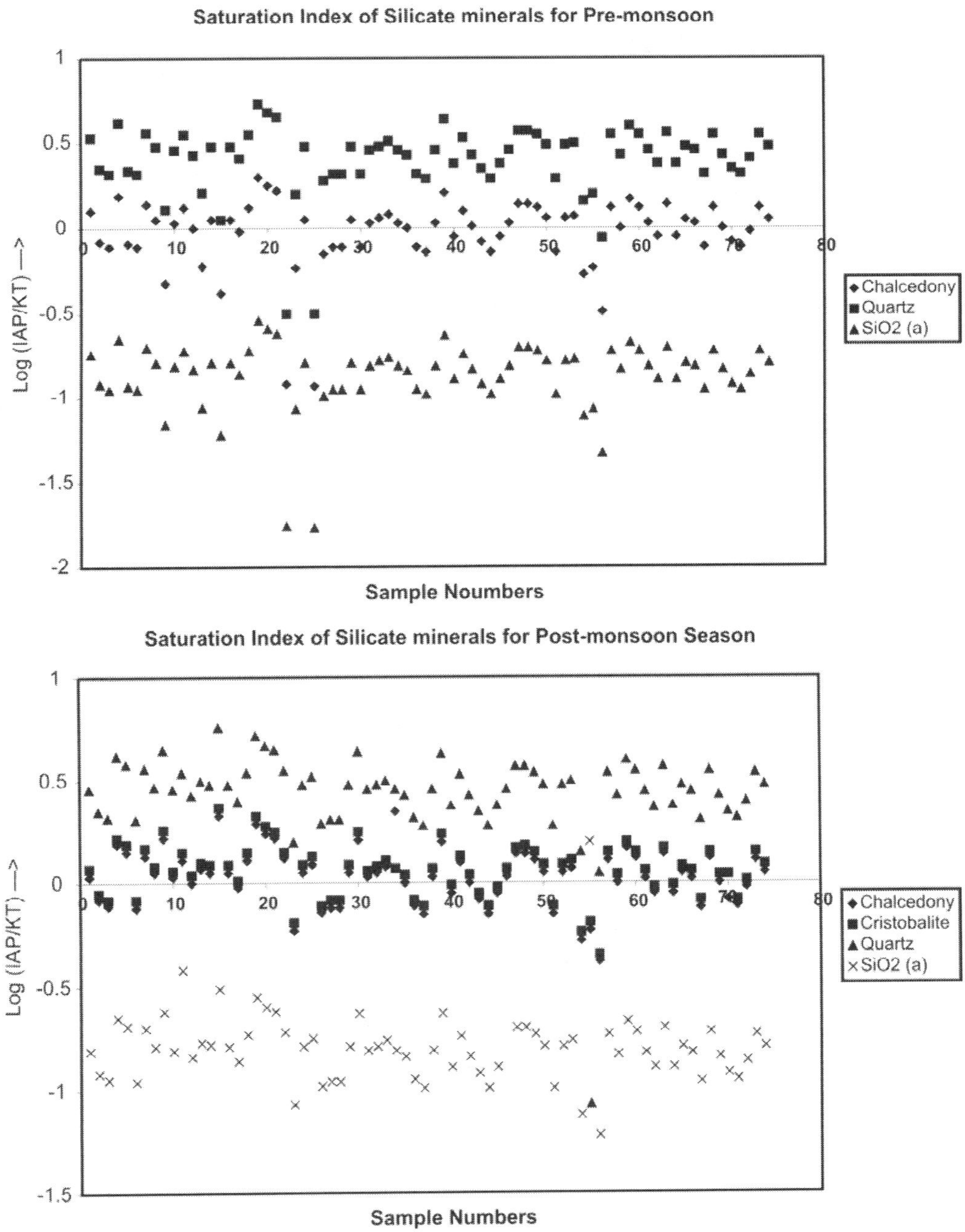

Fig. 6. a and b. Saturation index of silicate minerals for pre- and post-monsoon season.

The disequilibrium indices of sulphate minerals (Fig. 7a and b) Anhydrite, Epsomite, Thenandrite, Miralbite and Gypsum were represented and studied for dissociation factors in the study area. During both the seasons all the minerals falls in the state of undersaturation indicating lesser breakdown of sulfate minerals into the groundwater.

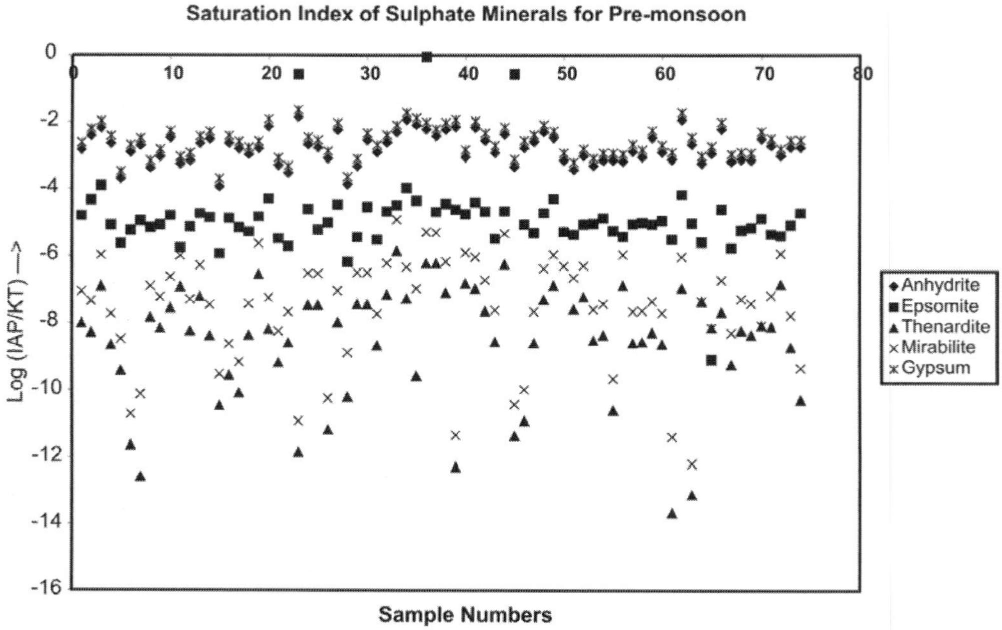

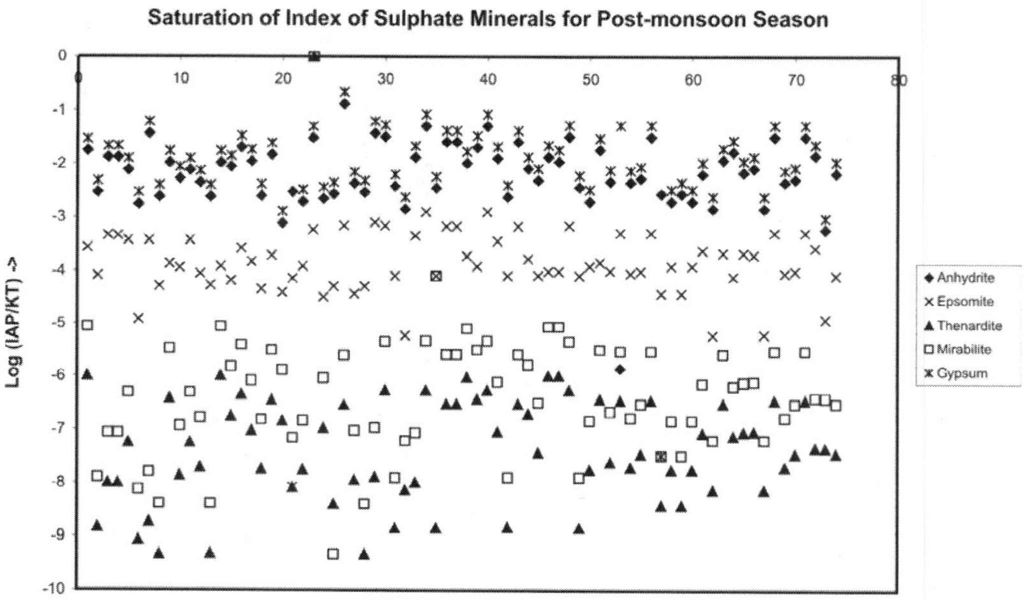

Fig. 7. a and b. Saturation index of sulphate minerals for pre- and post-monsoon season.

CONCLUSION

The hydrochemical investigation indicates that groundwater salinity is highly variable with significant zonation along groundwater flow direction from recharge area to discharge area. The ion exchange process controls the groundwater chemistry of the area. In POM season, percentage of water quality index rating is decreased; it may be due to over-exploitation of groundwater. Water quality in the study area is slowly reaching alarming stage so that proper planning is essential in this venture to preserve the fragile ecosystem. In general, in the recharge and discharge area groundwater quality is good, but when we enter Salem district it gets polluted due to the urbanization, industrialization and direct discharge of effluent. Saturation index of minerals indicates oversaturation and equilibrium state of silicate, carbonate and fluoride minerals during both seasons. Sulphate minerals revealed undersaturated state during both seasons. In general, water chemistry is guided by lithological influences on water chemistry by complex weathering process and ion exchange along with influence of ions from anthropogenic impact.

REFERENCES

APHA, (1995), Standard methods for the examination of water and waste water; (APHA).

BIS, (Bureau of Indian Standards) 10500 (1991). Indian Standard drinking water-specification. First revision, pp. 1–8.

Datta, P.S. and Tyagi, S.K. (1996). Major ion chemistry of groundwater in Delhi area: Chemical weathering processes and groundwater flow regime: *J. Geo. Soc. India*, **Vol. 47**, pp. 179-188.

Deutsch, W.J. (1997), Groundwater geochemistry: fundamentals and application to contamination. CRC, Boca Raton.

Edmunds, W.M., Carrillo-Rivera, J.J. and Cardona. A. (2002). geochemical evolution of groundwater beneath Mexico City. *J. Hydrol.* 258: 1-24.

Freeze, R.A. and Cherry, J.A. (1979). Groundwater: New Jersey, Prentice Hall.

Garrels, R.M. (1976). A survey of low temperature water mineral relations, in Interpretation of environmental isotope and hydrogeochemical data in groundwater hydrology: Vienna, International Atomic Energy Agency, pp. 65-84.

Hem, J.D. (1985). Study and interpretation of the chemical characteristics of natural water (3rd ed.): U.S. Geological Survey, Water Supply Paper 2254.

Lakshmanan, E., Kannan, K. and Senthil Kumar, M. (2003). Major ion chemistry and Identification of hydrogeochemical process of groundwater in part of Kancheepuram district, Tamilnadu, India, *J. Environ. Geosci.*, **Vol. 10**, pp. 157-166.

Loyd, J.W. and Heathcode, J.A. (1985). Nature inorganic hydrochemistry in relation to groundwater. Oxford University Press, New York

Milovanovic, M. (2007). Water quality assessment and determination of pollution sources along the Axios/Vardar River, Southeastern Europe. *Desalination* **213:** 159-173.

Mitra, B.K. and ASABE Member. (1998). Spatial and Temporal Variation of Groundwater Quality in Sand Dune Area of Aomori Prefecture in Japan.

Paulami Sahu, and Sikdar, P.K. (2008). Hydrochemical framework of the aquifer in and around East Kolkata Wetlands, West Bengal, *India Environ. Geol.,* **55:** 823-835.

Saxena, V.K. (2004). Geothermal Resources of India, Allied Publishers Pvt. Ltd., pp. 48-70.

Smith, K. (2001). Environment hazards: assessing risk and reducing disaster, 3rd ed. London: Routlege, p. 324.

Srinivasamoorthy, K., Chidambaram, M., Prasanna, M.V., Vasanthavigar, M., John Peter, A. and Anandhan, P. (2008). Identification of major sources controlling Groundwater Chemistry from a hard rock terrain – A case study from Mettur taluk, Salem district, Tamilnadu, India. *J. Earth Syst. Sci.* **117:** No.1. pp. 49-58.

Subba Rao, N. (1997). Studies on water quality index in hard rock terrain of Guntur district, Andhra Pradesh, India. National Seminar on Hydrology of Precambrian Terrains and hard rock areas, pp 129-134.

Tiwari, T.N. and Mishra, M.A. (1985). A preliminary assignment of water quality index of major Indian rivers. *Indian J. Environ. Proc.,* **5:** 276-279.

Trusdell, A.H. and Jones, B.F. (1973). Wateq: A computer program for calculating chemical equilibria of natural waters; *J. Research USGS* **2(2):** pp. 233-248.

WHO (World Health Organization), (1993). Guideline for drinking water quality, 2nd ed., **Vol 1:** p. 188.

12

Hydrogeochemical Study of Groundwater for Rural Development — A Case Study from Ponnarampatty Area of Valapadi Block, Salem District, Tamil Nadu

Asim Amitabh Pradhan*, Loganathan, R., Vanaroja, K. and Sivakumar, R.
Department of Geology, Presidency College (Autonomous), Chennai–600005
**Email: tinu.pradhan@gmail.com*

ABSTRACT

Water is very essential for our life, so one should know the quality of the water being used for various purposes. Groundwater is one of the most important sources of water for drinking, irrigation and industrial uses. Therefore, it requires a proper monitoring of its quality. The quality of water depends upon various processes and reactions that start from the moment it condensed in the atmosphere upto the time it discharges from the springs or well. Thus, an attempt has been made to evaluate the quality of groundwater for rural development from the surrounding area of Ponnarampatty, which is situated at the southern part of Valapadi Block, Salem District, Tamil Nadu. Nearly 15 samples were collected from the area and are physico-chemically analyzed. The quality analysis is made through the estimation of Ca, Na, K, Mg, TDS, EC and other anions. Groundwater of this area is free from P, Mn, N and NH_3. Maximum TDS observed in this area is 1687 mg/L. This high concentration of TDS observed in this area may be due to infiltration of effluent to the groundwater from the pharmaceutical industry which is located near the study area. In Pudur Komarapalayam area the iron concentration also exceeds the permissible limit so the water is not suitable for drinking purpose. All the groundwater samples except the few are suitable for domestic, agriculture as well as industrial purposes.

Keywords: Geochemistry, hardness, salinity, corrosivity ratio.

INTRODUCTION

Water is essential for sustenance of life, next to air, and is one of the principal elements, which influence economic, industrial and agricultural growth of mankind. Groundwater occurs in various conditions depending not only on geological environment of the area, but also depending on other factors such as topography of the area, porosity, permeability of the formation and weathering and fracturing of rocks. Since the groundwater occurrence depends on the above conditions the appropriate combination of scientific methods is required for the proper exploration and management of groundwater resources. The groundwater is polluted by industrial waste, agricultural waste, seawater intrusion, sanitary sewage, marine pollution, nuclear plant waste and acid rain. This type of polluted water spreads disease to human and other biotic life. In some area, land is polluted by fertilizers and possible lithogenic sources. Since the water is being widely used for drinking, domestic, irrigational and industrial purposes the determination of its quality becomes more important, as the intake of excess of certain minerals may lead to health disorders in humans as well as livestock. Similarly, the use of unsuitable waters may lead to undesirable results in agriculture and industries (Todd, 1980). Thus, an attempt has been made in Ponnarampatty for hydrogeological studies to find out water quality.

STUDY AREA

The study area is located between 78°15’E and 78°27’E longitude and between 11°29’N and 11°40’N latitude and is covered by various geological formations ranging from Archaean to Recent. The crystalline rocks like hornblende gneiss, charnockite of Archaean age to recent formations occupy 73% of total geographical area of the study area.

Methodology

Field study had been carried out and water samples were collected from 15 locations in the study area, and hydrogeochemical analyses (Apha, 1985) had been carried out for those samples (Tables 1 and 2). Trace element study had been carried out for 5 samples by using Atomic Absorption Spectrometer (AAS GBC Avanta). Thematic maps had been prepared for various physico-chemical and geochemical parameters for the study area and interpreted.

Results and Discussion

Balasubramanian and Sastri (1986) has developed a computer program (HYCH) for classifying the groundwater (Tables 1 and 2). The output data for collected samples are analysed and the quality assessment has been done by employing HYCH program and the following information thus obtained from the analyzed results of groundwater are discussed in detail.

The pH value of a solution is the negative logarithm of the concentration of hydrogen ions in moles per litre. In the study area, the pH values for the water samples ranges from 6.81 (Somampatty) to 7.46 (Citampatty). So in the study area pH is not at all a problem because it is within the permissible limit.

The electrical conductance of a substance is its ability to conduct an electrical current. Current flows in ionized or mineralized water because the ions are electrically charged and

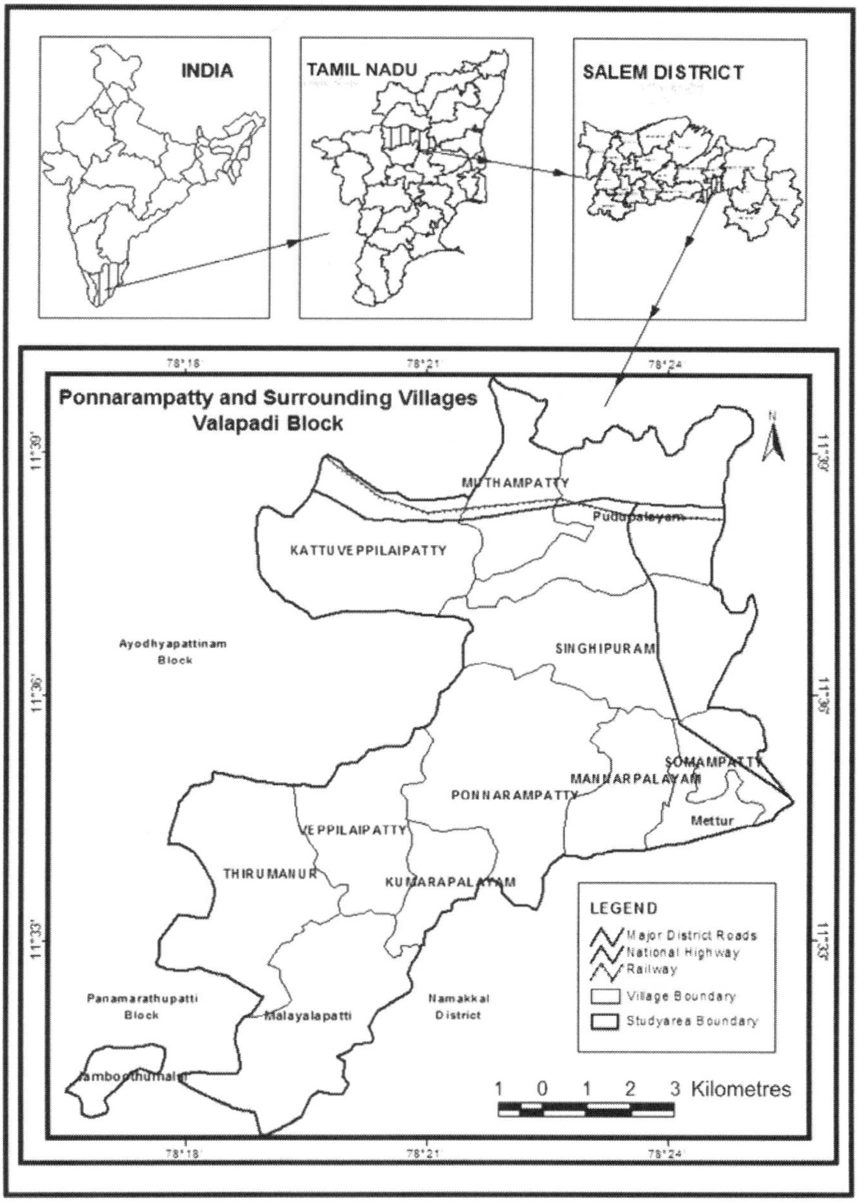

Fig. 1. Location map.

move towards a current source that will neutralize them. Electrical conductivity for the study area ranges from 755 mhos/cm in Ponnarampatty to 2410 mhos/cm in Komarapalayam. The electrical conductivity is higher at Komarapalayam as it is 2410 mho/cm while in all the other places the EC is within the permissible limit.

Table 1: Chemical parameters of the groundwater from Ponnarampatty area in ppm.

Sample no	Sample Location	pH	EC	TH	Ca	Mg	Na+K	HCO_3	Cl	NO_3	SO_4	TDS
1.	Thirumanur Kopangadu	6.96	859	312.1	74.0	31.0	29.0	348.0	32.0	13.0	37.0	601
2.	Thirumanur West	6.91	1215	322.8	75.0	33.0	95.0	276.0	168.0	34.0	43.0	850
3.	Annapuram	6.97	1826	568.7	134.0	57.0	168.0	304.0	360.0	42.0	95.0	1278
4.	Mudiyanur	6.92	1399	501.4	112.0	54.0	70.0	352.0	184.0	38.0	76.0	979
5.	Citampatty	7.46	955	382.4	89.0	39.0	18.0	348.0	64.0	15.0	37.0	668
6.	Mandakadu Malayapatty	6.82	1496	482.5	111.0	50.0	101.0	368.0	232.0	40.0	46.0	1047
7.	Tekkalpatty	7.02	1320	302.1	70.0	31.0	135.0	464.0	124.0	36.0	15.0	924
8.	V.Pudur Komarapalayam	7.01	1844	528.0	121.0	55.0	156.0	376.0	320.0	46.0	72.0	1290
9.	Komarapalayam	6.95	2410	783.5	174.0	85.0	165.0	456.0	480.0	73.0	33.0	1687
10.	Paravakadu	7.16	1507	438.8	105.0	43.0	113.0	360.0	212.0	40.0	62.0	1054
11.	Ponnarampatty	6.97	755	256.6	60.0	26.0	28.0	272.0	48.0	12.0	22.0	528
12.	Singipuram	7.02	1722	523.0	119.0	55.0	128.0	440.0	268.0	42.0	31.0	1205
13.	Mannarpalayam	7.53	986	340.1	77.0	36.0	77.0	392.0	84.0	15.0	42.0	690
14.	Somampatty	6.81	945	315.3	72.0	33.0	52.0	316.0	76.0	13.0	52.0	661
15.	Kanavaymedu	7.01	1000	340.1	77.0	36.0	45.0	360.0	68.0	21.0	35.0	700

Table 2: Geochemical parameters of the groundwater from Ponnarampatty.

Sample No.	Sample Location	Non-Carbonate Hardness	CR	SAR	RSC	Gibb's plot	Piper's Hydrogeochemical Facies
1.	Thirumanur Kopangadu	27.02	0.24	0.71	-0.54	Rock Interaction	Recent Dolomitic
2.	Thirumanur West	96.74	1.02	2.30	-1.93	Rock Interaction	Static And Disco-Ordinated
3.	Annapuram	319.75	1.99	3.06	-6.40	Rock Interaction	Contaminated With Gypsum
4.	Mudiyanur	213.19	0.96	1.36	-4.26	Rock Interaction	Static And Disco-Ordinated
5.	Citampatty	97.36	0.37	0.40	-1.95	Rock Interaction	Static And Disco-Ordinated
6.	Mandakadu Malayapatty	181.12	1.02	2.00	-3.62	Rock Interaction	Static And Disco-Ordinated
7.	Tekkalpatty	78.01	0.41	3.38	1.56	Rock Interaction	Dissolution And Mixing
8.	V.Pudur Komarapalayam	220.09	1.40	2.95	-4.40	Rock Interaction	Static And Disco-Ordinated
9.	Komarapalayam	410.21	1.56	2.56	-8.20	Rock Interaction	Static And Disco-Ordinated
10.	Paravakadu	143.91	1.01	2.35	-2.88	Rock Interaction	Static And Disco-Ordinated
11.	Ponnarampatty	33.80	0.33	0.76	-0.68	Rock Interaction	Static And Disco-Ordinated
12.	Singipuram	162.66	0.93	2.43	-3.25	Rock Interaction	Static And Disco-Ordinated
13.	Mannarpalayam	19.03	0.41	1.81	-0.38	Rock Interaction	Static And Disco-Ordinated
14.	Somampatty	56.48	0.51	1.27	-1.12	Rock Interaction	Static And Disco-Ordinated
15.	Kanavaymedu	45.25	0.37	1.06	-0.90	Rock Interaction	Static And Disco-Ordinated

The bulk of the Total Dissolved Solids include bicarbonate, sulphates and chlorides of calcium, magnesium, sodium and silica. In the study area, the total dissolved solid (TDS) of the groundwater is above 1000 ppm in six locations (Fig. 2) at the south-eastern side of the study area which are brackish (Carrol.D1962) in nature. Moderate quality water is seen at all other parts of the study area whereas not even a single place is noted in the study area having TDS less than 500 mg/lit.

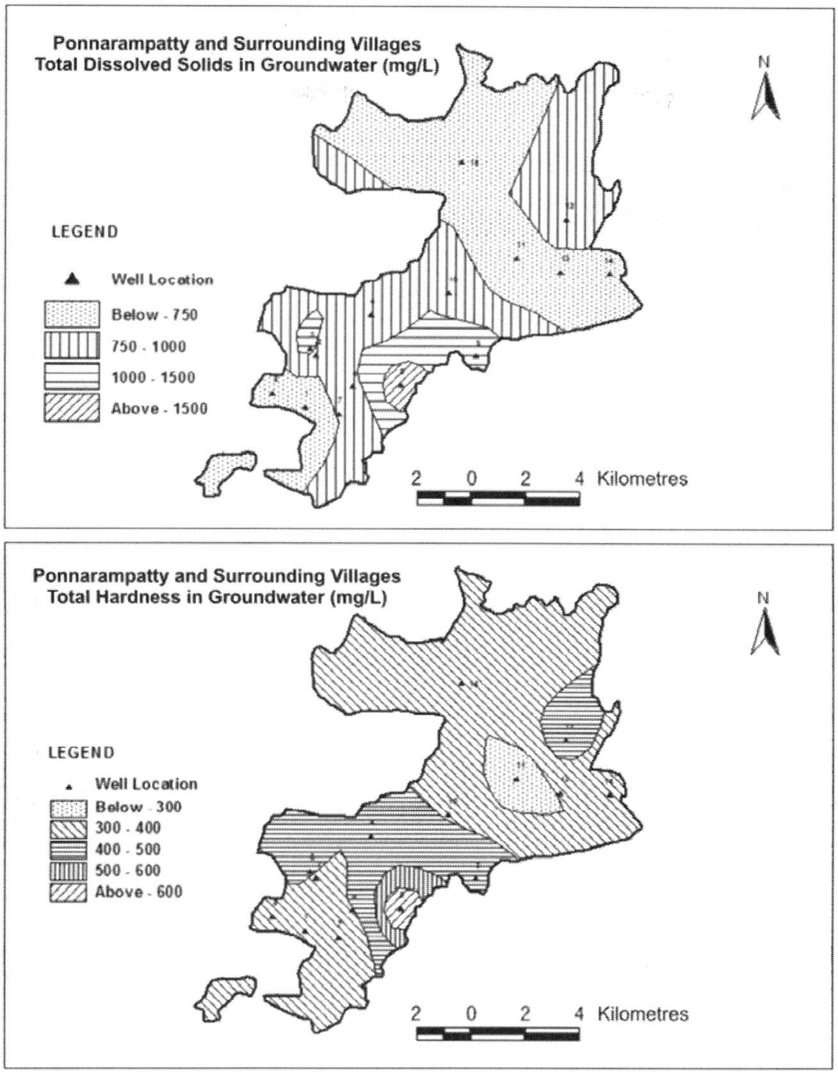

Fig. 2. TDS and total hardness map.

Hardness in water is primarily caused by calcium and magnesium cations although some heavy metals such as iron and manganese also consume soap.

$$\text{Total Hardness (TH)} = 2.497\ \text{Ca} + 4.115\ \text{Mg}\ \text{(Karanth, 1993).}$$

With reference to total hardness (Fig. 2), all the study areas except Ponnarampatty have extremely hard water while Ponnarampatty have hard water.

The principal sources of calcium in groundwater are some members of the silicate mineral group. Silicate minerals are not soluble in water, but weathering breaks them down into soluble calcium products and clay minerals. The carbonates and sulphates of calcium, however, are soluble in most natural water. According to the calcium content of the study area it exceeds the permissible limit of 110 ppm at the south-western places around Komarapalayam.

The concentration of magnesium in groundwater can attain a rather wide range. It is derived from the major natural sources like amphiboles, olivine, pyroxenes, dolomite and magnesite. However, most groundwater contain up to a few hundred ppm of magnesium. The magnesium content in the study area is within the permissible limit.

Sodium, Potassium Bicarbonate, Fluoride and Nickel content in the study area is found to be within the permissible limit. The southern and the eastern part of the study area contains

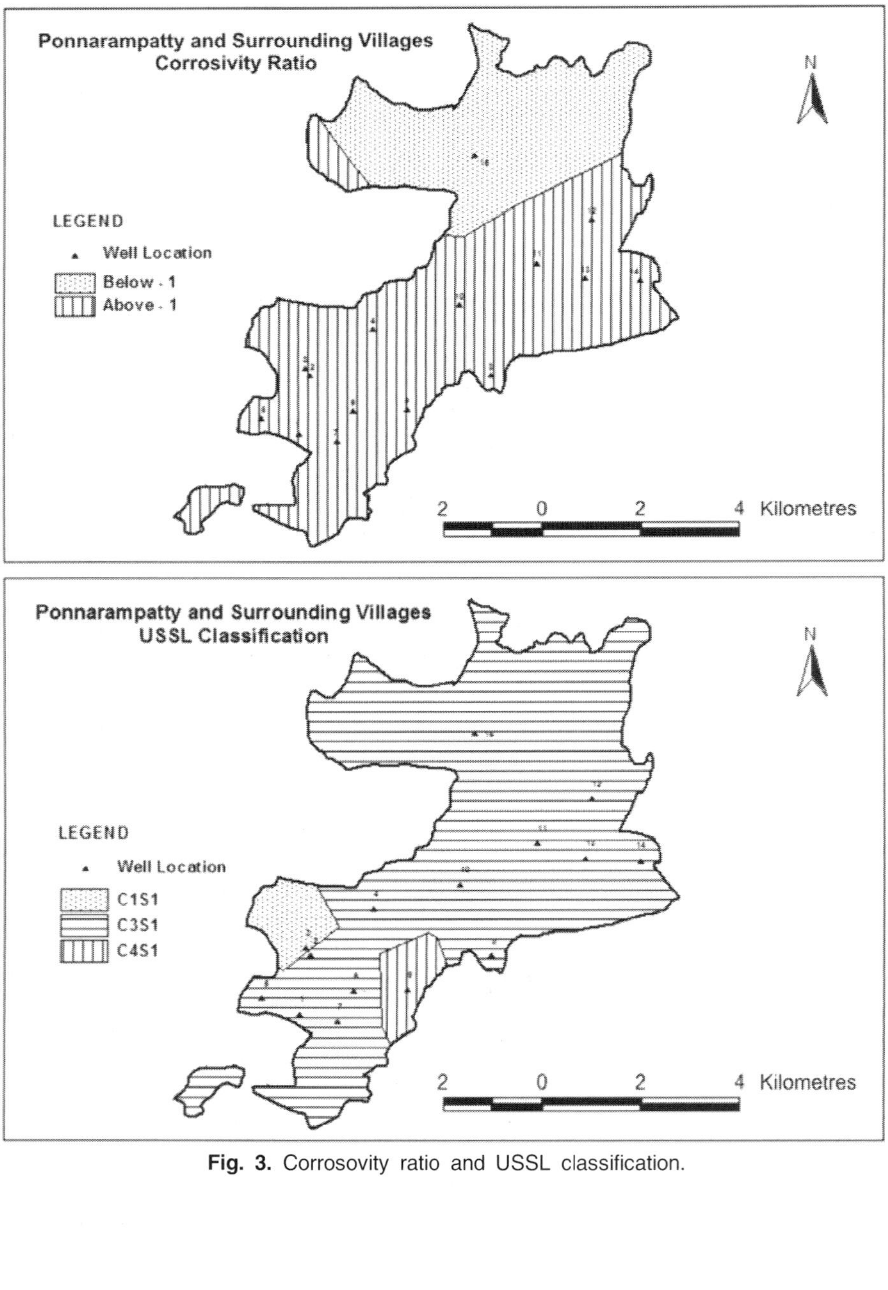

Fig. 3. Corrosovity ratio and USSL classification.

chlorine more than the permissible limit of 150 ppm, thus, they are of the class fresh brackish and at Komarapalayam it is very high of 480 ppm, thus, it falls under the class brackish.

The eastern part of the area has the corrosive ratio (Ryzner 1994) of more than one (Fig. 3) so in these areas one must use PVC pipes for supply of water whereas in all other places one can use metal pipe for the same purposes.

The United States Salinity Laboratory (USSL) has constructed a diagram (Fig. 3) for classification of irrigation water describing for classification of irrigation water describing 16 classes, with reference to SAR as an index for sodium hazards and specific conductance micromhos/cm as an index of salinity hazards. USSL classification shows three types of irrigation water in the study area. The majority of the samples occur in C3S1 type, where as the remaining samples falls C1S1 and C4S1 types.

In Gibb's classification (Gibb's 1970) the majority of the groundwater samples in the study area (Fig. 4) falls in the rock interaction area, which indicates the interaction between rock chemistry and the percolation waters in the subsurface.

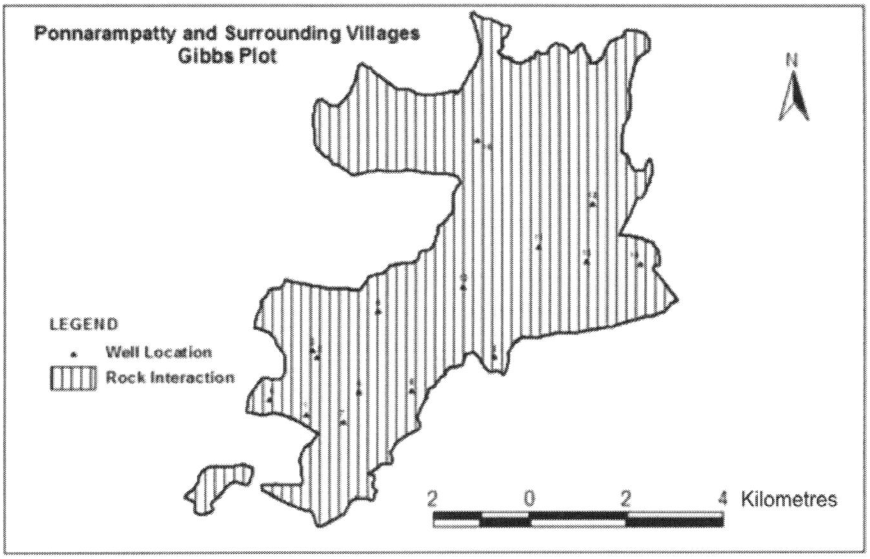

Fig. 4. Gibbs plot.

CONCLUSION

From the geochemical studies the following conclusions are derived;

1. The geology of the area constitutes Charnockite (Archaean) formations.
2. The predominant geomorphologic feature in the study area is pediment.
3. The pH of the groundwater is neutral in nature.
4. With reference to total hardness, all the study area except Ponnarampatty has very hard water.
5. The corrosivity ratio indicates that in the majority of the study area the water can be transported through metal pipes.

6. USSL classification shows that the samples reveal three types of irrigation water in the study area. The majority of the samples occur in C3S1 type, whereas the remaining samples falls C1S1 and C4S1 types.
7. The Gibb's diagram indicates that all the water samples in the study area fall in the rock interaction field.

ACKNOWLEDGEMENTS

Authors thank Prof. **Dr. G.B. Sukumaran, Head of Department of Geology** wholeheartedly for his meticulous guidance on the work and constant encouragement throughout the period of study. Trace element analysis of groundwater samples were carried out in the **Department of Geology, Presidency College (autonomous)**, Chennai–600 005 using the **GBC Avanta make AAS (Atomic Absorption Spectrophotometer)** procured from the **DST-FIST fund** which is sincerely acknowledged.

REFERENCES

Apha-Awwa (1985). Standard method for the examination of water and waste water, 16th edition-Washington D.C. APHA **1:** 1268.

Carrol, D. (1962). Rain water as a chemical agent of geologic processes. A review U.S. Geological survey water supply paper 520-f pp. 97-104.

Gibb's, R.J. (1970). Mechanism controlling worlds water chemistry science 170 pp. 1088-1090.

Karanth, K.R. (1993). Groundwater assessment development and management Tata McGraw-Hill Publishing Company Ltd. pp. 715.

Richards, C.A. (1954). Diagnoses and improvement of saline alkali soils USDA handbook, **60:** 160.

Ryzner, J.W. (1994). A New index for determining amount of calcium carbonate scale formed by water, *J. Ames W, Wass.* **36:** 472-486.

Todd, D.K. (1980). Groundwater hydrology John Wiley, pp. 535.

USSL, (1995). Classification of use of irrigation water, U.S. Department of Agriculture, Circu, 969, Washington.

13

Current Status of Nitrate Ion Contamination in Sopore Town, Kashmir

Imran A. Dar*, Mithas A. Dar and Sankar, A.

Department of Industries and Earth Science, Tamil University, Thanjavur–603001, Tamil Nadu, India

**Email: wonder_env@yahoo.com*

ABSTRACT

The prime objective of this research work is to evaluate the concentration of nitrate in groundwater of town and its environs. Systematic sampling was done, with a view to understand the source of nitrate concentration in the study area. Fifteen sample sites were selected and the samples were taken for a baseline study to understand the geochemistry of the study area and to assess its physico-chemical characteristics. The water quality parameters were investigated for summer (May, 2007) and winter (December, 2007) seasons and were compared with the standard values given by ICMR / WHO. The hydrochemical data of 15 samples indicates that the concentration of almost all parameters fall within the permissible limits except nitrate. Linear Trend Analysis on seasonal and annual basis clearly depicted that nitrate pollution in the study area is increasing significantly. About 85% of samples during summer season and 67% of the samples during winter season were showing the high concentration of nitrate, exceeding permissible limit of WHO (50 mg/L), which is due to the use of nitrogenous fertilizers in the study area. Appropriate methods for improving the water quality in the affected areas have been suggested.

Keywords: Hydrochemistry, fertilizers, nitrate, apple town.

INTRODUCTION

Water is one of the most essential requirements of all living beings. For a long time groundwater has been considered as a well protected resource. The reason for this was the belief in self-purification of the soil and as a rule the protection of groundwater by the covering layers. Man's influence on the quality of water is quite apparent and now a major concern. Groundwater has to be protected generally as it forms a principal source for drinking water and as it represents also a precious ecological part within the balance of water cycle. During recent years much of the emphasis is shown on in groundwater investigations in industrialized country. Rapid urbanization brings with it many problems as it places huge demands on land, water, housing, transport, health, education, etc. Environmental pollution has reached alarming levels in the last 5-6 years mainly due to industries and automobiles. This has shifted from problems of groundwater supply to considerations of groundwater quality. Freshwater being one of the basic necessities for sustenance of life, the human race through the ages has striven to locate and develop it. Water, a vital source of life in its natural state is free from pollution but when man tampers the water body it loses its natural conditions. Groundwater has become an essential resource over the past few decades due to the increase in its usage for drinking, irrigation and industrial uses, etc. The quality of groundwater is equally important as that of quantity. Groundwater is an essential natural resource for sustaining life and environment which is available in abundance and free gift of nature. Landuse practices have greatly polluted the groundwater quality. Nitrate is a widespread contaminant of ground and surface waters worldwide [1, 2]. The accumulation of nitrate in the environment results mainly from:

1. Non-point source runoff from the over-application of nitrogen fertilizers;
2. Point-sources such as Concentrated Animal Feeding Operations [3].
3. Point-sources from poorly or untreated human sewage.

Although nitrate and the other nitrogenous compounds are the essential elements in the life process of flora and fauna, nitrate is potentially hazardous when present in drinking water at sufficiently high concentrations [4]. As stated in the EPA publication "Is Your Drinking Water Safe?" [5] "Only two substances for which standards have been set pose an immediate threat to health whenever they are exceeded: bacteria and nitrate." In addition, high nitrate levels in water and feed lead to reduced vitality and increased stillbirth, low birth weight, and slow weight gain in livestock [6]. Nitrate level is monitored in municipal water supplies worldwide, and in foodstuffs, to prevent exposure of populations to harmful or toxic levels. Nitrate is contributed through biochemical activity by the microorganisms, freely and symbiotic species, such as nitrosomonas and nitrobactor [7]. Numerous studies have been done on nitrate contamination of groundwater in India and other countries [8]. Numerous sources in the environment contribute to the total nitrate content of natural waters [9] viz., atmosphere, geological sources, soils atmosphere, nitrogen fixation, human, animal wastes and agriculture. Lack of good sanitary practices and improper drainage systems may cause high rate of nitrate in the groundwater. The main thrust of this paper is to provide a methodological approach to explain high nitrate concentrations. The toxicity of nitrate to humans is due to body's reduction of nitrates to nitrite which is demonstrated by vascodilatory/cardiovascular effects at high dose levels and methemoglobinemia at lower dose levels [10]. Consumption of drinking water with nitrate, at concentrations greater than 50 mg/L causes Blue baby's syndrome, a disease where the skin becomes blue due to decreased efficiency of haemoglobin to carry the oxygen [11]. This phenomenon can occur in infants when approximately 70% of total haemoglobin has been converted to methaemoglobin [12].

High levels of nitrate in livestock feed and drinking water can result in reduced vitality and increased stillbirth, low birth weight, and slow weight gain and even death of the animals affected [6]. Chronic nitrate poisoning is correlated with abortions, stillbirths and stunted calves.

Abortion is attributed to maternal and foetal methemoglobinemia resulting in foetal anoxia (particularly in the last trimester of pregnancy). More recently, unpublished data indicates serum progesterone is reduced in chronic nitrate toxicosis in cattle and, possibly, horses. Chronic nitrate toxicosis causes loss of condition, loss of weight, reduced milk production and weakness. Deficiencies of vitamin A and E and hypothyroidism are reported in cattle and swine.

STUDY AREA

The study area consists of Apple (Sopore) town and its environs, of Baramulla District Kashmir, Consisting the total area of 61 km^2, and lies between Coordinates: 34°18'N 74°28'E to 34° 3'N 74°47'E, with an average elevation of 1,556 m (5,105 ft) from Mean Sea Level (Fig. 1). The Sopore town is famous because of dominance of apple trees and hence often called as an "Apple Town". The seasons are marked with sudden change and the year is divided into six seasons of two months each (Table 1). Annual rainfall of the valley recorded is about 75 cms. It rains in July and August and also in March and April. August is the warmest month. Temperature rises to 85^0 F. January is the coldest month. Temperature falls down to below zero. Longest sunshine hours are in September, October and November.

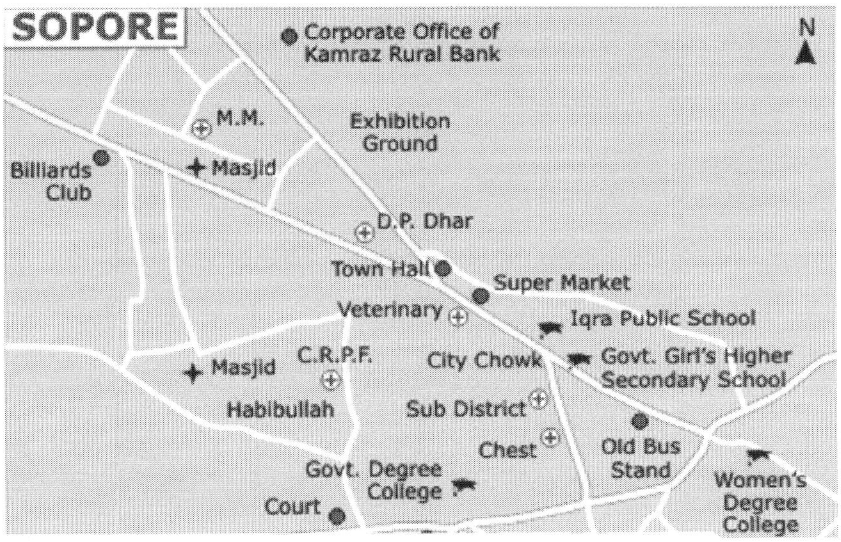

Fig. 1. Map of study area

Table 1: Seasons of study area with duration.

Seasons	Duration
Spring	March 15 to May 15
Summer	May 15 to July 15
Rainy	July 15 to September 15
Autumn	September 15 to November 15
Winter	November 15 to January 15
Ice-Cold	January 15 to March 15

Materials and Methods

A total of 15 samples from shallow wells and deep tube-wells were collected from various locations of the study area during summer (May, 2007) and winter (December, 2007) seasons. The details of the samples are given in Table 2. Samples were collected in 1-1 capacity polyethylene bottles. Prior to collection the bottles were thoroughly washed with diluted Nitric acid (HNO_3) and then with distilled water in the laboratory. The bottles were rinsed to avoid any possible contamination in bottling and every precautionary measure was taken. Methods of collection and analysis of water samples was adopted using standard protocols [13]. Then the samples were sealed, numbered and were carefully taken to the laboratory for the chemical analysis.

Table 2: Samples with mohalla name

Serial No.	Mohalla/ Colony	Sample No.
1.	Tawheed Bagh	S1
2.	Mumkak	S2
3.	Khwaja Gilgit	S3
4.	Colony Shalimar	S4
5.	Muslimpeer	S5
6.	Ashpeer	S6
7.	Sangrama	S7
8.	Duru	S8
9.	Shiva	S9
10.	Dangerpora	S10
11.	Nowpora	S11
12.	Alsafa Colony	S12
13.	Badam Bagh	S13
14.	New Colony	S14
15.	Shah Abad	S15

Results and Discussion

The different quality parameters of the study area were determined for 15 samples each for summer (May, 2007) and winter (December, 2007) and the results are presented in Tables 3 and 4. The values were compared with the standard values given by WHO and ICMR shown in Table 5. The nitrate concentration at 15 sampling stations in years 2003 to 2007 is shown in Table 6. Chemical analysis of nitrate shows that nitrate concentration at most places is exceeding the permissible limit; 85% during summer and 67% during winter season. From Fig. 2 it is clear that the value of nitrate concentration is found maximum in sample S14 (91mg/L), followed by S7 (81 mg/L) and S4 (78 mg/L) representing the drinking waters of Badambagh, Colony Shalimar, and Sangrama areas, respectively. The main source of this nitrate pollution during summer season was found to be the excessive use of nitrogenous fertilizers, as these areas are mainly agricultural areas. The water samples of S14, S7 and S4, presently threatening the health of people residing there was once called as an "ELIXIR" by the local people because of its purity.

Table 3: Analytical data of the groundwater during summer season (May, 2007).

S.No	pH	EC	TDS	Na	Ca	Mg	SO_4	Cl	F	NO_3^-
1	7.4	423	208	612	178	178	167	210	1.3	33
2	7.9	510	300	700	225	197	152	240	1.2	51
3	7.6	340	311	720	234	210	157	210	1.1	53
4	7.7	340	200	501	177	156	125	175	1.3	78
5	7.7	440	210	640	188	168	225	225	1.2	52
6	7.9	333	198	488	138	145	210	150	1.2	50
7	8.0	372	225	490	142	146	234	127	0.5	81
8	7.6	448	245	670	202	173	246	148	0.2	32
9	7.4	557	325	750	264	226	157	273	0.8	51
10	7.8	357	200	490	142	150	75	118	1.4	51
11	8.1	466	215	488	198	183	143	234	0.9	56
12	7.3	440	210	633	188	179	167	229	1.1	32
13	7.5	497	300	550	208	182	125	237	1.1	56
14	7.7	450	315	645	189	175	184	228	1.2	91
15	7.7	385	285	500	176	149	63	211	1.3	54

Except pH all parameters are measured in mg/L

Table 4: Analytical data of the groundwater during winter season (Dec., 2007).

S.No	pH	EC	TDS	Na	Ca	Mg	SO_4^-	Cl	F	NO_3^-
1	7.4	452	266	612	175	180	160	200	1.9	51
2	7.3	413	245	580	165	170	113	210	1.6	52
3	7.7	340	200	412	127	140	101	188	1.5	28
4	7.5	512	302	700	225	225	175	235	2.3	56
5	7.6	398	235	400	168	175	155	209	1.7	40
6	7.6	448	275	612	173	178	170	212	2.4	53
7	7.9	512	335	686	230	225	189	203	2.5	52
8	7.9	388	275	377	145	170	160	182	1.4	29
9	7.6	376	270	355	140	161	110	166	1.6	32
10	7.6	379	278	367	141	161	119	189	1.6	30
11	7.7	510	340	678	236	236	200	212	2.5	52
12	7.8	525	345	725	238	240	231	216	2.3	50
13	7.9	482	277	645	200	200	185	207	2.4	55
14	8.0	489	279	670	218	205	177	187	2.2	57
15	7.9	499	315	689	220	215	186	179	1.9	54

All parameters except pH are measured in mg/L

Table 5: Water Quality parameters. Their ICMR/ WHO Standards, and Assigned unit weights.

Parameter	Standard (Sn & Si)	Weightage (Wn)
pH	85	0.1420
Chloride	250	0.0048
Sulphate	250	0.0048
Alkalinity	20	0.0101
Nitrate	50	0.0242
TDS	1000	0.0012
Sodium	200	0.0060
Fluoride	1.5	0.8090

Table 6: Annual change in nitrate ion concentration (mg/L) in the study area.

2003	2004	2005	2006	2007
25	28	31	30	33
39	43	45	45	51
39	42	46	47	53
45	47	68	70	78
35	39	40	43	52
37	39	48	52	50
62	68	73	75	81
25	28	27	25	32
39	41	44	45	51
45	49	52	53	51
26	25	25	28	56
43	45	47	53	32
39	43	46	53	56
75	78	87	84	91
46	46	48	52	54

Graph showing nitrate concentrations in the groundwater of 15 sampling areas of Sopore Town and it's environs during the month of May, 2007. Peak concentrations were observed *in* Badambagh, Colony Shalimar, and Sangrama

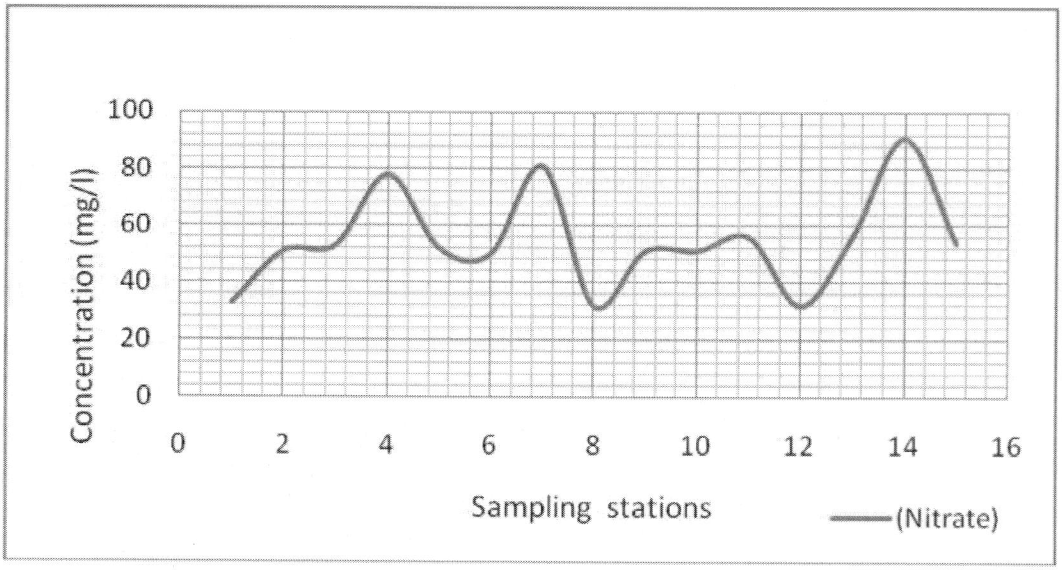

Fig. 2. Nitrate concentration of the study area, during summer season (May, 2007).

During winter season the same areas, i.e., S14, S7, and S4 are showing low nitrate concentrations as compared to their concentrations in summer season (Fig. 3) because of non-application of nitrogenous fertilizers during winter and hence strengthening the hypothesis that nitrate-pollution is caused due to use of nitrogenous fertilizers. Linear Trend Analysis (LTA) of nitrate showed that the Summer Linear Trend line is falling above the Winter Linear Trend line indicating that nitrate pollution is higher during summer season as compared to that of winter

Graph depicting nitrate concentrations in the groundwater of 15 sampling areas *of* Sopore Town and its environs during the month of Dec., 2007

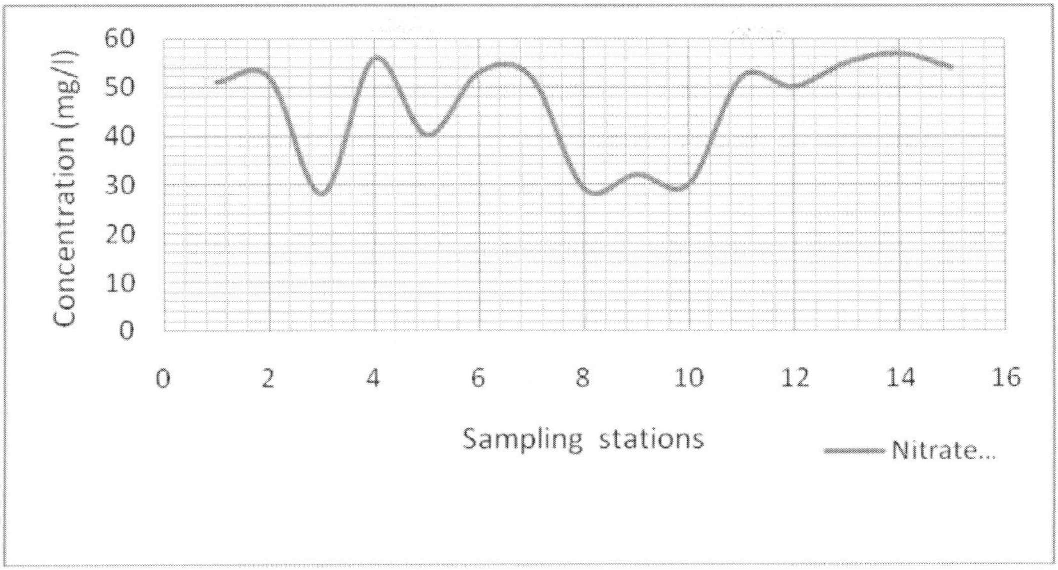

Fig. 3. Nitrate concentration of the study area, during winter season (Dec., 2007).

(Fig. 4). From the yearly LTA graph (Fig. 5) it is clear that Nitrate pollution goes on increasing with the passage of time. It is obvious from the graph that the LTA for 2007 is falling above the LTA for 2006. The LTA for 2006 is falling well above the LTA for 2005. Similarly, the LTA for 2005 is falling above the LTA for 2004 which inturn falls above to that of 2003 indicating the significant increase in nitrate ions during the years. Important source for this nitrate contamination in Tawheed Bagh (S1), Mumkak (S2), Shiva (S9), Nowpora (S11), Khwaja Gilgit (S3), Muslimpeer (S5), and Ashpeer (S6) areas is on-site sewage disposal in septic systems (Aravena R, Robertson WD 1998). Decomposition of soil organic matter, leaching of chemical fertilizers (Particularly N-fertilizer), human and animal excreta is potential sources of nitrate in groundwater [14].

The main culprit for the nitrate pollution in the groundwater of Duru, Dangerpora, and Alsafa Colony, New Colony, and Shah Abad areas was found to be irrigation with waste water. Moreover, plants absorb nitrate fertilizer through roots, which is then transformed into microorganisms [15]. The compensation of nitrate is insignificant in the soil environment [16], and hence nitrate percolates into groundwater. In present study the main sources of nitrate in groundwater are agriculture—85%, septic system-10%, and sewage and animal wastes 5% (Fig. 6).

CONCLUSION

Hydrochemical studies of the Sopore Town and its environs indicate that the concentration of nitrate is higher than permissible limit (50 mg/L) in most of groundwater collected from borewells.

The following graph depicts that Linear Trend Line of nitrate pollution for summer season falls well above that of winter linear trend line indicating high nitrate pollution during summer season as compared to that of winter season.

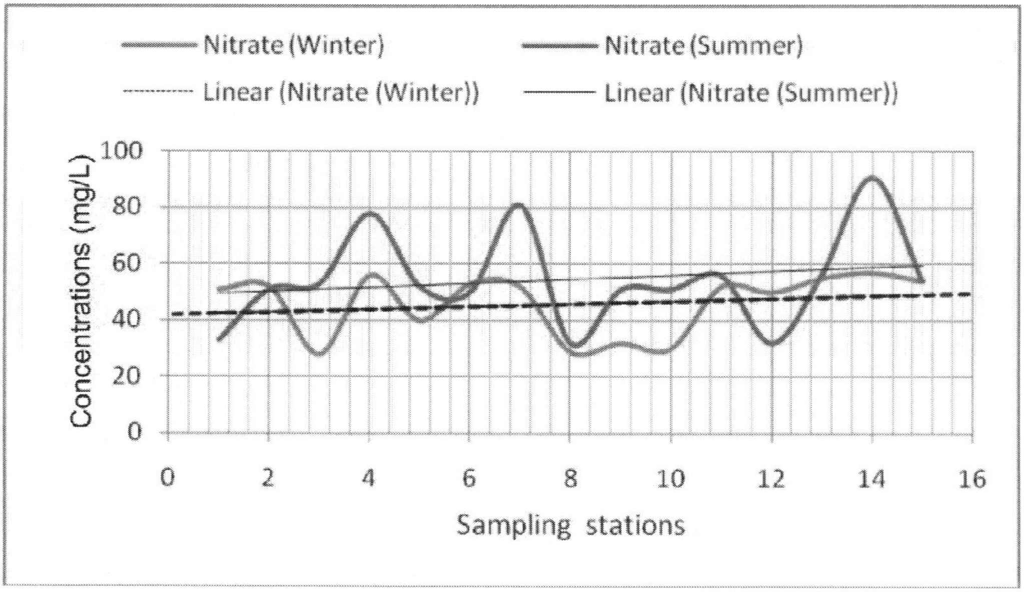

Fig. 4. showing the linear trend analysis of nitrate in summer and winter seasons, 2007.

The following LTA graph clearly indicates that the Linear Trend Line of Nitrate pollution for 2007. Falls above to that of 2006 which inturn falls above to that of 2005. The Linear Trend Line of 2005 falls above to that of 2004 which inturn falls above to that of 2003 indicating the Significant increase in nitrate pollution during the yrs.

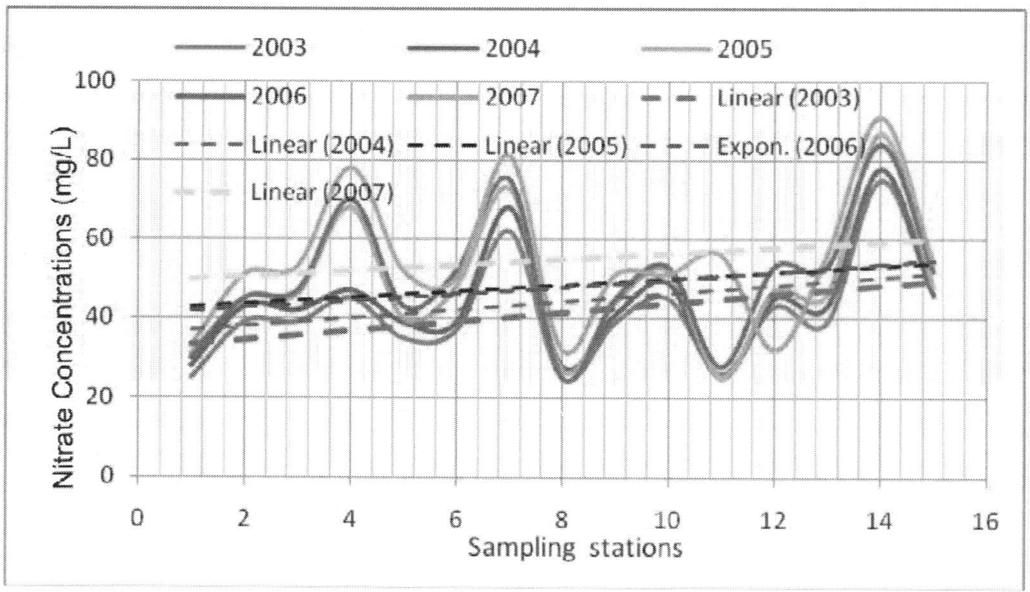

Fig. 5. showing LTA of nitrate in yrs. from 2003-07.

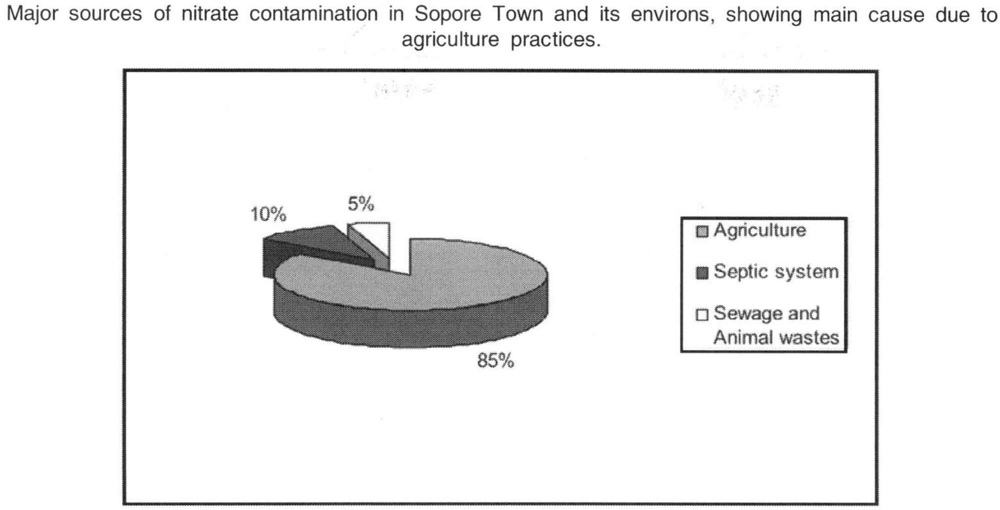

Fig. 6. Showing the sources of nitrate pollution in the study area, 2007.

The chief sources of nitrate pollution in the study area are agricultural activities, septic tanks and human and animal wastes. Irrigation with waste water was found the main source of nitrate pollution in Duru, Dangerpora, and Alsafa Colony, New Colony, and Shah Abad areas. Among the agricultural sources the common sources are inorganic fertilizer, urea and irrigation with waste water. Septic systems, animal waste and fertilizer are all potential sources of nitrate contamination. Water that comes into contact with a source of nitrate can carry that contamination through the soil and into the groundwater supply. Following remedial steps can reduce the risk of nitrate contamination:

1. Proper well location: Wells should be located uphill (upgradient) and at least 100 feet away from septic tanks, leachfields, animal confinement areas and fertilized areas.
2. Proper well construction: Make sure that the well casing extends above the ground, and construct an earth berm around the well to divert surface runoff away from the wellhead. Also, make sure your well has at least a 50 feet deep annular seal (grouting around the outside of the well casing) and an intact concrete slab on the wellhead.
3. Operate your septic system correctly: Maintaining the septic system regularly will help avoid system failure, which can lead to water contamination, the spread of disease, and the need for costly repairs. Follow these guidelines:
   (1) Don't run heavy vehicles over the area above your septic tank, drain pipes or leachfield.
   (2) Avoid planting trees or shrubs near drain pipes or the leachfield. Roots can clog the lines.
   (3) Don't dispose of chemicals or non-biodegradable materials in your toilet or drain.
   (4) Conserve water.
   (5) Install a lint trap on your washing machine.
   (6) Hire a reputable contractor to pump out your septic tank every two to three years.
   (7) If you have two leachfields, switch them every year.

4. Reduce your use of fertilizer: Use lawn and garden fertilizers only when necessary, and always according to the manufacturer's instructions. Farmers can use management practices that are both environmentally and economically sound. If these remedial measures are not taken immediately, then the water of Sopore Town, often called as an "Elixir", may turn into poison.

REFERENCES

1. Hallberg, G.R. (1989). Nitrate in groundwater in the United States. In: Nitrogen Management and Groundwater Protection, R.F. Follet, Ed., Elsevier, Amsterdam, pp. 35-74.
2. Puckett, L.J. (1995). Identifying the major sources of nutrient water pollution. Environmental Science & Technology: 408A-414A.
3. http://www.lpes.org/CAFO.html, http://epa.gov/guide/cafo, & http://www.factoryfarm.org).
4. Dissanayake, C.B. and Weerasooriya, S.V.R. (1987). Medical geochemistry of nitrates and human cancer in Sri Lanka. *Intern. J. Environmental studies.* **Vol. 30:** pp. 145-156.
5. US EPA Office of Water (WH-550) "Is Your Drinking Water Safe?" EPA 570 9-91-005, Sept 91. This document is no longer available on epa.gov website.
6. Committee on Nitrate Accumulation, Ag. Board, Div. of Biology & Agriculture, National Research Council. (1972). Hazards of Nitrate, Nitrite, and Nitrosoamines to Man and Livestock. In: Accumulation of Nitrate, National Academy of Sciences, Wash., DC, pp. 46-75.
7. Lunkad, S.K. (1994). Rising Nitrate levels in Groundwater and increasing N-Fertilizer consumption. *J. Bhu-Jal News.*
8. Lakshmanan, A.R., Krishna Rao, T. and Viswanathan, S. (1986). Nitrate and fluoride levels in drinking waters in the twin cities of Hyderabad and Secunderabad. *Indian J. Environ. Health*, **Vol. 28, No. 1:** pp. 39-47.
9. Handa, B.K., Geol, D.K. Kumar, A. and Sondhi, T.N. (1982). *Pollution of groundwater by nitrates in Uttar Pradesh.* IAWPC. *Tech. Annual.* **Vol. 9**, 95-103.
10. Federal Register, (1985). *National primary drinking water standards*, **Vol. 50 No. 219:** November 13, 1985. pp. 46880-47022.
11. Canter, L.W. (1987). *Nitrates in Groundwater from agriculture practices causes, prevention and clean-up. July.* Report to United Nations Development Program, University of Oklahoma, Norman, Oklahoma.
12. WHO, (1983). *Guidelines to drinking water quality*, World Health Organization Geneva.
13. APHA, AWWA, WPCF. (1998). *Standard Methods for the Examination of Water and Wastewater.* 20th edition American Public Health Association, Washington, DC, New York, USA.
14. Hill, A.R. (1982). *Nitrate distribution in the groundwater of Elliston region of Ontario, Canada. Groundwater,* **Vol. 20:** pp. 696-702.
15. Srinivasa Rao, Y., Reddy T.V.K. and Nayudu, P.T. (1997). Groundwater quality in the Niva River basin, Chittoor district, Andhra Pradesh, India. *Environmental Geology* (1), pp. 56-63.
16. Lance, J.C. (1972). Nitrogen removal by soil mechanism. *J. Water pollution control* Feb. **44:** 1352-1361.

14

Hydrogeochemical Studies in and around Kalpakkam, Kanchipuram District, Tamil Nadu

KARMEGAM, U*[1]., CHIDAMBARAM, S[1]., SASIDHAR, P[2]., RAMESH, R[1]., MANIKANDAN, S[1]., PARAMAGURU, P[1]., DHEIVANAYAGI, V[1]. AND AKILA, P[1].

[1]*Department of Earth Science, Annamalai University, Annamalai Nagar*

[2]*SRI-AERB, IGAR Kalpakkam*

**Email: megapriyan@gmail.com*

ABSTRACT

The coastal aquifer is similar to the continental aquifer except for the risk of water quality deterioration. This may be due to the natural or induced mixing with the seawater, it may also be due to the presence of old marine water in the aquifers. The study is taken up Kalpakkam area in and around area of Kanchipuram District. Tamilnadu. About 29 water samples from deep bore wells of different location. All the samples are slightly acidic to alkaline nature having pH ranging from 6.8 to 8.1. The total dissolved solids (TDS) range from 129.5 to 1643 mg/L. The order of abundance of cations and anions are Na > Ca > Mg > K and HCO_3 > Cl > So_4 respectively. The Hydrogeochemical Facies is generally Ca Mg Cl type. USSL classification shows that most of the samples fall in C3,S1category and suitable only for salt tolerant crops. The chemical composition of groundwater is controlled by rock water interaction. Coastal aquifer can be exploited to yield freshwater if the flow mechanisms are well-known wells are correctly constructed and placed. The present study deals with the geochemical process responsible for the change in water chemistry of the coastal aquifers and identification of susceptible zone of vulnerability/contamination.

Keywords: Hydrogeochemistry, classification, utility process.

INTRODUCTION

Water is the elixer of life, a precious gift of nature to mankind and millions of other species living on the earth. It is a fast becoming scare commodity in most part of the world. Water resources comprising surface water (river and lakes), groundwater and marine and coastal waters, support all living things including human beings. Though water is available in universe in huge quantity in the order of 1400 x 10^6 km^3, only 3% of the water in the universe is freshwater. Among the fresh waters, only about 5% of them or 0.15% of the total world water are readily available for beneficial use. The total water resource available in India is 1850 km^3, which is roughly 4% of the world's freshwater resources. The zone of influence immediately adjacent to the ion is known as hydration sheath (Dominico, 1990).

Groundwater is one of the important sources of water supply throughout the world. Groundwater constitutes one portion of earth's water circulatory system known as the hydrological cycle. Water bearing formation of the earth crust, act as conduits for transmission and as reservoirs for storage of water. Water entering this formation from the ground surface or from bodies of surface water, percolates gets transmitted and stored beneath the surface. At present nearly 1/5th of all the water used in the world are obtained from the groundwater (Sharma, 1998). Nearly 80% of the drinking water requirement and 50% of the municipal and industrial water requirement is met by groundwater. Water requirement in urban and rural areas throughout the world are increasingly at an alarming stage.

Though groundwater is an unnoticed and unacknowledged cornerstone in the foundation of many regional economical and environmental systems, it is far more reliable as a source of supply than surface water due to the long lack between changes in recharge, responses in groundwater level and well yields (Caster and Howsam, 1994). The present annual usage of water for India for various uses like agriculture, domestic, industrial, energy and other miscellaneous purposes is around 634 km^3 and expected to be doubled within next 50 years. Groundwater storage is essential due to lesser evaporation loss and lesser amount of direct pollution. Hence, it is essential to asses the groundwater quality and quantity.

Groundwater contains minerals carried in solution, the type and concentration of which depends upon several factors. The chemical composition of groundwater is depending upon the soluble products of rock weathering and decomposition in addition to external polluting agencies and changes in space and time. As the result of chemical and biochemical interactions between groundwater and the geological materials through which it flows, and to lesser extent because of contribution from the atmosphere and surface water bodies, groundwater contains a wide variety of dissolved inorganic chemical constituents in various concerts.

Pollution of groundwater occurs due to contaminants like urban, industrial and agricultural activities. WHO (1996) has reported that about 80% of the diseases in the world population are due to the consumption of polluted waters. The industrial, domestic effluents and leaching of solid waste dump result in groundwater contamination, which is quite often an irreversible process. Hence, it is essential to study all the major dissolved ions to understand the status of pollution in groundwater significant role in water quality in groundwater was also noted in many locations of the district.

STUDY AREA

The area proposed for the study extends from the Mahabalipuram in the north to Kuvathur in the south. The study area falls in between the geographical coordinates, 12°25' to12°37' N latitude and 80°00' to 80°12'E longitude (Fig. 1). The river Palar with its tributaries, drain into the sea along the region. The study area is well known due to the presence of the premier organization,

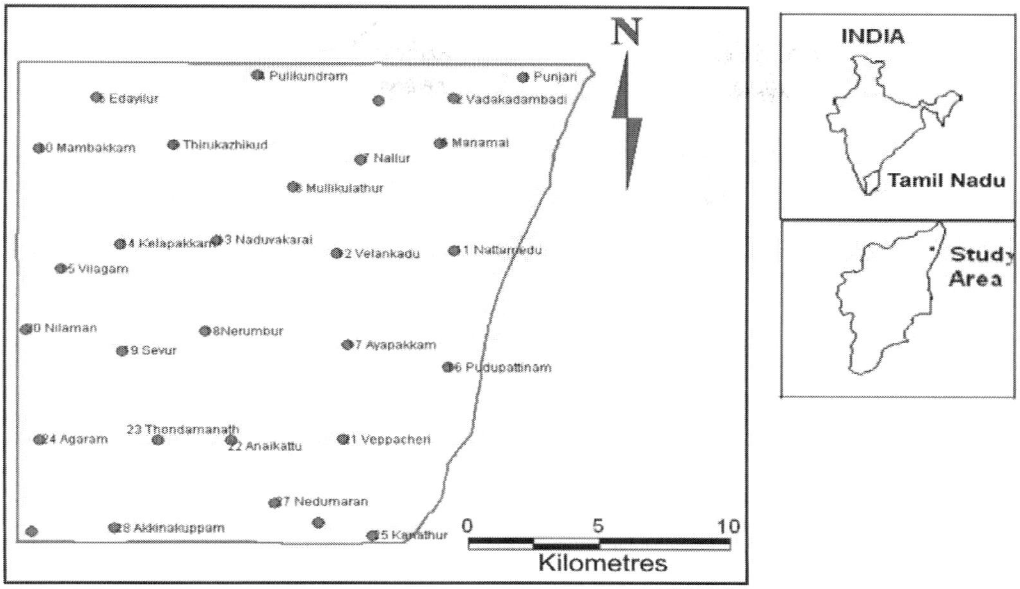

Fig. 1. Samples location map.

IGCAR at Kalpakkam. The coastal region is entirely compose of alluvium, which helps in easy infiltration of the surface water. The geomorphic features such as dunes and beach ridges are found along the coastal region. Apart from its importance as a coastal region it is also significant to note that the natural calamities like cyclones are frequently disturbing the normal ecosystem. This region was recently affected by the frequent rainfalls due to the depression in Bay of Bengal (BAAZ and FAANUZ).The aquiferal behaviour and its response to natural calamities in this region will help us to decipher the sensitive region and to identify the region for natural/ artificial recharge. In this paper, various groundwater geochemical parameters such as variation of salinity with seasons of Kalpakkam site have been reported. The average annual rainfall in the study area is 1237 mm and the water table fluctuation in 2008 has been found to be 2.43 ± 0.52 m. [5]

The subsurface lithology of the study area shows six types of geological formations including sand, clayey sand, clay, sandy clay, weathered rock and Charnockite. Earlier studies by Gurumoorthy *et al.* [6] has revealed that the Charnockite rock mass in this area is hard and compact and occurs at depths in the range greater than 15-20 m below land surface.

Methodology

Groundwater sample were collected along three tracts from north to south along 3 km from coastal (A), 6 km from coastal (B) and 12 km from coastal (C) (Fig. 1) systematically using clear acid washed polythene bottles. Sampling was carried during Feb. 2008 and analysis was carried out using standard procedures (APHA 1998). The parameter, such as temperature and pH were measured in the field. Electrical conductivity and total dissolved solids were measured in the laboratory. Salinity was determined by using refractometer. Na^+ and K^+ were determind by using flame photometer. Ca^{+} Mg^{+} Cl^{-} and HCO_3^{-} were determined by volumetric titration methods.

The data generated was processed to obtain Sodium absorption ratio (SAR), Residual carbon (R.S.P), Sodium percentage (Na%) and Total Hardness by using WATCLAST (Chidambaram, 2003). Different thematic layers were produced for SAR, RSC, Na%, and EC. Using MapInfo (GIS) Software V.8 and Vertical mapper. Water analyses were carried out by using standard procedures (APHA 1984; Ramanathan 1992 and Ramesh and Anbu 1996).

HYDROGEOCHEMISTRY

Groundwater is one of the replenishable resources on the earth. It contains minerals in solution; the type and the concentration depend on the surface and subsurface environment, rate of groundwater movement and source of groundwater. As a result of the solvent power of water, minerals are dissolved and carried into solution when the water moves along the aquifer from the recharge area to the discharge area, the chemistry is affected by various geochemical processes.

Water is formed by the union of two hydrogen atom with one oxygen atom. The bonding is asymmetrical with an angle of 105ï. Their asymmetrical arrangement gives rise to polar characteristic to the molecule. Water is effective in dissolving salts due to the higher dielectric constant and its molecules tend to combine with ions to form hydrated ions. The thermal agitations of ions in many minerals are greater to overcome the weaker charge of attraction that exists when surrounded by water, thus, allowing large number of ions to dissociate into aqueous solution. Groundwater can be viewed as electrolytic solution due to the presence of major and minor dissolved constituents present in ionic form. A charged species in water forces a local rearrangement in the structure and those nearby reorient themselves because of the charge. A random region separate the hydration sheath from the bulk solution, within the random region the water molecule are less oriented and they are in bulk solution or the hydration sheath.

WATER CHEMISTRY

Groundwater in the study area is generally alkaline in nature with pH ranging from 6.85 to 8.1 with an average of 7.55 (Table 1). EC is an indirect measure of ionic strength and mineralization of natural water. EC of pure water is around 0.05 μs/cm (Hem, 1991). EC ranges from 255.7 μs/cm to 3273 μs/cm with an average of 1185.56 μs/cm. Total dissolved solids (TDS) which is generally the sum of dissolved ionic concentration varies between 129.3 mgL^{-1} and 1643 mgL^{-1} with an average of 594.541 mgL^{-1}.

Anions

Bicarbonate represents the major sum of alkalinity. Alkalinity in water is the measure of its capacity of neutralization. It is formed mainly due to the action of atmospheric CO_2 and CO_2 released from organic decomposition. Groundwater samples bicarbonate ranges from 73.2 mgL^{-1} to 475.8 mgL^{-1} with an average of 261.8 mgL^{-1} $HCO_3 > Cl > SiO_4 > SO_4$. HCO^{-}_3 is higher in indicating the contribution from silicate and carbonate chemical weathering process. Cl^{-} is higher in indicating the impact of saline water and Base Exchange reactions (Allan freeze and Cherry 1979; Bower 1979). Silica is the third most abundant element in the earth crust and

Table 1: Chemical composition of groundwater (in mg/L) except EC and pH (EC in ìs/cm)

	Location	PH	EC	TDS	Ca	Mg	Na	K	Cl	HCO_3	PO_4	SO_4	SIO_4
1	Punjari	7.63	1408	706.5	45	75.5	197.3	44.8	85.3	395	0.3	9.4	64
2	Vadakadambadi	7.74	809.8	405.5	40	46.5	192.3	24.5	88.62	305	0.2	3.4	75.5
3	Kuchchikadu	7.81	1072	541.8	40	52	210	34	84.6	414.8	0.2	7	61.5
4	Pulikkudram	7.43	1009	510.5	60	62.5	197.7	24.5	123.6	292.8	0.25	3.4	52.5
5	Edayilur	7.59	1440	726.9	56	57.6	210	25	150	305	0.2	7.3	35
6	Manamai	7.97	742.7	373.8	48	52.8	252.9	4.5	88.62	317.2	0.15	2	60.5
7	Nallur	7.55	3273	1643	140	140.6	72.2	9.3	120	451.4	0.3	22.5	86
8	Mullikulottur	7.86	998.9	499.8	52	63.4	270	39.3	115	329.4	0.35	3.3	86
9	Therukazhikundram	7.42	1097	556.4	210	360.5	790.6	680.5	63.5	195	3.3	6	59.5
10	Mambakkam	7.71	352.6	176.2	12	24	250.1	4.5	53.17	146.4	0.15	0.2	90
11	Nanamedu	8.09	2992	1499	49	84.4	350.4	65.8	205.6	402.6	0.4	7.3	32.5
12	Velangadu	7.71	1192	607.6	36	52.8	390.5	4.5	177.2	427	0.3	6	36
13	Naduvakkarai	7.63	2260	1122	48	60.4	430	80.5	250.5	475.8	0.35	8.2	44
14	Kilappakkam	7.88	1012	509.1	32	36	220.5	50.8	70.9	439.2	0.15	5.8	34
15	Velagam	7.63	2140	1068	80	86.4	280	34.5	160.5	402.6	0.3	7.7	50
16	Kalpakkam	7.42	748.2	374.8	12	16.8	220.5	24	124.07	219.6	0.1	8.8	27
17	Ayappakkam	7.31	840.2	419.7	20	26.4	157.6	22.5	106.35	146.4	0.1	11	56
18	Nerumbur	7.28	367.4	184.1	16	16.8	240.1	4.4	17.72	158.6	0.3	2.2	44
19	Sevur	7.95	1735	867.5	60	130.5	780	410	85	130	2	9.4	66
20	Nilamangalam	7.63	587.4	294.4	32	39	235.5	80.5	53.17	341.6	0.35	1.1	68.5
21	Veppancheri	8.1	1028	513.7	230	380	790.5	650	55.5	120	3.4	1	94.5
22	Anaikattu	7.5	2798	1401	120	160	520	160.2	655.8	204	0.4	10.3	75
23	Surpilaiyarkuppam	7.41	315.5	158	40	50.4	240.1	25.5	35.45	109.8	0.4	2.5	95
24	Agaram	7.2	1573	787.4	44	55.2	320	120	148	183	0.45	7.8	68
25	Kanathur	6.88	265.6	133.7	80	150.9	640	310	95	73.2	1.8	2.5	63
26	Kuvathur	7.2	1354	677.2	190	210	710	460	95.5	192	2.35	17.5	77.5
27	Nedumaram	6.85	255.7	129.3	20	50	257	24	35.45	134.2	0.4	1.3	67.5
28	Akkinampattu	7.19	441.8	221.4	24	64	245.2	14	35.45	134.2	0.4	6.5	72.5
29	Tiruvadur	7.62	272.5	132.9	10	20	251.7	20	35.45	146.4	0.35	2.2	54
	Minimum	6.85	255.7	129.3	10	16.8	72.2	4.4	17.72	73.2	0.1	0.2	27
	Maximum	8.1	3273	1643	230	380	790.6	680.5	655.8	475.8	3.4	22.5	95
	Avarage	7.55	1185.5	594.52	63.65	90.53	342.16	119.03	117.75	261.8	0.67	6.33	61.91

essential component of almost all minerals. Silica concentration ranges from 27 mgL^{-1} to 95 mgL^{-1} with an average of 61.91 mgL^{-1} Silicate is also higher that indicates weathering of silicate minerals. Sulfate is found in water due to its lesser breaking down of organic substances from weathered soil/water and due to the influence of saline waters (Miller 1979; Craig and Anderson 1979 and Singh and others 1994).

Cations

Sodium is the important and most abundant alkali metal which is highly mobile and soluble in groundwater. The dominance of cations are as follows $K^+ > Na^+ > Mg^{2+} > Ca^{2+}$. K^+ ranges from 4.4 mgL^{-1} to 680.5 mgL^{-1} with an average of 119.03 mgL^{-1}. Na^+ concentration vary from 72.2 mgL^{-1} to 790.6 mgL^{-1}, with an average of 342.16 mgL^{-1} groundwater samples. The major source of sodium and potassium is from weathering of potash and sodic plagioclase in waters due to its abundance in silicate rocks.

Calcium is an important element in many geochemical processes. Minerals like Calcite, Plagioclase and Hornblende are the primary sources for calcium in groundwater. Calcium concentration is ranging from 10 mgL^{-1} to 230 mgL^{-1} with an average of 63.65 mgL^{-1}. Mg^{2+} was ranging from 16.8 mgL^{-1} to 380 mgL^{-1} with an average of 90.53 mgL^{-1}.

Residual Sodium Carbonate

All values expressed in epm. The water having excess of Carbonate and Bicarbonate Co cover the alkaline earth mainly Ca and Mg in excess of allowable limits affects agriculture unfavourably (Richards 1954).

$$RSC = (CO_3+HCO_3) - (Ca+Mg)$$

The tendency of alkaline earth influences the suitability of water for irrigation purpose. The water gets more precipitated with Ca and Mg and as a result Na in water gets increased in the form of sodium carbonate.

The distribution of RSC was drawn using Richards (1954) as good, medium and bad categories. Good category falls in 93% of the study area, medium category falls in 3.5% of the study area and the bad category is represented 3.5%.

Sodium Absorption Ratio

Salinity of groundwater and SAR determines its utility for agricultural purposes. Salinity originates in groundwater due to weathering of rocks and leaching from top soil, anthropogenic sources along with minor influence on climate. The level of Na^+ and HCO_3^- in irrigation Kelly (1951) and Tijani (1994) groundwater affects permeability of soil and drainage of the area.

SAR is an estimate of the extent to which sodium ion present in the water would be adsorbed by the soil and is expressed as follow.

$$SAR = Na\sqrt{Ca+Mg/2}$$

Were concentration are expressed in equivalent per million (epm).

The distribution of SAR, was plotted, using **(Richard 1954)** Classification was made as excellent and good categories. More than 90% of samples the falls under excellent class and remaining 10% percentage falls under good category.

Sodium Percentage

Na% (Eaton, 1950) classification of groundwater for irrigation purpose was calculated as

$$Na\ \% = \frac{(Na+K)\ *100}{(Ca+Mg+Na+K)}$$

The ratio of sodium and potassium in the sum of cation is an the important factor in considering water for agriculture uses, for example, relatively affected soil structure and permeability, resulting in this alkaline soil. The sodium concentration of irrigation water is of prime importance and plays a significant part in determining the permeability of soil Na absorbed on clay surface, as a substitute for Ca and Mg may damage the soil structure making it compact and impervious.

The distribution of Na% classification (Richard 1954) was used as excellent good, Permissible, Doubtful, and Unsuitable categories show in all natural waters % of Na content is a parameter to assess its suitability for agriculture purpose (Wilcox 1948) sodium combining with CO_3 can add to the formation of alkaline soils and sodium combining with Cl form saline soils. Both these soils do not helping growth of plants. A maximum of 60% of Na is allowed for agriculture purpose (Ramakrishna 1998). The excellent category falls in 3% of the study area. The permissible category is represented in 24% of the study area. The doubtful region falls in 59% part of the study area and undoubtful in 14% of the study area.

Hardness

Hardness of water is related to its reaction with soap and to the scale of incrustation accumulating in containers (or) conducts where water is heated or transported. Since soap is precipitated by Ca and Mg ion. Hardness is defined as the sum of concentration of their ion expressed an Mg/L of $CaCO_3$. Hardness increases by metallic ion dissolved in water. Hardness is used as an indicator of the rate of scale formation in hot water heater in low pressure boiler. USGS Hardness four type soft, slightly hard, moderately hard and very hard, Scaling and deposition problems in air-conditioning plants are associated with total hardness of the water. Total hardness of more than 180 mg CaCO3/L can be treated as very hard water and it can lead to scaling problems in air conditioning plants (Hem 1970). The distribution of TH, was plotted using USGS classification as slightly hard, moderately hard and very hard categories. The slightly hard is represented 14% in the study area, 17% moderately and 69% hard category of the study area.

Spatial Distribution of EC

The spatial distribution of Electrical conductivity (EC), was drawn, using Wilcox (1995)

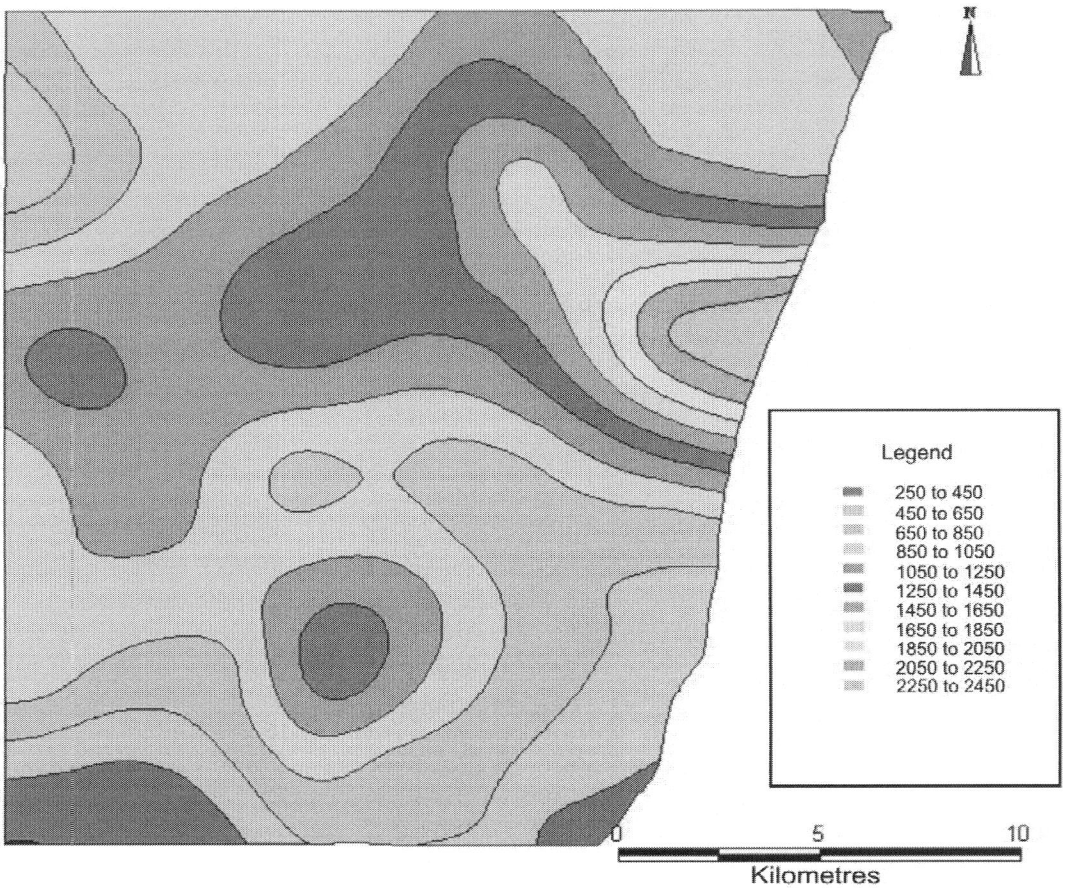

Fig. 2. Spatial distribution of electrical conductivity in μs/cm.

classification as excellent, good, permissible, doubtful and poor categories (Fig. 2) EC ranged from 255.7 μs/cm to3273 μs/cm average of 1185.56. EC of rainwater is 0.05 μs/cm (Hem, 1991). The most of the samples falls in south-west part range from 250–1250 due to the infiltration of present Palar river basin in coastal aquifer of the study area and few samples were falls above 1250 ranges in north-east part of the study area, due to the saline water intrusion of the study area and because of the tourist area the water is contamination in Mamallapuram.

GEOCHEMICAL CLASSIFICATION

The geochemical evolution of groundwater can be understood by plotting the concentrations of major cations and anions in the Piper trilinear diagram. The plot shows that most of the samples fall in the field of mixed $CaNaHCO_3$ type. Some samples are also representing NaCl type (Fig. 3). It indicates the dominance of Ca in the cations and interplay of HCO_3 and Cl in anions. From the plot alkali (Na) exceeds the alkaline earths (Ca and Mg) and strong acid Cl exceeds the week acids (HCO_3 and SO_4).

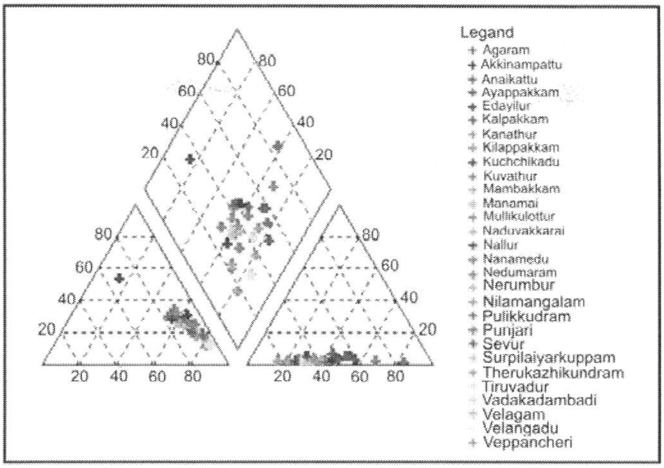

Fig. 3. Piper plot.

The suitability of groundwater for irrigation purpose is mainly based upon estimation of parameters like SAR, Na%, RSC (Table 2). Total Na^+ concentration and EC is important in classifying the irrigation water (Wilcox 1955). Salinity of groundwater and SAR determines its utility for agricultural purposes. Salinity originates in groundwater due to weathering of rocks and leaching from topsoil, anthropogenic sources along with minor influence on climate. The level of Na^+ and HCO^-_3 in irrigation groundwater affects permeability of soil and drainage of the area. Na^+ is an important cation which in excess, deteriorates soil structure and reduces crop yield. Scholler (1965) proposed a measure called "Index of Base Exchange" (IBE) to describe metamorphic reactions taking place in groundwater. There are substances which absorb and exchange their cations with cations present in groundwater. Those substances are called Permutolites, e.g. Clay minerals like Kaolinite, Illite, Chlorite, Halloysite, Glauconite, Zeolites and organic substances. In this Kaolinite, Illite, Chlorite and Halloysite are the clay minerals in which ions are held at edges and their ionic exchange capacity is low. This case is reverse in Montmorllianite and Vermiculite, the exchange capacity is higher when the number of ions held on the surface is more. Chloro-alkaline indices, CAI1 and CAI2 are used to measure extent of base exchange during rock water interaction. Where there is an exchange of Na^+ and K^+ in groundwater with Mg^{2+} or Ca^{2+} in rock/alluvium both the indices are positive. All ionic concentration is expressed in epm. The indices to base exchange indicate that there is a significant exchange of Na+K in rock to the Ca+Mg in groundwater (Chidambaram 2000).

There is also less notable introducing of Na+K in groundwater into the Ca+Mg in rock from the matrix. The Scholler water type indicates that all the samples fall in class Type I.

In SAR classification Richards (1954), majority of the samples grouped in excellent to good category. In Wilcox (1955) classification of electrical conductivity most of the samples fall within "permissible" to "good" limit with minor representations in doubtful category. In Na% Eaton (1950) classification of groundwater for irrigation purposes, majority of the samples grouped in unsafe zone and minor representations also falls in safe zone. In Wilcox (1955) classification for Na% most of the samples fall in "unsuitable" to "permissible" range. The

Table 2: Result of computer program WATCLAST.

Category	Grade	N = 29	Category	Grade	N = 29	Category	N = 29
Na% Wilcox (1955)			**USGS Hardness**			**TDS Classification (USSL,1954)**	
Excellent	0-20	1	Soft	<75	0	<200	6
Good	20-40	0	Slightly Hard	75-150	4	200-500	7
Permissible	40-60	7	Moderately Hard	150-300	5	500-1500	15
Doubtful	60-80	17	Very Hard	>300	20	1500-3000	1
Unsuitable	>80	4	**IBE Schoeller (1965)**			**Cation Facies**	
Na% Eaton (1950)			(Na+k)rock->Mg/Ca g.w.		28	Ca-Mg Facies	0
Safe	<60	8	(Na+k)g.w.->Mg/Ca rock		1	Ca-Na Facies	15
Unsafe	>60	21	**Schoeller Classification (1967)**			Na-Ca Facies	14
S.A.R. Richards (1954)			Type I		29	Na Facies	0
Excellent	0-10	26	Type II		0	**Anion facies**	
Good	10 18	3	Type III		0	HCO_3 Facies	0
Fair	18-26	0	Type IV		0	HCO_3-Cl-SO_4 Facies	0
Poor	>26	0	**Corrosivity Ratio (1990)**			Cl-SO_4-HCO_3 Facies	21
R.S.C. Richards(1954)			Safe	<1	25	Cl- Facies	8
Good	<1.25	27	Unsafe	>1	4	**Hardness Classification (Handa,1964)**	
Medium	1.25-2.5	1	**Chloride Classification (Stuyfzand,1989)**			Permanent Hardness (NCH)	
Bad	>2.5	1	Extremely fresh	<0.14	0	A1	1
EC Wilcox (1955)			Very fresh	0.14-0.84	1	A2	0
Excellent	<250	0	Fresh	0.84-4.23	22	A3	5
Good	250-750	10	Fresh Brackish	4.23-8.46	5	Temprorary Hardness (CH)	
Permissible	750-2250	15	Brackish	8.46-28.21	1	B1	0
Doubtful	2250-5000	4	Brackish-salt	28.21-282.1	0	B2	7
Unsuitable	>5000	0	Salt	282.1-564.1	0	B3	1
			Hyperhaline	>564.3	0		

dominant cation facies is the Ca-Na and Na-Ca facies and the Anion facies is dominated by chloride, sulphate and bicarbonate ion. According to chloride classification by Stuyfzand (1989) majority of samples grouped in fresh to fresh brackish category, indicating the usage of this water agricultural activity. In USGS hardness classification most of the samples grouped in very hard to moderately hard and few representations fall in the slightly hard category. The majority of the groundwater in this region is having Corrosivity Ratio CR < 1, safe category. Few samples having CR > 1, fall in unsafe category.

GEOCHEMICAL CLASSIFICATION

In the USSL diagram, SAR (alkali hazard) and specific conductance, majority of groundwater samples fall in C3S1 zone (Fig. 4) indicating high salinity and low sodium hazard. Gibbs (1970) proposed a method for identifying relationship between water composition and mechanism controlling chemistry of groundwater. Most of the groundwater samples in cation plot fall away from the zone indicate apart from the weathering other factors also play a dominant role in the

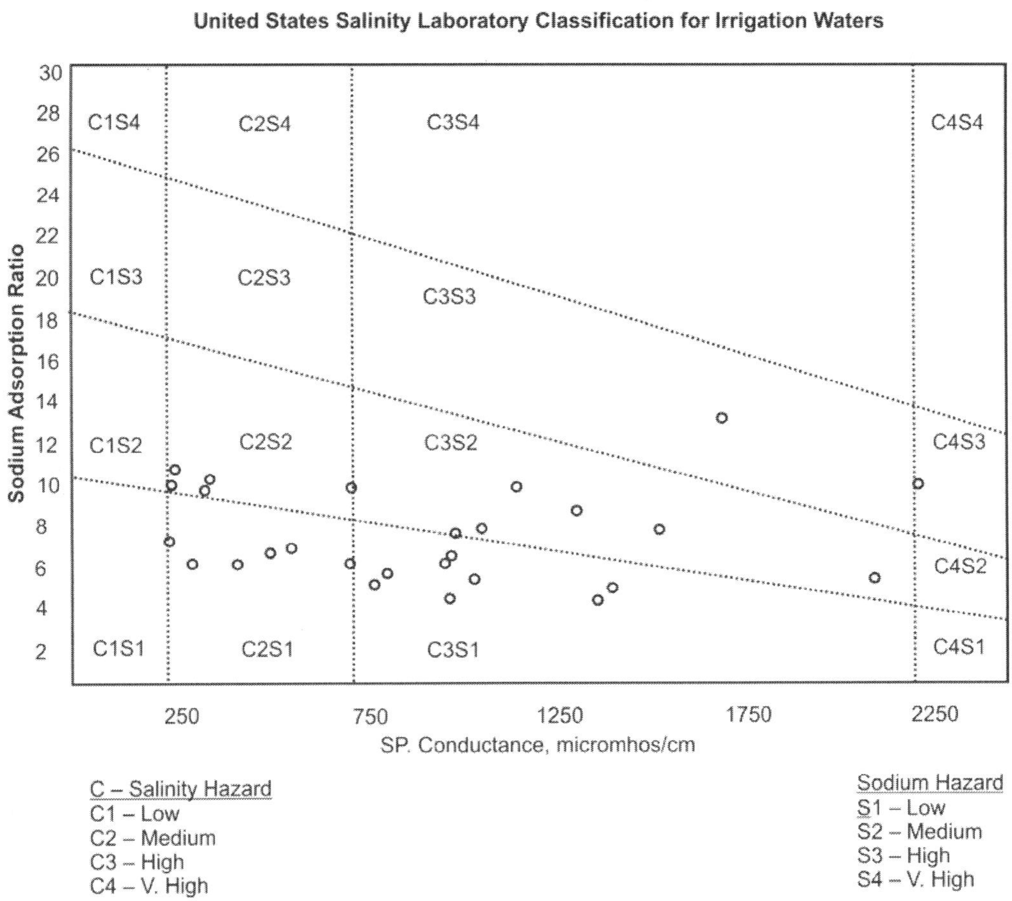

Fig. 4. USSL diagram.

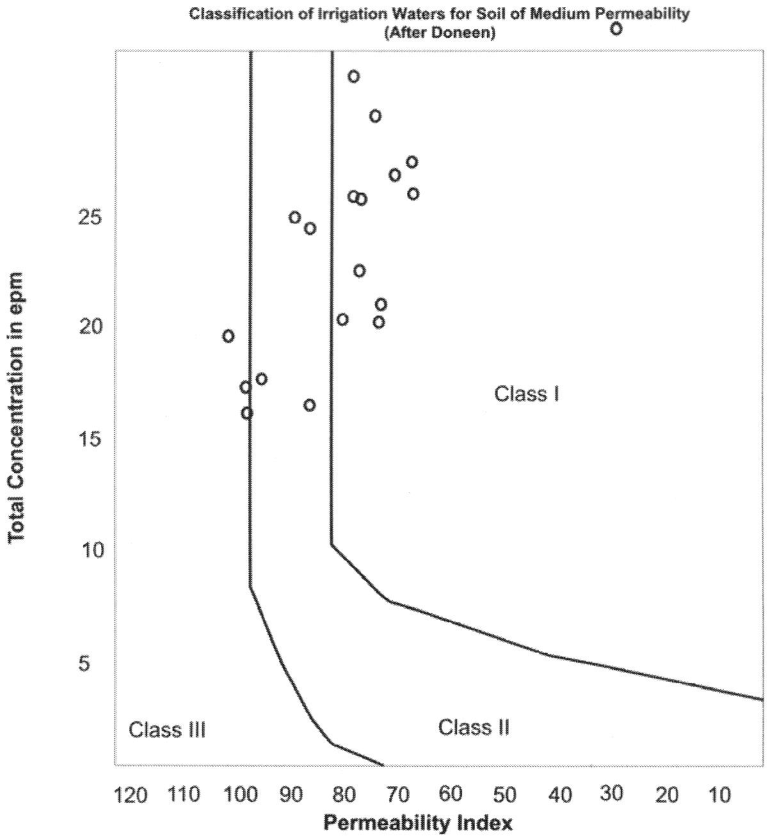

Fig. 5. Doneen plot.

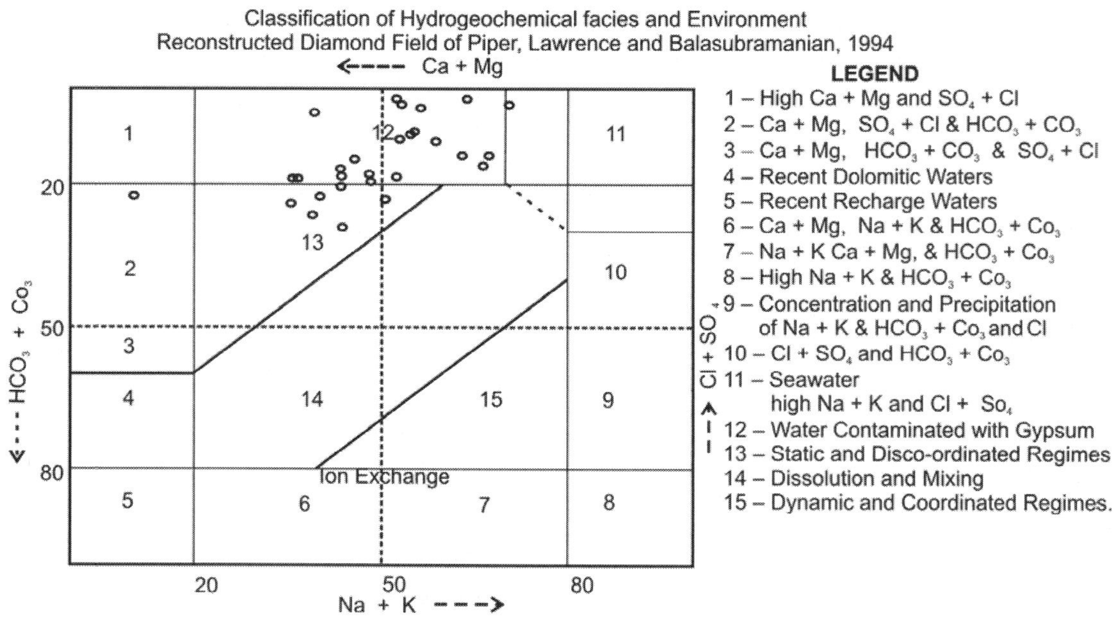

Fig. 6. Modified hill poper plot (Johnson).

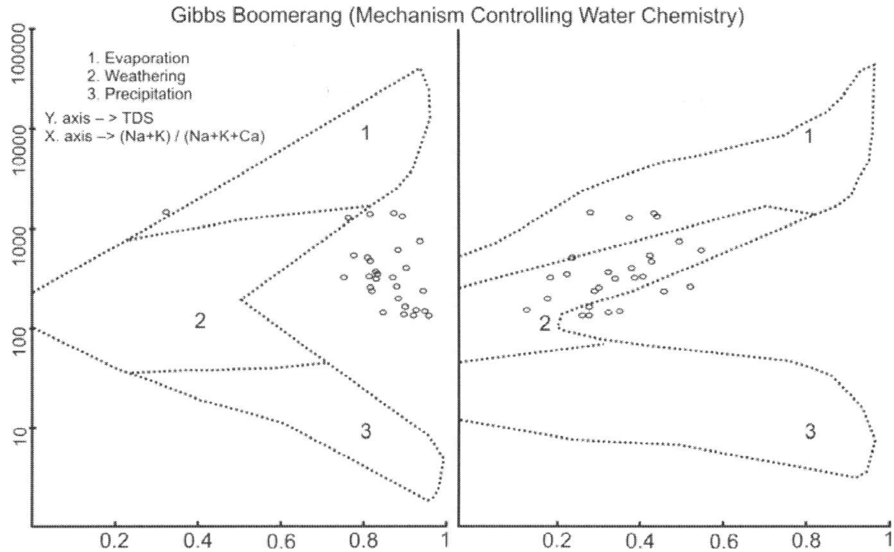

Fig. 7. Gibbs plot.

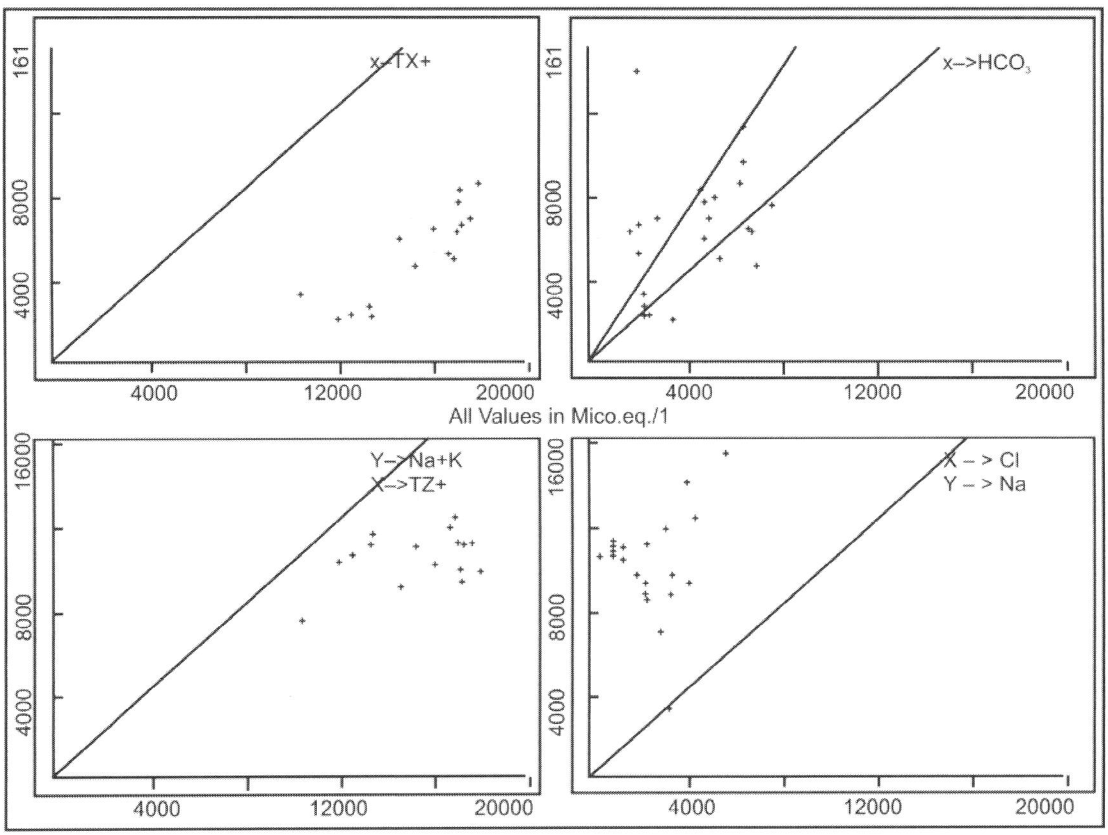

Fig. 8. Relationships between ions.

evaporation system. In the anion plot most of the groundwater samples fall in weathering zone may be due to the weathering of feldspar and also minor representations fall in evaporation zone (Fig. 5). Permeability index is an important factor which influences quality of irrigation water, in relation to soil for development in agriculture. Based on permeability index, Doneen (1948) classified groundwater as Class I, Class II and Class III to find out suitability of groundwater for irrigation purpose. In the study area (Fig. 6) majority of groundwater samples fall in Class I and Class II. In Johnson plot (Fig. 7), majority of groundwater samples fall in water contaminated with Gypsum zone, few samples fall in static and disco–ordinated regions zone. The graphical plot for certain ions are plotted to determine the dominance of certain ions. The plot (Fig. 8) for $(Ca+Mg)/TZ^+$ the majority of groundwater samples fall in TZ^+ side of the equiline indicating the higher concentration of TZ^+. The plot for $Ca+Mg/HCO_3$, majority of samples fall in Ca+Mg side indicating predominance of calcium and magnesium over alkaline. The plot for $(Na+K)/TZ^+$ indicates that all samples fall in cation zone indicating their dominance which might have been derived form feldspar weathering or by alkali soils. In the plot for Cl and Na majority of the samples fall in Cl side due to leaching process.

REFERENCES

Doneen, L.D. (1948). The quality of irrigation water. California agriculture dept. 4-**11:** 6-14.

Gibbs, R.J. (1970). Mechanisms controlling world's water chemistry, *Science* **170:** 1088-99.

Hem, J.D. (1970). Study and interpretation of the chemical characteristics of natural waters. USGS water supply **1473:** 363.

Hem, J.D. (1991). Study and interpretation of the chemical characteristic of natural waters, 3rd edn. US Geol survey, water supply paper 2254. Scientific pub, Jodhpur.

Stuyfzand, P.J. (1989). Nonpoint sources of trace elements in potable groundwaters in the Netherlands. proceedings 18th TWSA water workings testing and research institute Kiwa.

USSL, (1954). diagnosis and improvement of saline and alkali soils, USDA handbook **60:** 147.

WHO, (1971). International standards for drinking water 3rd edition Geneva, pp. 70.

Wilcox, L.V. (1955). Classification and use of irrigation water. U.S. *Geological Department Agri. Circ.* **969:** 19.

15

Physical and Chemical Analysis of Drinking Water Samples in Nagamalai Pudukkottai, Madurai

BAKKIALAKSHMI, S[1*]., SHANTHI, B[a]., PREMA, V[1]. AND MENAKA, T[1].

[1]*Department of Physics, Annamalai University, Annamalai Nagar–608002*

[a]*Centralized Instrumentation and Services Laboratory (CISL), Annamalai University, Annamalai Nagar–608002, Tamil Nadu, India*

**Email: bakkialashmi66@yahoo.com*

ABSTRACT

Physical and chemical studies of drinking water samples in Nagamalai Pudukkottai area in Madurai District was carried out in 2008. From an overall view of water quality parameters determined in our experimental area, it is found that in general, water is hard and it has high TDS, calcium and chloride. The presence of fluoride at about 1 mg/L in drinking water is known to protect against dental caries but high levels of fluoride cause mottling of the dental enamel (1). The water from sampling stations S_1, S_2, S_3, S_4, and S_5 have above 1 mg/L of fluoride. The water from these stations are not suitable for drinking purpose. Hardness of water appears to have a beneficial affect against cardiovascular disease. From the analysis we conclude that the water in this area is not good for drinking and it may lead to some diseases.

Keywords: Physical parameters, chemical parameters and water pollution.

INTRODUCTION

In industrial zone, different types of industries emits different types of pollutants. The concentration of each pollutant depends upon the situation of point with respect to the industries that emits the pollutants. Air and water pollution combined with mismanagement of solid wastes have caused serious damage to the quality of human life.

Dasgupta Adak and Purohit (1) described correlation coefficient of some physico-chemical characteristics of surface and groundwater of Rajganpur, Rourkela. Their corresponding water quality index and analysis of results showed that experimental waters are suitable for drinking purpose except pond water. Study of groundwater for irrigation near industrial area in Rajasthan has been made by many workers, Hussain *et al.* Ojha, (2,3).

Indirabai and Seenamma George (4) have analysed the drinking water quality in selected areas of Trichy town, Tamil Nadu. Their results show water quality involving physical, chemical and biological parameters. They suggested purification of water before use for drinking (4).

Santhi *et al.*, (5) have conducted investigation about the fluorides content in groundwaters of Villioor Union, Tirunelveli district, Tamil Nadu. Their survey reveals the fact that majority of the residents of these villages were affected by fluorosis.

Materials and Methods

Waste water samples were collected in pre-cleaned, acid-washed and dried polyethylene containers of 2-litre capacity. Cleaning procedure of these containers involves soaking them in a 2% nitric acid solution for about 24 hours, followed by multiple rinses with doubly distilled water.

In the present study, water samples were collected at five different locations.

Results and Discussion

Physical Parameters

Various physical examinations such as appearance, odour, turbidity and electrical conductivity are done and the results are given in Table 1.

Table 1: Physical parameters.

Sample No.	S1	S2	S3	S4	S5	Maximum Permissible Limit	
Appearance	Clear	Clear	Clear	Clear	Clear	Colourless	Colourless
Odour	None	None	None	None	None	Unobjectionable	Unobjectionable
Turbidity NTU	0	0	0	0	0	1	10
Total Dissolved Solids (mg/L)	1064	119	511	1200	1150	500	2000
Electrical Conductivity (mho/cm)	1520	170	730	1750	1600	-	-

A – CPHEEO Standards desirable limit

B – CPHEEO Standards cause for rejection when exceeds

Appearance, Odour, Turbidity and TDS

The appearance of the water samples were analysed visually all of our samples S1, S2, S3, S4 and S5 are clear. All of our samples are found to be odourless. The turbidity values are below the desirable limit (i.e. 0 (zero)). TDS values for the water samples at S1 is below the desirable limit. All other samples S2, S3, S4, and S5, exceeds the desirable limit but below the limit of cause for rejection when exceeds. It is shown in Fig. 1.

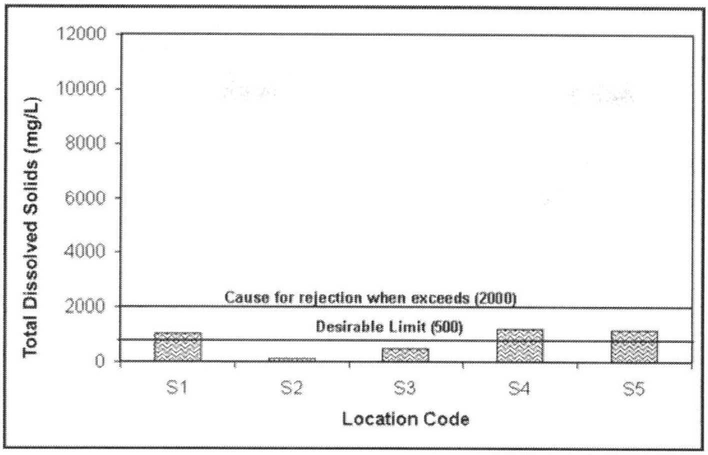

Fig. 1. Variation of TDS with location code.

Electrical Conductivity

The Electrical Conductivity values ranges from 1710 to 170 mho/cm. for our location. According to the classification made by the United State salinity Laboratory, EC greater than 2250 category of water is considered poor. This water is not suitable for irrigation under ordinary condition but may be used occasionally under very special circumstances. (5) In the present study, the EC values at all the light sports are lower than the permissible limit. It is shown in Fig. 2.

Chemical Parameters

The observed chemical parameters are given in Table 2.

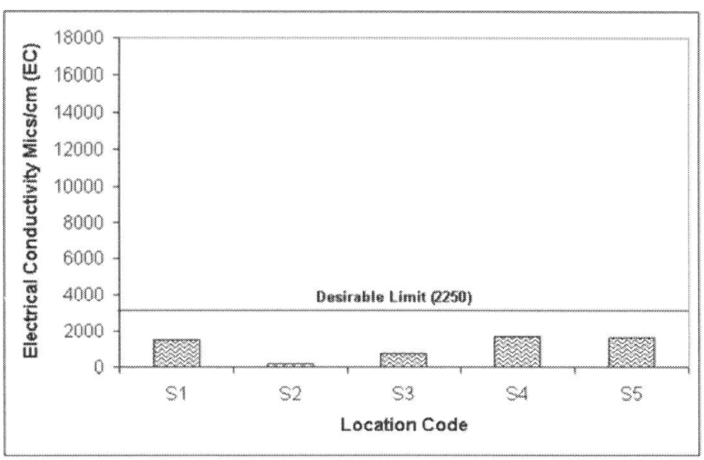

Fig. 2. Variation of electrical conductivity with location code.

Table 2: Chemical parameters.

Sample No.	S1	S2	S3	S4	S5	Maximum Permissible Limit	
Distance (km)	0.0	0.25	0.50	0.75	1.0	**A**	**B**
pH	7.0	6.6	8.0	8.5	8.2	7.0	6.5
Alkalinity pH ($CaCo_3$)	0	0	0	0	0	-	-
Total alkalinity ($CaCo_3$)	284	40	196	300	290	200	600
Total Hardness ($CaCo_3$)	432	40	252	442	430	200	600

A – CPHEEO Standards desirable limit
B – CPHEEO Standards cause for rejection when exceeds
Results are expressed in mg/L

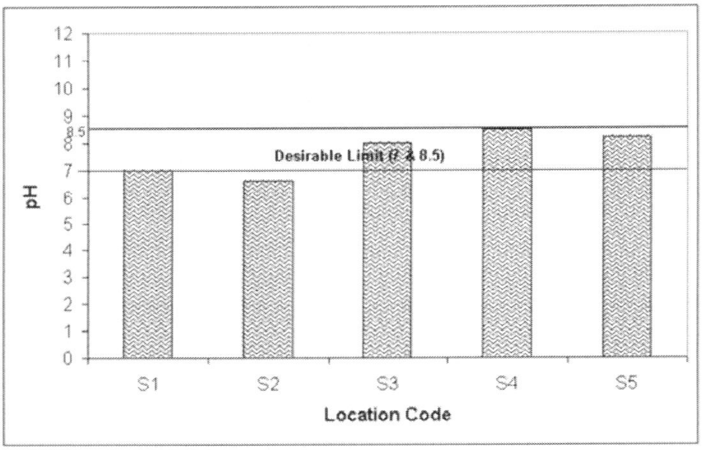

Fig. 3. Variation of pH with location code.

pH, Alkalinity Total and Total Hardness

It can be observed that the pH of the water varies from 6.6 to 8.5 in our location. The lowest value 6.6 occurs at the sample station S2. The highest value 8.5 (within the permissible limit) occurs at S4. The variation of pH with location code is shown in Fig. 3. The samples collected for the present study in S1, S4 and S5 are found that exceeds the desirable limit for alkalinity total value. Desirable limit for the alkalinity total is 200. The water from any of our sampling stations S1, S4 and S5 cannot be used for irrigational or industrial purposes. The hardness in water is mainly due to its calcium and magnesium contents (6). The ranges of hardness value is 40 to 442. The sampling station S2 have very low value compared with the permissible limits of literature (WHO). All other sampling stations exceeds the permissible limit and below the cause for rejection limit.

Srinivasa Rao and Venkateswaralu (7, 8) have reported that the hardness values 186-920 mg/L. According to them few samples have very high hardness values, which are due to the presence of high salt content. This variation is shown in Fig. 4.

Concentration of Heavy Metals in Water Collected at Various Sampling Stations

The concentration of heavy metals like Calcium, Magnesium, Iron, Manganese. Free ammonia, Nitrite, Nitrate, Chloride, Fluoride and Sulphate were observed from Table 3.

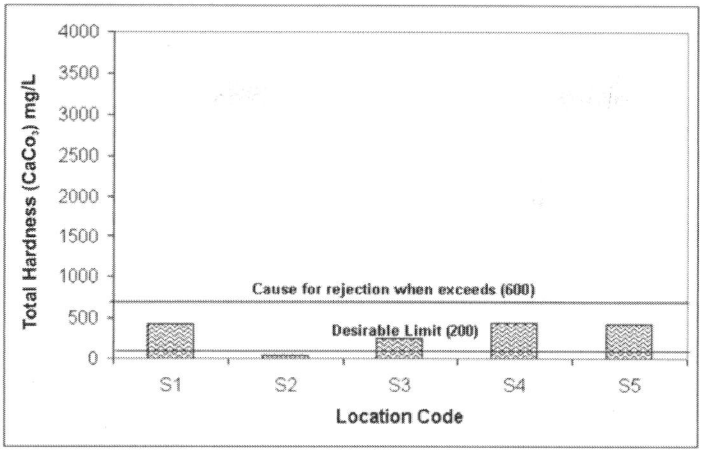

Fig. 4. Variation of total hardness with location code.

Table 3: Concentration (ppm) of heavy metals in water collected at various locations.

Sample No.	S1	S2	S3	S4	S5	Maximum Permissible Limit	
Calcium (Ca)	120	8	64	130	120	75	200
Magnesium (Mg)	32	5	22	40	38	30	150
Iron (Fe)	0	0	0	0	0	0.1	1.0
Manganese (Mn)	0	0	0	0	0	0.05	0.5
Free Ammonia (NH3)	0	0	0	0	0	-	-
Nitrite (NO_2)	0	0	0	0	0	-	-
Nitrate (NO_3)	46	0	23	50	45	45	100
Chloride (Cl_2)	212	34	72	252	231	200	1000
Fluoride (F_2)	2	1	1.0	1.4	1.3	1.0	1.5
Sulphate (SO_4)	66	0	33	83	72	200	400

Calcium (Ca) and Magnesium (Mg)

Calcium is one of the most abundant substances of natural water. The quantities of calcium in natural water generally vary from 10 to 100 mg/L depending upon the disposal of the industrial waste. The calcium values vary from 8 to 130. These results are tabulated is Table 3 and the values are represented by Fig. 5. The water in sampling station S2 contains minimum calcium. The permissible limits of magnesium prescribed by ICMR and ISI are 50-150 mg/L (Fig. 5). The concentration of Mg is higher than the WHO limits. Hence, it may not be suitable for irrigation purpose. It is shown in Fig. 6.

Iron (Fe), Manganese (Mn), Free Ammonia (NH_3), Nitrite (NO_2) and Nitrate (NO_3)

Iron is one of the most abundant substances of natural water. The quantities of iron in natural water generally vary from 0.1 to 1.0 mg/L. The iron content in all samples is nil. The desirable limit for manganese is 0.05. All of our water samples fall within the desirable limit. No Mn is observed in any samples. All the sampling stations S1, S2, S3, S4 and S5 have no free ammonia.

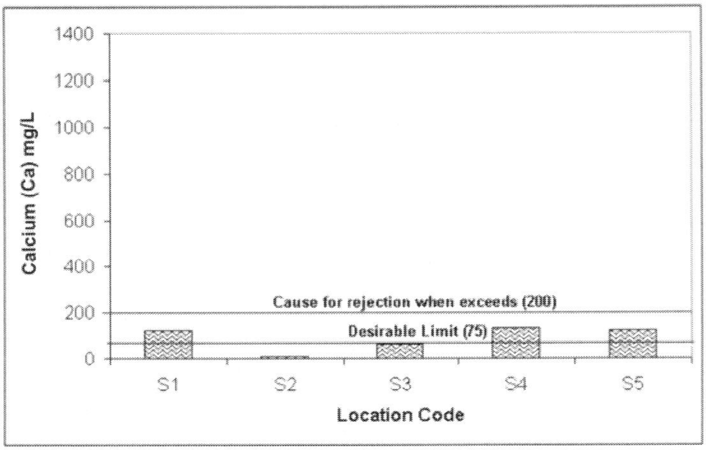

Fig. 5. Variation of calcium with location code.

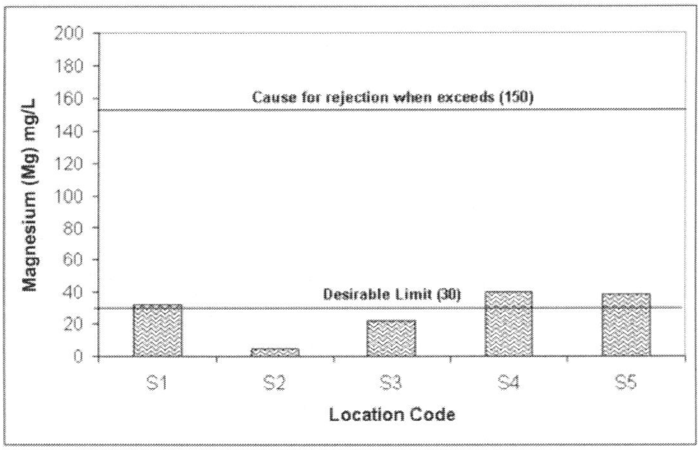

Fig. 6. Variation of magnesium with location code.

All the sampling stations have no nitrite. The sampling station S2 has no nitrate and all others have the Nitrate within the prescribed limit (S1= 46, S3 = 23, S4 = 5.0, S5 = 45).

Chloride (Cl_2), Fluoride (F_2), Sulphate (SO_4) and Phosphate (PO_4)

Chloride is highly soluble with most of the naturally occurring cations and do not precipitate with sediments and cannot be removed biologically in the treatment of waste. The amount of chloride present in well water is found to be in the range of 34 to 252 mg/L. The fluoride value in our sampling station S2 fall within the permissible limit. All other stations slightly exceeds the limit. Sulphate occurs in water largely in oxidized form. It may be also present as sulphides. Discharge of industrial wastes and domestic sewage in water tends to increase its concentration. Water containing more than 500 mg/L sulphate content tastes bitter and beyond 1000 mg/L it has purgative effect.

Discussion

From the results it is noted that the quality of well water varies from place to place. TDS, hardness, calcium and chloride values are higher than the permissible limit.

The pH of all sampling stations are within the prescribed limit. The electrical conductivity values are all above 250 mmhos/cm. This suggests the requirement of some treatment of water before used for drinking and other purposes. The concentration of some heavy metals like Ca, Cl_2 in wastewater are higher than the prescribed limits except S2 given by FAO, ISI and WHO. The higher concentration of these heavy metals present in the water samples may be toxic to plants. As such the water is unsuitable for irrigation purpose unless it is treated properly. If this is untreated, the habitants using this water for drinking purpose would be much suffered of heavy metals dose.

The presence of fluoride at about 1 mg/L in drinking water is known to protect against dental caries but high levels of fluoride cause mottling of the dental enamel (1). The water from sampling stations S1, S2, S4, and S5 have above 1 mg/L of fluoride. The water from these stations are not suitable for drinking purpose.

Hardness of water appears to have a beneficial affect against cardiovascular disease.

CONCLUSION

Keeping in view of all these results, it can be conduced that the water from this area needs efficient treatment in reducing or neutralizing the toxic elements presents in it, thus making it suitable for irrigational and drinking purpose except S2. Hence, it is recommended that water must be thoroughly analysed and studied before being used for domestic applications.

REFERENCES

1. Dasgupta Adak, *et al.*, **20(9):** (2000). *Indian J. Environ. Production*, pp. 681-687.
2. Hussain, J., Hussain, I. and Ojha, K.G. **13(3):** (2001), *Asian J. chemistry*, pp. 1114-1122.
3. Hussain, J., Hussain, I. and Sharma, K.C. **13(3):** (2001). *Asian J. Chemistry*, 1123-1126.
4. Indirabai, W.P.S. and Seenamma George, **21(2):** (2002). Pollution Research, pp. 209-214.
5. Hussain. J., *et al.*, 13 (3); (1995), *Asian J. Chemistry*, pp. 1114-1122.
6. Trivedy and Goel, J.K., (1986).
7. Water Pollution 2 edition, Srinivasa Rao and Venkateswaralu, **20(3):** (2000). *Indian J. Environ. Protection*, pp. 161-164.
8. Raman, N. and Murugesan, A. **13(3):** (2001). *Asian J. Chemi.*, pp. 995-1000.

16

Defluoridation of Groundwater by Aquatic Macrophyte *Hydrilla verticillata*

RAMANATHAN, AL.,[1*] SADAF NAZNEEN[1], CHIDHAMBARAM, S.,[2] AND SALLAUDIN, M.[3]

[1]*School of Environmental Sciences, JNU*

[2]*Dept. of Earth Sciences, Annmalai University*

[3]*Green Technology Division, MOEN and F, New Delhi*

**Email: alrjnu@gmail.com*

ABSTRACT

In this paper, we outline the fluoride removal from groundwater techniques using aquatic macrophyte *Hydrilla verticillata* at various fluoride concentrations with respect field conditions at various parts of our country. It was found that plants absorb more to from lower F concentration waters after one week. Time factor seems detailed to be an important to remediate groundwater rich in fluoride. We have to undertake detail study to maximize the adsorption of F by using other aquatic plants and we have to a carry out study for longer periods as well.

Keywords: Defluoridation, *Hydrilla verticillata*, cation, anion.

INTRODUCTION

It is estimated that nearly one-third of the world's population is dependent on groundwater for drinking (UNEP, 1999). Intensive use of natural resources and anthropogenic impacts are posing a great threat to water quality (Foster 1995; Mor *et al.*, 2006). Nearly 12 million of the total 85 million tonnes of fluoride deposits on the earth's crust are found in India. This is one of the reasons why fluorosis is endemic in India (UNICEF 1999). In India, at present 20 states have become endemic to fluorosis (Gupta *et al.*, 2006). An estimation shows that around 62 million people including 6 million people suffer from fluorosis because of consumption of fluoride rich water (Raju *et al.*, 2009). It is estimated that around 200 million people among 25 nations world over are affected by fluoride pollution (Suther *et al.*, 2008).

Fluoride the ionic form of fluorine is ubiquitous in nature. It is most reactive of all elements therefore does not exist in free form in nature (Jha and Jha, 1981). Fluoride enters groundwater mainly by leaching and weathering of bedrocks containing high fluoride content. According to WHO optimum fluoride level in drinking water should be between 0.5 – 1.2 mg/litre. Excessive ingestion of fluoride can damage tooth forming enamel causing dental fluorosis and skeletal fluorosis in later stages. Many study reports suggest high fluoride content in various water bodies in India. In India, at present, it has been estimated that fluorosis is prevalent in 17 states of India (Table 1). The safe limit of fluoride in drinking water is 1.0 mg/L. The endemic fluorosis in India is generally largely of geochemical origin.

District known to be endemic for fluoride in various states of India (18)

States	Districts	Range of fluoride concentration (mg/L)
Assam	Karbianglong, Nagaon	0.1–18.1
Andhra Pradesh	All districts except Adilabad, Nizamabad, West Godhavari, Visakhapatnam, Vijzianagaram, Srikakulam	0.11–20.0
Bihar	Palamu, Daltonganj, Gridh, Gaya, Rohtas, Gopalganj, Paschim, Champaran	0.6–8.0
Delhi	Kanjhwala, Nijafgarh, Alipur	0.4–10.0
Gujarat	All districts except Dang	1.58–31.0
Haryana	Rewari, Faridabad, Karnal, Sonipat, Jind, Gurgaon, Mohindergarh Rohtak, Kurukshetra, Kaithal, Bhiwani, Sirsa, Hisar	0.17–24.7
Jammu and Kashmir	Doda	0.05–4.21
Karnataka	Dharwad, Gadag, Bellary, Belgam, Raichur, Bijapur, Gulbarga, Chitradurga, Tumkur, Chikmagalur, Manya, Bangalore, Mysore	0.2–18.0
Kerala	Palghat, Allepy, Vamanapuram, Alappuzha	0.2–2.5
Maharashtra	Chandrapur, Bhandara, Nagpur, Jalgaon, Bukduna, Amravati Akola, Yavatmal, Nanded, Sholapur	0.11–10.2
Madhya Pradesh	Shivpuri, Jabua, Mandla, Dindori, Chhindwara, Dhar Vidhisha Seoni, Sehore, Raisen and Bhopal	0.08–4.2
Orissa	Phulbani, Koraput, Dhenkanal	0.6–5.7
Punjab	Mansa, Faridkot, Bhatinda, Muktsar, Moga, Sangrur, Freozpur, Ludhiana, Amritsar, Patila, Ropar, Jallandhar, Fatehgarh Sahib	0.44–6.0
Rajasthan	All the 32 districts	0.2–37.0
Tamil Nadu	Salem, Periyar, Dharampuri, Coimbatore, Tiruchirapalli, Vellore, Madurai, Virudunagar	1.5–5.0
Uttar Pradesh	Unnao, Agra, Meerut, Mathura, Aligarh, Raibareli, Allahabad	0.12–8.9
West Bengal	Birbhum, Bhardaman, Bankura	1.5–13.0

(Meenakshi and Maheshwari, 2006)

In the past studies, various natural materials like red soil, untreated charcoal, serpentine, flyash have been used for fluoride removal (Chidambaram *et al.*, 2003). The ability of soil to absorb fluorine from solution has been studied by earlier researchers (Bower and Hatcher, 1967; Fluhler *et al.*, 1982; Gupta *et al.*, 1982). The present study is based at finding low cost, environment-friendly method for removal of fluoride. In the past few years, plant based phytoremediation is fast gaining importance because of its environment-friendly and cost-effective approach. Many aquatic macrophytes along with *Hydrilla verticillata* have been found to effectively remove heavy metals from wetlands by accumulating metals in their bodies (Kumar *et al.*, 2008). In certain studies it has also been found that macrophytes are efficient in removing chemical pollutants like petrochemicals, ammonia nitrates, etc (Ratushnyak, 2008) (Anna A, 2008). Not much study has been carried out with aquatic plants which might prove effective in removing fluoride directly from the water. In few studies aquatic macrophyte *Hydrilla verticillata* has been found to remove fluoride to some extent (Sinha *et al.*, 2000). In Tamil Nadu where many cases, the water sources have been rendered unsafe not only for human consumption but also for other activities such as irrigation and industrial needs. Therefore, now there is a need to focus greater attention on fluoride, which is the major inorganic pollutant of natural origin found in groundwater. In this paper aquatic macrophyte *Hydrilla verticillata* has been investigated for fluoride removal under field conditions to replicate it in the flourosis affected regions of Tamil Nadu.

MATERIALS AND METHODS

The plants of *Hydrilla verticillata* were collected from Bhatkal Lake, Faridabad. The water samples were collected from the lake to test the presence of fluoride which show negative results. The plants were grown for some period in cement tanks in the experimental garden of School of Environmental Sciences, JNU for acclimatization. Different concentrations of fluoride (1, 2, 3, 4 and 5 ppm) were prepared in tap water which had tested and found no prior fluoride presence. The use of the tap water was subjected to fluoride had been tested removal efficiency had no prior of these plants in field conditions. The tap water also served as the minimal nutrient medium for plants for the period of seven days. Plants with same biomass, i.e. nearly 8 grams of fresh weight were kept in each pot. Plants kept in plain tap water served as control. The pots were placed in field conditions during October/November 2008. The plants were harvested after seven days and water samples were collected in plastic bottles. The plants were grown in plastic pots to avoid fluoride loss due to sorption by the walls of the glass apparatus.

The fluoride uptake in the treatments was estimated by consort fluoride selective ion electrodes (model C933). Na and K and Ca in the treated samples were measured by Atomic Absorption Spectrophotometer (AAS). Bicarbonate, Chloride and Phosphate were measured by titrimetric method. Nitrate and Sulphate were analysed by Spectrophotometer (Hach).

RESULTS AND DISCUSSION

Figs. 1, 2 and 3 show the fluoride removed by absorption as well as fluoride remaining in water after harvesting *Hydrilla verticillata* on 4th day and 7th day. The figures shows that plant uptake fluoride both on 4th day as well as 7th day treatments. It is observed that maximum removal of Fluoride is observed in the 1 ppm after 4th and 7th day. But removal percentage is highest on 7th day compared to 4th day, but in general in both the days the removal capacity gradually decreased with increase in fluoride concentration in water. However, at least 0.5 to 1 mg/L of fluoride is removed from groundwater by this plant in each case. Table 3 gives shows the percentage of removal of fluoride after 7th day which clearly shows that 60 uptake is there for 1 ppm, whereas for 2-5 ppm they vary from 20-28% and it is a good indication that the higher concentration of fluoride can be removed using this aquatic plants.

Tables 1 and 2 indicate the dissolve ions after four days of various fluoride concentration water treated with this aquatic plants shows that at 0 (control) ppm F, Ca is less and Na and K is high followed by high HCO_3 with low Cl, and NO_3. After treating with 1-5 ppm F water Ca increased and Na and K show slight decrease, followed by increase of HCO_3, SO_4, NO_3 and Cl. After seven days of treatment Ca increases followed by decrease of K and Na with increase

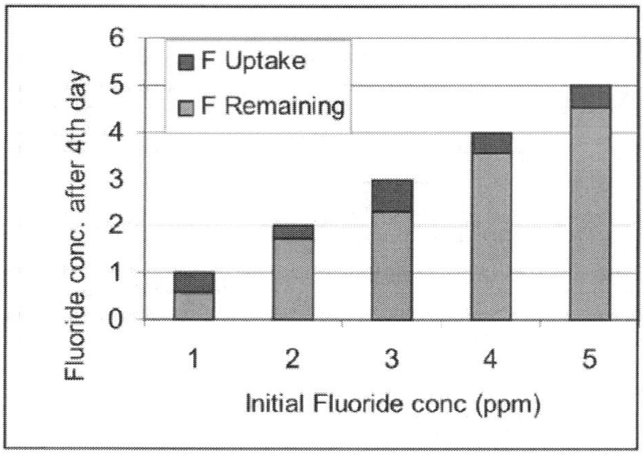

Fig. 1. Fluoride absorbed as well as fluoride remaining in water after harvesting *Hydrilla verticillata* on 4th day.

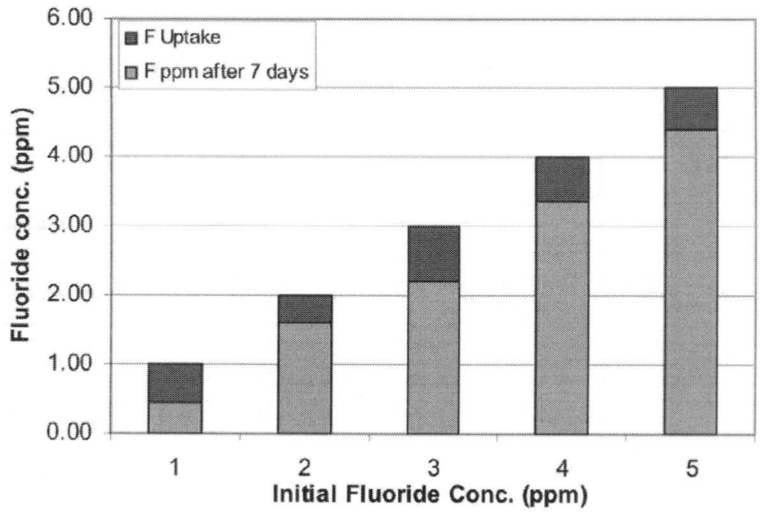

Fig. 2. Fluoride absorbed as well as remaining fluoride in water after harvesting *Hydrilla verticillata* on 7th day.

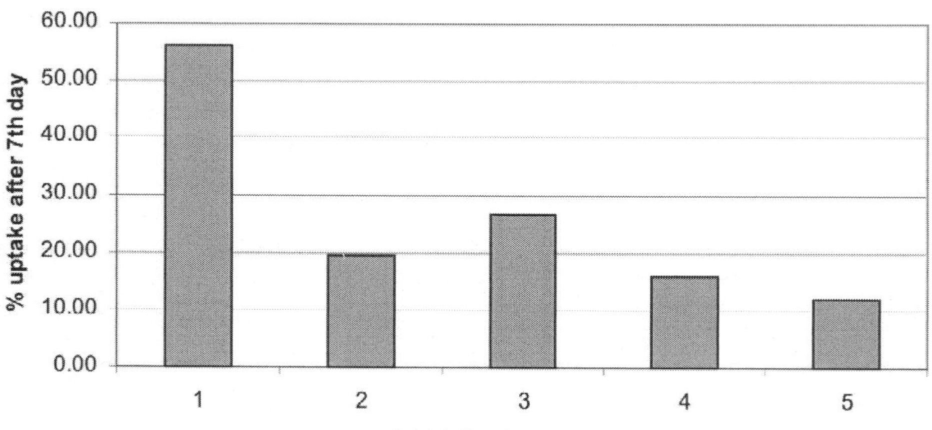

Fig. 3. Percentage removal of fluoride by plants of *Hydrilla verticillata* after 7th day.

in HCO_3, NO_3 and SO_4. The two observations indicate that low calcium and high bicarbonate alkalinity favor high fluoride content in groundwater which has been altered after the plants where grown. The Ca increased and HCO_3 reduced and hence the F is absorbed by the plants and these changes are reflected in the groundwater hydrochemistry. But we observed that NO_3, Cl and SO_4 increased in groundwater during this study and it becomes very high after 7 days indicating that it might have come from the plant which we brought from lakes and that has enhanced these ions in water by the leaching from these plants. In both the case PO_4 is more in controlled and less in others which shows that plans uptake PO_4 for their growth.

Table 1: Major cations in the solution treated with fluoride after 4 days.

Treatments (mg/L) control	Na^+ (mg/L) 9.95	K^+ (mg/L) 1.58	Ca^+ (mg/L) 47.21
1	8.76	1.45	54.04
2	6.22	2.62	49.67
3	5.01	0.78	51.51
4	4.24	0.54	57.86
5	7.63	0.22	53.92

Table 2: Major anions in the solution treated with fluoride after 4 days.

Treatments (mg/L)	Cl^-(mg/L)	HCO_3^- (mg/L)	PO_4^{-3} (mg/L)	NO_3^- (mg/L)	SO_4^{-2} (mg/L)
control	2.41	122	0.148	00.00	57.83
1	9.026	129.2	0.125	0.71	70.28
2	8.17	136.4	0.095	1.19	58.21
3	8.01	142.6	0.072	0.25	90.00
4	7.94	134.3	0.050	0.61	80.62
5	8.21	144.5	0.035	0.46	66.22

Table 3: Major cations in the solution treated with fluoride after 7 days.

Treatments (mg/L)	Na^+ (mg/L)	K^+ (mg/L)	Ca^+ (mg/L)
Control (TW)	9.95	1.58	47.21
1	7.23	2.59	60.87
2	4.02	0.89	52.62
3	6.12	0.45	60.67
4	4.02	0.96	69.42
5	7.93	3.14	65.32

Table 4: Major anions in the solution treated with fluoride after 7 days.

Treatments (mg/L)	Cl^-(mg/L)	HCO_3^- (mg/L)	PO_4^{-3} (mg/L)	NO_3^- (mg/L)	SO_4^{-2} (mg/L)
Control (TW)	2.41	122.0	0.148	0.00	57.83
1	2.31	146.4	0.123	2.71	66.41
2	0.71	140.3	0.074	5.44	64.36
3	1.28	115.9	0.050	2.60	101.49
4	0.71	122.0	0.045	3.39	94.28
5	1.06	146.4	0.026	2.19	61.21

Tables 5-8 indicate that after 4 days Ca show good correlation with treatments and it become very positive and strong after 7 days. This strong positive correlation with bicarbonate (r = + 0.85) is in good agreement with earlier work (Jha *et al.*, 2008). Good correlation is also observed between K and Ca, K, Ca, HCO_3 and treatments after 4th day. Whereas the HCO_3 and treatment has become positive after 7 days. Among the cations calcium shows a positive correlation (r = + 0.66) with increase of the fluoride treatments. Among the anions phosphate shows a significant negative correlation on the 4th day and in the 7th day treatments. Results also show a positive correlation with the calcium.

Table 5: Correlation matrix for cations after 4th day.

	Treatments (mg/L)	Na+ (mg/L)	K+ (mg/L)	Ca+ (mg/L)
Treatments (mg/L)	1			
Na+ (mg/L)	-0.63837	1		
K+ (mg/L)	-0.69035	0.250613	1	
Ca+ (mg/L)	0.669349	-0.54816	-0.61558	1

Table 6: Correlation matrix for anions after 4th day.

	Treatments (mg/L)	Cl-(mg/L)	HCO_3^- (mg/L)	PO_4^{-3} (mg/L)	NO_3^- (mg/L)	SO_4^{-2} (mg/L)
Treatments (mg/L)	1					
Cl-(mg/L)	-0.10962	1				
HCO_3^- (mg/L)	0.85253	0.00197	1			
PO_4^{-3} (mg/L)	-0.99563	0.020526	-0.86379	1		
NO_3^- (mg/L)	0.138372	0.499129	0.228339	-0.17177	1	
SO_4^{-2} (mg/L)	0.439246	0.123419	0.465449	-0.47589	-0.22175	1

Table 7: Correlation matrix for 7th day in cations.

	Treatments (mg/L)	Na+ (mg/L)	K+ (mg/L)	Ca+ (mg/L)
Treatments (mg/L)	1			
Na+ (mg/L)	-0.40618	1		
K+ (mg/L)	0.124885	0.54147	1	
Ca+ (mg/L)	0.812874	-0.43282	0.179607	1

Table 8: Correlation matrix for 7th day for anions.

	Treatments (mg/L)	Cl-(mg/L)	HCO_3^- (mg/L)	PO_4^{-3} (mg/L)	NO_3^- (mg/L)	SO_4^{-2} (mg/L)
Treatments (mg/L)	1					
Cl-(mg/L)	-0.53734	1				
HCO_3^- (mg/L)	0.094987	0.278317	1			
PO_4^{-3} (mg/L)	-0.96671	0.615849	-0.01604	1		
NO_3^- (mg/L)	0.307735	-0.41559	0.307596	-0.45599	1	
SO_4^{-2} (mg/L)	0.394142	-0.23628	-0.66909	-0.49298	0.212115	1

Following are the limitations of this study:

- The absorption is highly dependent on pH and not checked for wide range of pH.
- High concentration of total dissolved salts (TDS) can make water non-potable.
- Presence of sulphate, phosphate results in ionic competition.
- The process has low adsorption capacity, poor integrity and needs detail study.

The process removes only a smaller portion of fluoride (18–33%) at higher concentration from 2-5 ppm, and therefore need more attention to augment its adsorption capacity.

- Large pots, space, continuous monitoring, etc., is required for this experiment.
- Silicates may have adverse effect on defluoridation along with temperature, which also may affect the defluoridation capacity.

CONCLUSION

The present study indicates that *Hydrilla verticillata* plants are effective in removing fluoride from contaminated water under field conditions. Removal efficiency decreased with increase in concentration of fluoride which shows that there is no effect of increase in concentration on fluoride removal capacity of *Hydrilla verticillata*. The nutrient quality analysis along with the fluoride treatments shows no significant trend. The plants remove fluoride under specific environmental conditions. The fluoride removal efficiency varies according to the quality of the groundwaters, so actual applications may vary from the generalizations made here. Therefore, this method should be tested using the various natural waters to be treated before implementation in the specific region. Hence, this study needs further detailed investigation.

REFERENCES

Bower, C.A. and Hatcher, J.T. (1967). Adsorption of fluoride by soils and minerals. *Soil Sci.* **103:** 209-218.

Chidambaram, S., Ramanathan, A.L. and Vasudevan, S. (2003). Fluoride removal studies in water using natural materials. *Water. SA* **29(3):** 339-343.

Fluhler, H., Polomski, J. and Balser, P. (1982). Retention and movement of fluoride in soils. *J. Environ. Qual.* **11:** 461-468.

Foster, S.S.D. (1995). Groundwater for development – An overview of quality constraints. In Nash, H. and McCall, G.J.H. (Eds.), Groundwater quality. 17th Special Report (pp. 1-3). London United: Chapman and Hall.

Gupta, S., Banerjee, S., Saha, R., Datta, J.K. and Mondal, N. (2006). Fluoride geochemistry of groundwater in Nalhati-1 Block of Birbhum District, West Bengal *Fluoride,* **39(4):** 318-320.

Jha, L.B. and Jha Mohan, (1981). Fluoride pollution in India. *Intern. J. Environ. Studies,* **19:** 225-230.

Soni Hiren, J.I., Kumar, Rita., Kumar, Nirmal N. and Bhatt, Ira. (2008). Macrophytes in Phytoremediation of Heavy Metal contaminated water and Sediments Pariyej community reserve, Gujarat, India. *Turkish journal of Fisheries and Aquatic sciences,* **8:** 193-200.

Meenakshi, Maheshwari, R.C. (2006). Fluoride in drinking water and its removal, *J. Hazardous Materials,* **B137:** 456-463.

Janardhan Raju, N., Sangita Dey, and Kaushik Das, (2009). Fluoride Contamination in groundwaters of Sonbhadra District, Uttar Pradesh, India. *Current Science,* **96(7):**

Ratushnyak Anna, A. (2008). The role of Aquatic macrophytes in Hydroecosystems of Kuibyshev Reservoir (Russia) American-Eurasian. *J. Agric. and Environ. Sci.* **4(1):** 01-08.

Jha, S.K., Nayak, A.K. and Sharma, Y.K. (2008). Fluoride occurrence and Assessment of Exposure dose of Fluoride in Shallow Aquifers of Makur, Unnao District, and Uttar Pradesh India. *Environ Monit Assess.*

Sinha, S., Saxena R. and Singh, S. (2000). Fluoride removal by *Hydrilla verticillata* (l.f.) Royle and its Toxic effects **65:** 683-690.

Surinder Suther., Vinod K. Garg., Shushant Jangir., Simranjeet Kaur., Nidhi Goswami, and Singh Sushma, (2008). Fluoride contamination in drinking water in rural habitation of Northern Rajasthan, India. *Environmental Monitoring Assessment* **(145):** 1-6.

UNICEF, (1999). State of the Art Report on the Extent of Fluoride In Drinking Water and the Resulting Endemicity in India. Report by fluorosis and Rural Development Foundation for UNICEF, New Delhi.

17

Speciation Analysis of Trace Elements in Water by Atomic Absorption Spectrometry

RAMESH KUMAR, A[1]., AND RIYAZUDDIN, P.[2]

[1]*Central Groundwater Board, South Eastern Coastal Region, E1, Rajaji Bhavan, Besant Nagar, Chennai–600090*

[2]*Department of Analytical Chemistry, University of Madras, Guindy Gampus, Chennai–600025*

**Email: riyazdr@yahoo.co.uk*

ABSTRACT

The determination of specific form of trace element in water is important as the toxicity, geochemistry and other characteristics depend on speciation of the element. However, the determination of chemical species is a challenging task due to lack of sensitivity of analytical instruments, stability of species, matrix effects, etc. Various separation techniques such as co-precipitation, solvent extraction, solid-phase extraction, etc. can be effectively utilized to separate and pre-concentrate the elemental species, which subsequently can be determined by atomic absorption spectrometry (AAS). In addition, suitable modification of reaction conditions such as pH, concentration and type of acid and tetrahydroborate (THB) results in the determination of different oxidation states of arsenic, selenium, antimony and tellurium using hydride generation AAS. This paper discusses the various approaches for the determination of different species of arsenic, chromium, selenium, etc. using AAS. Problems related to stability of species, reliability of results are also discussed.

Keywords: Trace element, speciation, water, AAS.

INTRODUCTION

Metals and metalloids are important contaminants of the atmosphere, aquatic systems and terrestrial environment. It is well known that the toxicity of elements depends upon their physico-chemical forms, rather than their total concentrations. For example, Cr(III) is essential, whereas Cr(VI) is toxic. For humans, the more assimilable form of iron is haem iron, whereas the inorganic form of iron salts and simple organic complexes are far less available. Organometallic compounds are usually much more toxic than the ions of the corresponding inorganic compounds. Mercury, tin and lead obey this general rule, whereas arsenic is an exception, because most arsenicals are less toxic than inorganic arsenic species.

In addition, the hydrogeochemistry of elements depend on the specific form in which the elements occur, for example, under the natural water pH (6-8), Cr(III) is precipitated as $Cr(OH)_3$, whereas Cr(VI), because of its anionic nature (CrO_4^{2-}) shows high mobility. Similarly, As(III) is more mobile than As(VI). Cu(I) tends to be associated with particulate matter, whereas Cu(II) exist as ionic form, hence more toxic. Therefore, the determination of specific form of an element, i.e., speciation analysis is required to understand the role of an element in living organisms, and in the environment.

Speciation Analysis–Evolution and the Present Scenario

In the early 1950s, aquatic geochemists first made the distinction between 'dissolved metal' and 'particulate metal' using 0.45 ìm membrane filter for filtration. The simultaneous rapid development of electrochemistry enabled the analyst to distinguish between 'free' and 'complexed metal' species, thus highlighting the first analytical speciation schemes. Subsequently, several computer programs have been developed based on equilibrium calculation models for speciation. The work of Florence[1] carried out during 1980s using stripping voltametric techniques can be considered as a milestone in speciation studies. The metal fractionation techniques of Tessier *et al.*[2] for soils and sediments samples were also considered important contribution in understanding the metal partition between various mineral phases.

The advancement of various atomic spectrometric techniques such as AAS, AFS, ICP-AES, in combination with various separation procedures allowed the separation of specific forms of metals and its determination. Examples include the various species specific co-precipitation techniques, ion-exchange and complexation and liquid-liquid extraction techniques. Speciation analysis took a major turning point since the coupling of gas chromatography with various element specific detectors. This has resulted in the redefinition of the concept of metal speciation, which was previously rather vague. In 2000, IUPAC[3] published a report defining the various terms related to 'speciation'.

i. *Chemical species.* Chemical elements: specific form of an element defined as to isotopic composition, electronic or oxidation state, and/or complex or molecular structure.
ii. *Speciation analysis.* Analytical chemistry: analytical activities of identifying and/or measuring the quantities of one or more individual chemical species in a sample.
iii. *Speciation of an element; speciation.* Distribution of an element amongst defined chemical species in a system.
iv. *Fractionation.* Processes of classification of an analyte or a group of analytes from a certain sample according to physical (for example; size, solubility) or chemical (for example; bonding, reactivity) properties.

Hyphenated Techniques in Speciation Analysis

Hyphenated techniques or coupled techniques are a combination of two dissimilar techniques, mostly a combination of separation (chromatography) and detection technique (AAS, AFS, ICP-AES and ICP-MS).[4] The most common separation technique used is gas, and liquid chromatography. Often liquid chromatography is preferred, because the former requires tedious derivatization and extraction steps. HPLC is frequently used for species separation and the most commonly used are ion pair chromatography (IP-HPLC), reversed phased chromatography (RP-HPLC) and size exclusion chromatography (SE-HPLC). The detectors mostly used are HGAAS, ICP-MS and ES-MS-MS.

The advantages of the hyphenated techniques are good separation, minimized sample handling and the ability to work with aqueous and non-aqueous samples. However, due to the high acquisition and operating cost of these instruments, their availability is limited. In addition, a high degree of operator skill is required.

Atomic Spectrometry in Speciation Analysis

Atomic spectrometric techniques like atomic absorption spectrometry (AAS), atomic fluorescence spectrometry (AFS) and inductively coupled plasma-atomic emission spectrometry (ICP-AES) are basically total element techniques. These techniques could well be used as species specific detectors, if the target analyte species are separated by suitable separation techniques like co-precipitation, liquid-liquid extraction (LLE), liquid-solid extraction (LSE), etc. or ion-exchange methods.[5, 6]

Since most trace elements in natural and contaminated water occur in the mg L^{-1} range, a preconcentration step is generally required. Flame AAS generally sensitive in the mg L^{-1} level obviously requires a preconcentration step. Graphite furnace (GF) AAS and hydride generation (HG) AAS are sensitive in the mg L^{-1} range, hence, generally it does not require preconcentration.

Speciation Analysis of Arsenic Using HGAAS

Hydride generation AAS (HGAAS) technique has widely been used for the inorganic speciation analysis of As, Bi, Sb, Se and Te.[7] The trivalent states of As, Bi and Sb readily forms the respective hydrides, whereas the pentavalent state forms the hydride only after its conversion to the trivalent state. The reduction rate of pentavalent to trivalent state is very slow compared to the formation of hydrides, and most often it requires a reducing agent such as KI or L-cysteine. The formation of hydrides from As, Bi and Sb is a pH dependant reaction, as the oxyions of the elements should be fully hydrogenated before its reaction with tetrahydroborate (THB). As pK_1 for arsenic acid is 2.3, the reaction must therefore be carried out at very low pH, that is, 1-2 mol L^{-1} HCl is generally employed. Arsenite (pK_1 9.2), on the other hand is protonated under natural water conditions, and reacts with THB under mild acidic conditions. Thus, differentiation of As(III) and As(V) could be achieved simply by exploiting the pH dependency of THB reduction.

Another approach for As speciation analysis is based on the use of appropriate concentration of acid and THB for selective hydride generation and is limited to the determination of inorganic As species only. This approach is based on the redox property of THB, which act as a reductant and hydride source. Most procedures use low concentration of acid/THB to generate AsH_3 form As(III), and higher concentration of acid/THB for total As determination.

A major drawback of the above procedures is the interference of methylated species such as monomethylarsonate (MMA) and dimethylarsinate (DMA).[8] It can be eliminated by collecting the evolved arsines in a 'U' tube immersed in liquid nitrogen, warmed and sequentially released according to their boiling point (arsine; –55 °C, methylarsine; 2 °C, dimethylarsine; 55°C).[9] Alternatively, specific reaction mediums can be used with different pretreatment techniques.[10] However, methylated forms of arsenic rarely occur in natural waters.[11] Some of the reported methods and their salient features are given in Table 1.

Speciation Analysis of Selenium Using HGAAS

Selenium, exists in the environment in several oxidation states and as well as a variety of inorganic and organic compounds. Dissolved inorganic selenium can be found in natural waters as selenide (-II), as colloidal elemental Se(0), as selenite ion Se(IV) and as selenate anion Se(VI). Organic forms of Se that may be found in aqueous environment are volatile (methyl selenides) or non-volatile (tri methyl selenonium ion, seleno amino acids and their derivatives). As only Se(IV) is reduced to SeH_2 with THB, the other species should be converted into Se(IV) in order to generate SeH_2; performing analysis using three separate aliquots, i.e. a) with no further treatment–direct determination of Se(IV), b) after reduction of Se(VI) to Se(IV) with HCl or HBr/KBr-Br_2–the sum of Se(IV) and Se(VI), thus, Se(VI) is determined by difference, c) mineralization of organic matrix by UV-irradiation or wet digestion followed by reduction to Se(IV)- all Se species. The difference between the total Se and the sum of Se(IV) and Se(VI) is attributed to organic selenium compounds.[7]

Many workers determined inorganic selenium species, i.e. Se(IV) and Se(VI), utilizing the formation of SeH_2 only from the +IV state. Se (VI) is reduced to the +IV state by heating/boiling with 4-6 mol l^{-1} of HCl or with KBr/HBr. The KBr reduction has been reported to be very efficient.[7]

In addition to the offline-reduction, several methods have been reported based on the online-reduction using FI manifold. The online-reduction is mostly achieved by heating with KI/HCl, microwave, UV irradiation, etc. Typical examples of selenium speciation analysis are given in Table 2.

Speciation Analysis of Chromium

Speciation analysis of chromium has been studied by several workers.[19] In aquatic environment, the most stable oxidation states of chromium are Cr(III) and Cr(VI). Under reducing conditions, Cr(III) is the more thermodynamically stable form, as cations (e.g., Cr^{3+}, $Cr(OH)_2^+$ and $Cr(OH)_4^-$) with high particle affinities, and as precipitates (e.g., $Cr(OH)_3$ and $FeCr(OH)_3$) at pH ranges commonly found in natural waters. Under oxidizing conditions, Cr(VI) is thermodynamically favored and exists as soluble oxyanions.

Several analytical procedures have been proposed for the determination of chromium species using element specific detectors such as AAS for quantification (Rubio 1999).[20] The separation of chromium species is generally carried out using co-precipitation or complexation-solvent extraction techniques. The ammonium pyrolidinedithiocarbamate (APDC) ligand has widely been utilized in several chromium speciation analysis procedures.[21]

Coprecipitation of Cr(III) species with $Fe(OH)_3$ serves as an effective technique for several elemental species. Cr(III) can be selectively separated from Cr(VI) with co-precipitation with $Fe(OH)_3$,[22] $Al(OH)_3$[23] and $PbSO_4$[24]; $BaSO_4$[25] for Cr(VI) and $Fe(OH)_2$,[26] $Bi(OH)_3$[27] and $Pb(PDC)_2$[28] for both oxidation states. Also several methods are proposed for the separation of Cr(VI) using different resin such as AG1-X4 [29] AG50W-X8[30], Amerlite LA-2.[31]

Table 1: Selected literature references for arsenic speciation analysis in natural waters by HGAAS.

Species	Matrix	Preservation	Reaction conditions	DL ($\mu g\ L^{-1}$)	Technique	Reference
As(III), As(V)	-	-	As(III): pH: 4-5, As_T: 5 mol L^{-1} HCl	-	HGAAS	[12]
As(III), As(V), MMA, DMA	River and Seawaters	-	As(III): pH: 4-5	-	HG-CT-AES	[13]
As(III), As(V)	Groundwater	1 mL HCl+ 500 mg L^{-1} Fe(II)	As(III): 2 mol L^{-1} HCl+0.035% THB As_T : KI prereduction	1.4	FI-HGAAS	[14]
As(III), As(V), MMA, DMA	Groundwater, Seawaters	0.1 mol L^{-1} HCl, 4ºC	As(III)+As(V)+MMA: 1 mol L^{-1} HCl+ KI prereduction As(III)+DMA: 6 mol L^{-1} AcOH As(III)+As(V)+DMA: 6 mol L^{-1} AcOH+ KI prereduction Total As: 1 mol L^{-1} tartaric acid	0.1	CF-HGAAS	[15]
As(III), As(V), MMA, DMA	Surface water	-	Total As: 0.01 mol L^{-1} HCl+ 2% THB 5% L-Cys, 10 min reaction time DMA: 1 mol L^{-1} HCl, 0.3% THB 4% L-Cys, <5 min reaction time As(III): 4 mol L^{-1} HCl+0.05% THB MMA: 4 mol L^{-1} HCl+0.03% THB 0.4% L-Cys, 30 min reaction time As(V): by difference	0.6, 0.5, 0.6	CF-HGAAS	[10]
As(III), As(V)	Groundwater	-30 ºC	As(III): 5% citric acid, As_T: KI prereduction	0.03, 0.02	HG-GFAAS	[16]
As(III), As(V), MMA, DMA	-	-	As(III): 4 mol L^{-1} HCl MMA+DMA: 0.165 mol L^{-1} HCl+$KMnO_4$ As_T: 0.85 mol L^{-1} Tartaric acid	0.2, 0.5, 0.3, 0.3	FIHGAAS	[17]
As(III), Total As	Seawater	-	As(III): pH 7, As_T: alkaline persulpahte digestion, As_T(all species): UV irradiation	0.015	FI-HG-GFAAS	[18]

DL: detection limit, CT: cryogenic trap, UV: ultraviolet, AES: atomic emmission spectrometry, FI: flow injection, AcOH: acetic acid, As_t : total arsenic

Table 2: Selected literature references for selenium speciation analysis in natural waters by HGAAS.

Species	Matrix	Preservation	Reaction conditions	DL (mg L^{-1})	Technique	Reference
Se(IV), Se(VI) DMeSe, DMeDSe	Seawater	HCl	U-tube at liq. N_2 was used to separate methylated species	-	HG-CT-AAS	[41]
Se(IV), Se(VI)	-	-	Heating with HCl	0.003	HGAAS	[42]
Se(IV), Se(VI)	River water	-	Se(IV): Extraction with DAN and HG	0.03	Solvent extraction-HGAAS	[43]
Se(IV), Se(VI)	Mineral water	-	Se(IV): Extraction with DAN and HG	0.011	Solvent extraction-HGAAS	[44]
Se(IV), Se(VI)	Seawater	-	Se(IV) retained on mercapto-cilica	-	SPE-HGAAS	[45]
Se(IV), Se(VI)	Seawater	-	Se(IV) retained on anion exchange resin	-	SPE-HGAAS	[46]
Se(IV), Se(VI)	Seawater	-	Se(IV) retained on anion exchange resin	0.006	SPE-HGAAS	[47]
Se(IV), Se(VI)	Seawater	-	Online microwave prereduction	0.025	HGAAS	[48]
Se(IV), Se(VI)	-	-	Online-heating with HCl	0.05	HGAAS	[49]
Se(IV), Se(VI)	Seawater	-	Offline-heating with HCl	0.70	FI-HGAAS	[50]
Se(IV), Se(VI)	-	-	Online-MW reduction	1.0	FI-HGAAS	[51]

Table 3: Selected literature references for chromium speciation analysis in natural waters by AAS.

Species	Matrix	Reaction conditions	DL (mg L^{-1})	Reference
Cr(III), Cr(VI)	Lake water	$Pb(PDC)_2$ co-pption	0.02	[28]
Cr(III), Cr(VI)	River water	APDC-DIBK sequential extraction	10	[34]
Cr(III), Cr(VI)	Mineral water	Selective retention of Cr(VI)-PDC Complex on solid phase; total Cr after HNO_3 oxidation	0.03, 0.3	[52]
Cr(III), Cr(VI)	Seawater	Cr(III): $Fe(OH)_3$ co-pption and APDC-MIBK Extraction, Cr(VI) determined by direct extraction	0.040	[22]
Cr(III), Cr(VI)	-	Ion exchange of Cr(VI)	0.1	[29]
Cr(III), Cr(VI), Organic-Cr(III)	Seawater	Cr(III): Copption with $Fe(OH)_3$ Cr(III)+Cr(VI): Copption with $Bi(OH)_2$ Organic-Cr: persulphate digestion		[27]
Cr(III), Cr(VI)	Natural waters	APDC-MIBK ectraction at pH 3.5 without oxidation	0.3	[33]

Table 4: Selected literature references for the stability of arsenic, selenium and chromium species.

Matrix	Species	Conditions	Stability(Days)	Reference
Natural water	As(III), As(V)	0.2% H_2SO_4	125	[53]
Mine water	As(III), As(V)	10 m mol L^{-1} H_3PO_4, pH <2, 6 ºC dark	90	[54]
Groundwater	As(III), As(V)	0.45 mm filtration + EDTA	14	[55]
River water	As(III), As(V)	5 ºC	30	[56]
Acid mine water	As(III), As(V)	6 mol L^{-1} + sample cleanup by cation exchange and 4 ºC	2	[57]
Groundwater	As(III), As(V)	0.45 mm filtration, pH <2 with HCl + EDTA, storage in dark	45	[38]
Groundwater	As(III), As(V) Se(IV), Se(VI)	pH <2 with HCl, 4 ºC dark	180, 90(As(III))	[40]
Seawater	Se(IV), Se(VI)	pH <2 with HCl, 4 ºC	180	[58]
Seawater	Cr(III), Cr(VI)	None	30	[59]
Groundwater	Cr(III), Cr(VI)	None	21	60]
Groundwater	Cr(III), Cr(VI)	None	14	[61]
Natural water	Cr(VI)	pH >8	-	[62]
Natural water	Cr(VI)	pH 9.2: 0.5 mol L^{-1} Na_2CO_3, 0.5 mol L^{-1} $NaHCO_3$ 0.1 mol L^{-1} EDTA	7	[63]

Selective determination of chromium species by chelation-solvent extraction using sodium diethyldithiocarbamate (NaDDC) and ammonium pyrolidine-dithiocarbamate (APDC) has been reported by several authors.[32,33] The Cr-PDC or Cr-DDC complex is extracted with different solvents; the choice often depends on the subsequent quantitation technique used. For example, methylisobutyl ketone (MIBK) and diisobutyle ketone (DIBK) are the preferred solvents for FAAS,[33, 34] whereas, chloroform, and ethyl acetate are used for HPLC.[35, 36]

As the APDC ligand reacts readily with Cr(VI), the determination of total chromium requires oxidation of Cr(III) by $KMnO_4$.[37] The inertness of Cr(III) was ascribed to the difficulty of displacing the water of coordination from the strongly hydrated Cr(III) ion by APDC.[36] However, using high concentration of APDC, Cr(III)+Cr(VI) could be determined.[33] Honma described a procedure in which, Cr(VI) and Cr(III) were extracted successively after 5 and 90 minutes of reaction time with APDC at pH 1.9-2.1.[34]

Cr(III) and Cr(VI) forms different complexes with PDC ligand, that is, Cr(III) $Cr(PDC)_3$, and Cr(VI) forms $Cr(PDC)_2(OPDC)$ and a minor amount of $Cr(PDC)_3$. This property has widely been utilized for the speciation analysis of Cr(III) and Cr(VI) by HPLC.[35, 36]

Though several studies have reported the iron hydroxides co-precipitation procedure, there are some drawbacks are observed. Iron introduces interferes in the detection method either spectrophotometry or AAS. In addition, the coprecipitation of Cr(III) with $Fe(OH)_3$ presents the drawback that Cr(VI) may be partially adsorbed by the precipitate. In order to circumvent these drawbacks other coprecipitating agents such as lead and barium sulphates were proposed.[24]

Stability of Chemical Species

Species stability is an important issue, since it is not always possible to analyze samples immediately after collection. Instability may be due to chemical reaction between species, interaction with other matrix components/container, microbial activity, temperature, pH, light, etc. Often, the preservative technique used for a particular matrix may or may not be effective for other matrixes. This is evidenced by the contradictory resulted reported in the literature. Hence, stability studies reported based on synthetic samples are seldom valid for real samples. In general, preservation of samples at 4°C and dark conditions improve species stability.[38-40] Table 4 summarizes important preservation methods adapted to the preservation of arsenic, selenium and chromium in water samples.

The reliability of speciation analysis data depends on several factors right from sample collection to analysis. Strict quality control practices are essential to improve the reliability of analytical results. Till date no species specific certified reference materials (CRMs) are available for natural water samples. Hence, spiking, i.e. adding a known amount of the analyte to the sample appears to be an alternate tool for checking the accuracy of the data. Field spikes are more appropriate, because any likely changes due to sampling to analysis can be checked.

CONCLUSION

Speciation analysis of elements found in environmental samples is essential for meaningful assessment of toxicity and for studying the geochemistry. Due to lack of awareness, interest and instrumental facilities, many laboratories still determine the total elemental concentrations. This article summarizes some of the simple procedures, which could easily be carried out using AAS available in most of the laboratories. Sample preservation still remains an area which poses the major difficulty in speciation studies. The reliability of analytical results depends on strict quality control measures practiced from sample collection to analysis.

ACKNOWLEDGEMENTS

The first author thanks Mr. B.M. Jha, Chairman, CGWB, Faridabad and Dr. N. Varadaraj, Regional Director, and E. Sampath Kumar, Supt. Hydrogeologist, CGWB, SECR, Chennai for their encouragement and permission to publish this paper.

REFERENCES

1. Florence, T.M. (1983). *Trends Anal. Chem.,* **2:** 162.
2. Tessier, P.G., Campbell, C. and Bisson, M. (1979). *Anal. Chem.,* **51:** 844.
3. Templeton, D.M., Ariese, F., Cornelis, R., Danielsson, L-G., Muntau, H., Van Leeuwen, H.P. and Lobinski, R. (2000). *Pure Appl. Chem.,* **72:** 1453.
4. Lobinski, R., Pereiro, I.R., Chassaigne, H., Wasik, A. and Szpunar, J. (1998). *Anal. At. Spectrom.* **13:** 859.
5. Morabito, R. (1995). *Fresenius J. Anal. Chem.,* **351:** 378, 1995.
6. Havezov, I. (1996). *Fresenius J. Anal. Chem.,* **355:** 452.
7. Ramesh Kumar, A. and Riyazuddin, P. (2007). *Int. J. Environ. Anal. Chem.,* **87:** 469.
8. Hinners, T.A. (1980). *Analyst,* **105:** 751.
9. Howard, A.G. (1997). *J. Anal. At. Spectrom.,* **12:** 267.
10. Shraim, B., Chiswell, B. and Olszowy, H. (1999). *Talanta,* **50:** 1109.
11. Shraim, B., Chandrasekaran, N., Anuradha, C.D. and Hirano, S. (2002). *Appl. Organomet. Chem.,* **16:** 202.
12. Aggett, J. and Aspell, A.C. (1976). *Analyst,* **101:** 341.
13. Braman, R.S. and Foreback, C.C. (1973). *Science,* **182:** 1247.
14. Sigrist, M.E. and Beldomenico, H.R. (2005). *Spectrochim. Acta, Part B,* **59:** 1041.
15. Bundaleska, J.M., Stafilov, T. and Arpadjan, S. (2005). *Int. J. Environ. Anal. Chem.,* **85:** 199.
16. Niedzielski, P. and Siepak, M. (2005). *Centrt. Eur. J. Chem,* **3:** 82.
17. Rude, T.R. and Puchelt, H. (1994). *Fresenius J. Anal. Chem.,* **350:** 44.
18. Cabon, J.Y. and Cabon, N. (2000). *Anal. Chim. Acta,* **418:** 19.
19. Kotas, J. and Stasicka, Z. (2000). *Environ. Pollut.,* **107:** 263.
20. Rubio, R., Sahuquillo, A., Rauret, G. and Quevauviller, P. (1992). *Int. J. Environ. Anal. Chem.,* **47:** 99.
21. Marques, M.J., Salvador, A. and Morales-Rubio, A. (2000). *Fresenius J. Anal. Chem.,* **367:** 601.
22. Mullins, T.L. (1984). *Anal. Chim. Acta,* **165:** 97.
23. Hiraide, M., Yoshida, Y. and Mizuike, A. (1976). *Anal. Chim. Acta,* **81:** 185.
24. Obiols, J., Devesa, R., Garcia-Berro, J. and Serra, J. (1987). *Int. J. Environ. Anal. Chem.,* **30:** 197.
25. Yamazaki, H. (1980). *Anal. Chim. Acta,* **113:** 131.
26. Cranston, R.E. and Murray, J.W. (1978). *Anal. Chim. Acta,* **99:** 275.
27. Nakayama, E. (1981). *Anal. Chim. Acta,* **131:** 247.
28. Wang, L., Hu, B., Jiang, Z. and Li, Z. (2002). *Int. J. Environ. Anal. Chem.,* **82:** 387.
29. Pankow, J.F. and Janauer, G.E. (1974). *Anal. Chim. Acta,* **69:** 97.
30. Ball, J.W. and McCleskey, R.B. (2003). *Talanta,* **61:** 305.
31. Melaku, S., Cornelis, R., Vanhaecke, F., Dams, R. and Moens, L. (2005). *Microchim. Acta,* **150:** 225.
32. Wai, C.M., Tsay, L.M. and Yu, J.C. (1987). *Microchim. Acta,* **II:** 73.
33. Subramanian, K. (1988). Anal. *Chem.,* **60:** 11.
34. Honma, Y. (2003). *Bull. Chem. Soc. Jpn.,* **76:** 1397.
35. Andrle, C.M. and Broekaert, J.A.C. (1993). *Fresenius J. Anal. Chem.,* **346:** 653.
36. Wang, J.S. and Chiu, K. (2004). *Anal. Sci.,* **20:** 841.
37. USEPA, (1983). Methods for chemical analysis of waters and wastes, Cincinnati, Ohio,
38. McCleskey, R.B., Nordstrom, D.K. and Maest, A.S. (2004). *Appl. Geochem.,* **19:** 995.
39. Gomez-Ariza, J.L., Pozas, J.A., Giraldez, I. and Morales, E. (1999) *Int. J. Environ. Anal. Chem.,* **74:** 215.
40. Rassler, M., Michalke, B., Schramel, P., Schulte-Hostede, S. and Kettrup, A. (1998). *Int. J. Environ. Anal. Chem.,* **72:** 195.
41. Cutter, G.A. (1983) *Anal. Chim. Acta,* **149:** 391.

42. Apte, S.C. and Howard, A.G. (1986). *J. Anal. At. Spectrom.* **1:** 379.
43. Narasaki, H. and Mayumi, K. (2000). *Anal. Sci.,* **16:** 65.
44. Tamari, Y. and Ogura, H. (1997). *Bunseki Kagaku,* **46:** 313.
45. Sahin, F., Volkan, M., Howard, A.G. and Ataman, O.Y. (2003). *Talanta,* **60:** 1003.
46. Itoh, K., Chikuma, M., Nishimura, M., Tanaka, T., Tanaka, M., Nakayama, M. and Tanaka, H. (1989). *Fresenius Z. Anal. Chem.,* **333:** 102.
47. Larraya, M.G., Cobo-Fernandez, Palacios, M.A. and Camara, C. (1994). *Fresenius J. Anal. Chem.,* **350:** 667.
48. Gallignaani, M., Valero, M., Brunetto, M.R., Burguera, J.L., Burguera, M. and de Pena, Y.P. (2000). *Talanta,* **52:** 1015.
49. Stripeikis, J., Costa, P., Tudino, M. and Troccoli, O. (2000). *Anal. Chim. Acta,* **408:** 191.
50. Cobo Fernandez, M.G., Palacios, M.A. and Camara, C. (1993). *Anal. Chim. Acta,* **283:** 386.
51. Burguera, J.L., Carrero, P., Burguera, M., Rondon, C., Brunetto, M.R. and Gallignani, M. (1996). *Spectrochim. Acta, Part B,* **51:** 1837.
52. Chwastowska, J., Skwara, W., Sterlinska, E. and Pszonicki, L. (2005). *Talanta,* **66:** 1345.
53. Cheam, V. and Agemian, H. (1980). *Analyst,* **105:** 737.
54. Daus, B., Weiss, H., Mattusch, J. and Wennrich, R. (2006). *Talanta,* **69:** 430.
55. Gallagher, P.A., Schwegel, C.A., Wei, X. and Creed, J.T. (2001). *J. Environ. Monit,* **3:** 371.
56. Hall, G.E.M., Pelchat, J.C. and Gauthier, G. (1999). *J. Anal. At. Spectrom,* **14:** 205.
57. Oliviera, V., Sarmiento, A.M., Gomez-Ariza, A., Nieto, J.M. and Sanchez-Rodas, G. (2006). *Talanta,* **69:** 1182.
58. Gomez-Ariaz, J.L., Pozas, J.A., Giraldez, I. and Moralez, E. (1999). *Int. J. Environ. Anal. Chem.,* **74:** 215.
59. Sirinawin, W. and Westerlund, S. (1997) *Anal. Chim. Acta,* **356:** 35.
60. Whalley, A., Hursthouse, S., Rowlatt, P., Iqbal-Zahid, H. and Vaughan, R. Durant. (1999) *Wat. Air soil pollut.* **112:** 389.
61. Sacher, F., Raue, B., Klinger, J. and Brauch, H-J. (1999). *Int. J. Environ. Anal. Chem.,* **74:** 191.
62. Parks, J.L., McNeill, L., Frey, M., Eaton, A.D., Haghani, A., Ramierez, L. and Edwards, M. (2004). *Wat. Res.,* **38:** 2827.
63. Pavel, J., Kliment, J., Stoerk, S. and Suter, O. (1985). *Fresenius Z Anal. Chem.,* **321:** 587.

18

Procurement of Pentachlorophenol Degrading Bacteria from Tannery Effluent for their Waste Water Treatment

KUSHAGRA[1*], SHASHI PRABHA[1], SUJEET KUMAR[2], THAKUR, I.S.[1] AND RAMANATHAN, AL.[1]

[1]*School of Environmental Sciences, Jawaharlal Nehru University, New Delhi, India*

[2]*School of Biotechnology, jawaharlal Nehru University, New Delhi, India*

**Email: mekushagra@gmail.com*

ABSTRACT

Water resources deterioration includes qualitative as well as quantitative aspects especially in developing countries like India. Being the reservoir for all shorts of dumps generated from point and non-point sources, water bodies gets perturbed very potentially. Industries are one of the major consumers of water having 3,200 major industrial units in India, along with small industries, mostly discharging their effluent directly to water bodies without appropriate treatment. Tannery/leather industry is amongst the seventeen most polluting industries, listed by Ministry of Environment and Forest (MOEF), processing 7,00,000 tonnes of hides and skins per year, releasing 75,000 m^3/day of effluents. Pentachlorophenol (PCP) is used as a biocide in tanning and is a major contaminant in the tannery effluent. It poses toxicity by inactivating respiratory enzymes, uncoupling oxidative phosphorylation and inhibiting of glycolytic phosphorylation. Microbes have great ability for horizontal gene transfer and their high growth rates, allows them to evolve quickly and to degrade persistent toxic contaminants. Proper identification of potential microbes and their hidden genetic resources should be deciphered and utilized. Microbial biodegradation of pentachlorophenol can be tool for tannery effluent treatment.

Two bacterial strains were isolated from sediment of leather tanning effluent of Kanpur Tannery hub and enriched against PCP in continuous enrichment in the chemostat, had shown potential to utilize it. The strains were screened on the basis of growth, utilization of PCP and its ability to release chloride. Study of bioconversion of PCP into intermediates was confirmed by Gas Chromatography (GC) and X-ray diffraction analysis of metabolites. The member strains were *Enterobacter* sp. (PCP1) and *Pseudomonas* sp. (PCP3) on the basis of 16S rDNA sequence comparison. Formation of intermediates like DCHQ, hyroquinone, 1,4-benzoquinone which is substrate for ring cleavage catalyzed by pcpA gene products indicates the possible mineralization of PCP.

Keywords: Pentachlorophenol, tannery-effluent, chemostat, biodegradation, X-ray diffraction, *Enterobacter* sp., *Pseudomonas* sp., ring cleavage.

INTRODUCTION

Tannery/leather industry is amongst the seventeen most polluting industries, listed by Ministry of Environment and Forest (MOEF), processing 700000 tonnes of hides and skins per year, releasing 75000 m^3/day of effluents. The leather processing requires a large quantity of chemicals which releases huge quantity of chemicals in waste water and pentachlorophenol (PCP) is one of the major contaminant of this industry. PCP is a biocide, used in many industries including tanning, paper and pulp and released into effluent. Its large amount finally gets onto sediments (up to 7000 mg kg^{-1} of soil) Shiu *et al.* (1994) Thakur *et al.* (2001) and Kringstad K P(1984). PCP is highly toxic and recalcitrant compound and placed as priority pollutant by U.S. Environmental Protection Agency and European Community. Bacterial communities can adopt in such high PCP exposure and may develop potential to utilize and degrade it due to their abundance, greater ability of conjugative and horizontal gene transfer among themselves in sediments. Lovley (2003) and Neilson *et al.* (1994). Bacteria with the ability to degrade PCP are widespread in soil and sediment cores; however, most studies of PCP biodegradation have been carried out under laboratory conditions with selected PCP-degrading bacteria (Haggblom *et al.*1989). Pentachlorophenol is biodegraded and mineralized by only a limited number of bacteria and also by bacterial consortium including *Pseudomonas* sp. (Radehaus and Schmidt 1992; Lee *et al.*1998). PCP degradation in aerobic condition has been extensively studied. PCP 4-monooxygenase (PcpB) oxidizes PCP to 2,3, 5,6-tetrachloro*p*- hydroquinone (TeCH) Thakur *et al.*(2001). TeCH reductive dehalogenase (PcpC) converts TECH to 2,3, 6-trichloro-*p*-hydroquinone and then to 2,6-dichloro-*p*-hydroquinone (DCHQ) by reductive dechlorination (Xun 1992). DCHQ is subject to ring cleavage by DCHQ 1, 2-dioxygenase (PcpA), producing 2-chloromaleylacetate (2-CMA) (Ohtsubo 1999).

The present study is carried with a view to exploit the potential of indigenous microbial community of PCP contaminated sediment, by bioaugumentation through continuous enrichment in the chemostat against PCP as a sole carbon source and tracing the possible metabolites by X ray diffraction. We report here two bacteria (*Enterobactor* sp. and *Pseudomonas* sp.) potentially degrade PCP and forming metabolites including DCHQ, hydroquinone, chlorohydroquinone by X ray diffraction analysis which act as substrate for DCHQ 1,2-dioxygenase (PcpA) causing ring cleavage, which suggests possible mineralization of PCP by bacterial strains isolated from PCP contaminated sediments (Ohtsubo 1999).

Materials and Methods

Sediment samples and culture conditions in chemostat. Sediment samples together with liquid effluent (1:10 w/v) were collected from three sites of the main channel of tanneries located at Kanpur, India. Bacterial cells were extracted from sediment core containing bacterial community by centrifugation of sediments at 8000 rpm for 10 min and streaked on mineral salt medium containing (gm/L): $Na_2HPO_4.2H_2O$, 7.8; KH_2PO_4, 6.8; $MgSO_4$, 0.2; Fe $(CH_3COO)_4NH_4$, 0.01; Ca $(NO_3)_2.4H_2O$, 0.05; $NaNO_3$, 0.085, trace element solution 1 ml/L, PCP (100 ppm) as a sole source of carbon and bromothymol blue (0.1%) (Thakur 1995). The bacterial community capable of indicating utilization of PCP in MSM-PCP agar plates with bromothymol (changes blue to yellow) was enriched in a chemostat, and continuous enrichment was performed using mineral salt medium and PCP (0.1 g/L) in 2-litre glass vessel provided by stirring, 250 rev/min, temperature, 29ºC; pH, 7.2– 7.4; an airflow of 500 ml/min and medium flow rate of 10 ml/h., as described earlier (Shah and Thakur 2002). The growth of the bacterial community was determined by measuring A_{560}, with MSM-PCP taken as control.

Isolation and identification of PCP-degrading bacteria. Bacterial community obtained from stabilized chemostat after continuous enrichment was spread over MSM-PCP agar plate and purified by repeated streaking and restreaking on nutrient agar plate (Thakur *et al.* 2001). The bacterial colonies appeared on nutrient agar plates were morphologically characterized and purified by repeated culturing. The morphologically distinct isolates able to utilize PCP were identified biochemically and by comparing the 16S rDNA sequences among other bacterial sequences.

Isolation of genomic DNA. Fresh overnight grown bacterial cultures were taken for genomic DNA isolation. Genomic DNA purification Kit (QIAamp DNA Mini Kit, QUIGEN) was used for this purpose. Genomic DNA was extracted according to the manufacturer's instructions.

PCR amplification and sequencing of 16S rDNA. Two set of universal primers ($P_{0,}$ P_4 and P_{4A}, P_6) were used for 16S rDNA amplification and were synthesized from Microsynth, Switzerland (Studholme 1999). Primer P_0 (27f) 5'-GAGAGTTTGATCCTGGCTCAG-3' and P_4 (561r) 5'-CTTTACGCCCAGTAAT-3' were used for amplify 27-561 fragment of 16S rDNA in an Eppendorf thermal cycler with a temperature profile of 94°C for 10 min, 94°C for 1 min, 45°C for 1 min, 72°C for 1 min, and 72°C for 10 min; steps 2 to 4 were repeated for 30 cycles. Primer P_{4A} (561f) 5'-AATTACTGGGCGTAAAG-3' and $P_{6\ (1495r)}$ 5'-CTACGGCTACCTTGTTACGA-3' were used for amplify 561-1495 fragment of 16S rDNA in an Eppendorf thermal cycler with a temperature profile of 95ºC for 2 min, 95ºC for 1 min, 50ºC for 2 min, 72ºC for 3 min and 72ºC for 7 min; step 2 to 4 were repeated for 35 cycles. The 50 µl of the reaction mixture contains PCR buffer $(NH_4)_2SO_4$, 2 mM $MgCl_2$, 0.2 mM each dNTP, 0.25 mM primer, 1.0 U TaqDNA polymerase, and 10 ng genomic DNA (Kuisiene 2002).

The two sets of amplicons were purified and eluted through 1% agarose. QIAquick Gel Extraction Kit was used for that purpose. Purified both set of 16S rDNA amplicons of each members of consortium were sequenced from Microsynth, Switzerland. The sequences were used for finding out the taxonomic status of bacterial strains with the help of PHYLIP and CLUSTALW software and with the help of tree tool, phylogenetic tree were prepared.

PCP degradation by the isolates. The bacterial isolates were grown in minimal salt medium containing PCP (100 mg/L) as source of carbon and energy. The growth of the bacterial cells was determined by measuring optical density at 560 nm (A_{560}) and utilization of carbon source was analyzed by extraction and determination of PCP by spectrophotometer at 320 nm and Gas Chromatography and X-ray diffraction (Radehaus and Schmidt 1992). PCP was estimated spectrophotometrically by extracting the organic phase, once in 5 ml NaOH (0.5N). The PCP was measured by taking the absorbance at 319 nm and comparing with standard solution (Edge Hill and Finn 1983).

Gas Chromatography. The biodegradation of PCP was analyzed by Gas Chromatography of metabolites extracted from MSM-PCP culture medium. After pelleting the bacterial cells, the aqueous phase was removed and acidified with HCl to a pH~2, and extracted with ethylacetate with vigorous shacking for 1 hour with the repeating the process thrice and the organic phase collected and passed through Na_2SO_4 column to remove air and moisture. The organic phase was concentrated by Rotavapour. Finally, concentrate was dissolved in acetone: hexane (2:1) mixture.

A HP 6890 series GC system and electron capture detector (ECD) fitted with a fused silica chemically bonded capillary column (EQUITY ™ 5,28092 U, 0.25 mm in inside diameter by 30 m in length; 0.25 mm film thickness) was used for GC. Each sample (1 ml) was injected at 60°C in splitless mode. After 2 min at 60°C, the column temperature was increased by 45°C/min to 100°C, 8°C/min to 200°C and finally 20°C/min up to 325°C. Final temperature was maintained for 7.5 minute. Nitrogen gas was used as carrier gas. Detector and injector temperature were maintained at 325°C and 280°C respectively.

X-ray diffraction. Metabolites extracted were analyzed for their characteristic X-ray diffraction pattern. Extracted metabolites were converted into powder form over a glass slide. For diffraction analysis X-pert system; Philips equipped X-pert Highscore softwere, were used in which Cu electrode (K-α) was meant for X-ray generation. Diffractogram was expanded form 5°θ-85°θ value for each sample. Each sample gives a specific Peack list in diffractogram. 'X-pert Highscore' was used for analysis of diffractogram which gives suggested compounds in samples.

Results and Discussion

Bioaugmentation by chemostat. Indigenous population, capable of degrading PCP, able to change bromothymol MSM-PCP agar plate yellow was enriched in a chemostat to put required

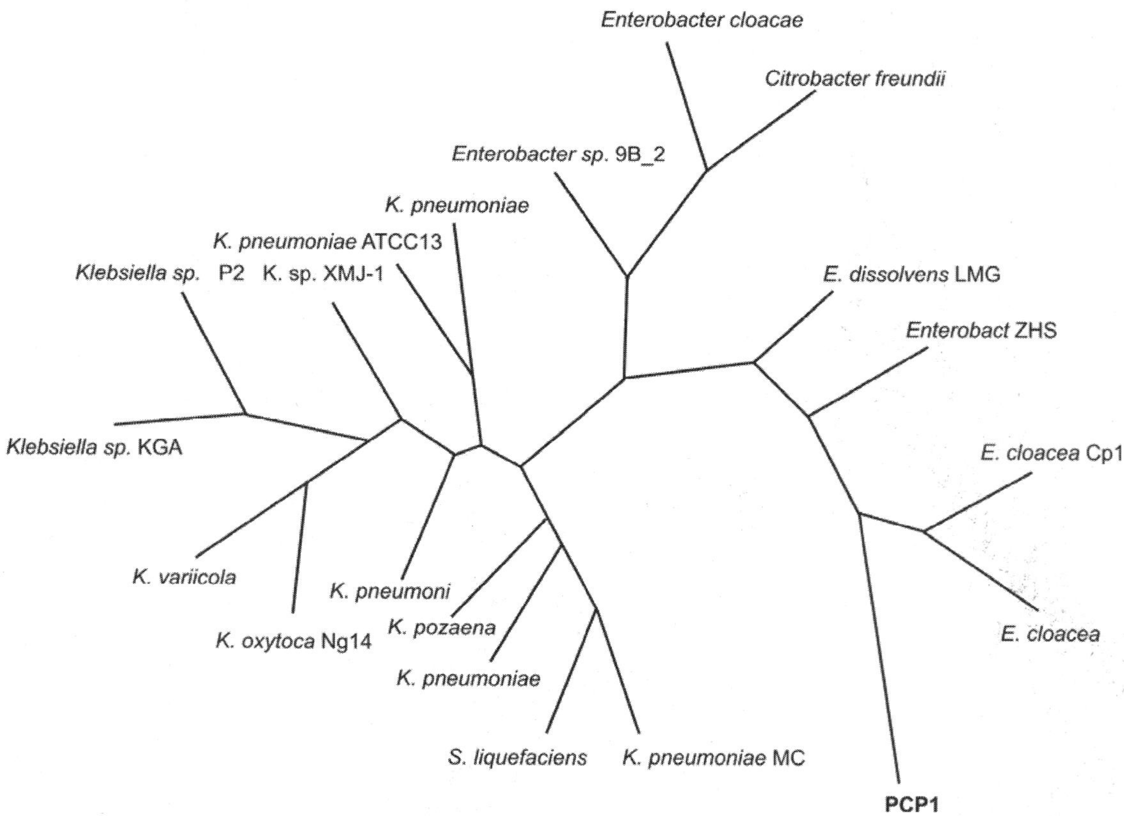

Fig. 1. Phylogenetic tree of PCP1 (where, K: *Klebsiella;* E: *Enterobacter;* Enterobact: *Enterobacteraceae*).

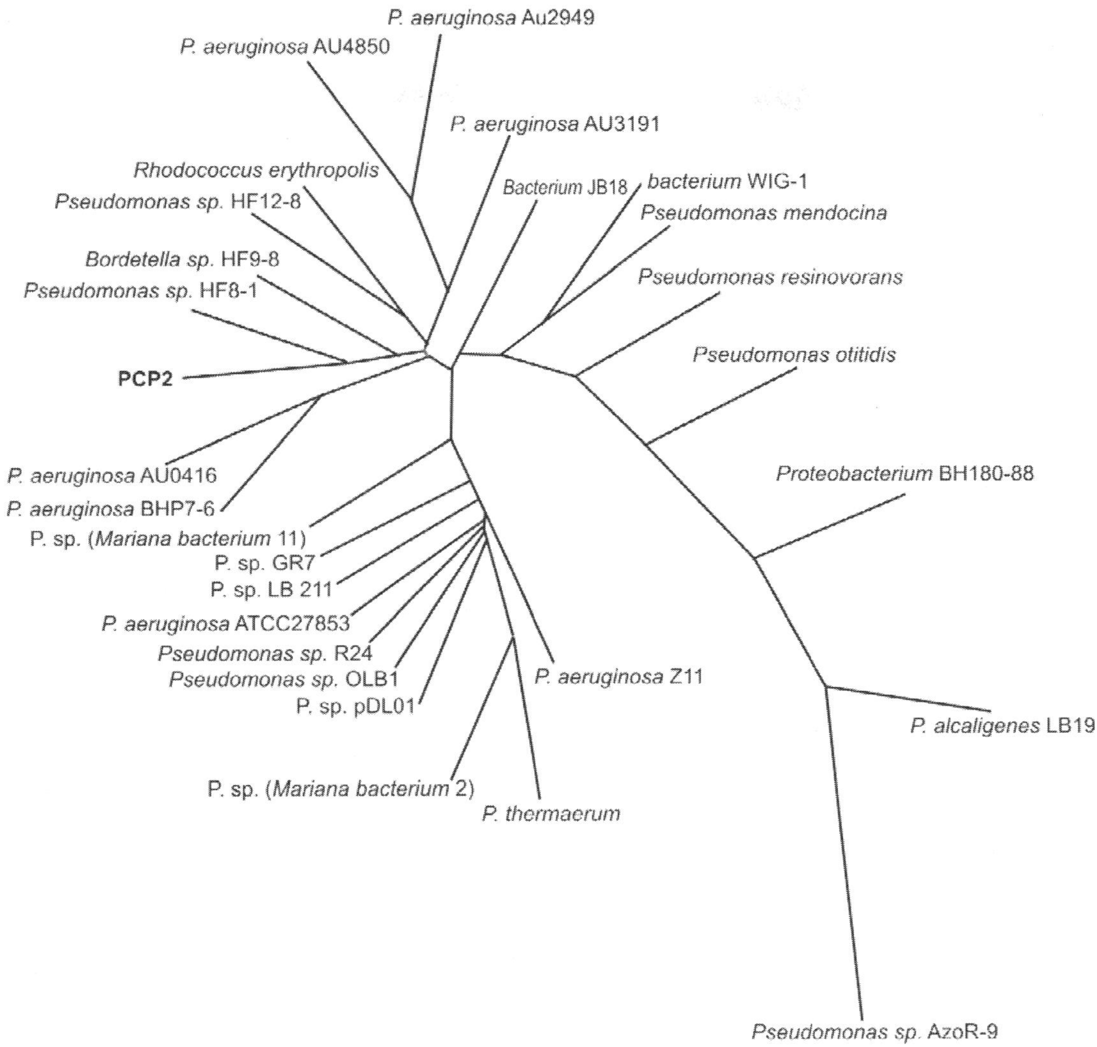

Fig. 2. Phylogenetic tree of PCP2 (Where **P**: *Pseudomonas*).

selective pressure for PCP utilization (Thakur 1995, Bouwer and Zehnder, 1993). A mixed stable community was obtained from a sediment core after continuous enrichment in MSM-PCP a sole source of carbon and energy in the chemostat after some initial fluctuations the culture turbidity increased, and stabilized. The culture medium removed under aseptic condition in ten-fold dilution, and platted on MSM-PCP agar. Two different isolates (PCP1 and PCP2) were obtained on nutrient agar plates and were taken for further analysis.

Identification of bacterial isolates. Morphological and biochemical characterization reveals that both isolates were Gram –ve. PCP1 was cocci, and PCP2 was short rod shaped appeared in the microscope. Both these strains were highly sensitive to Rifampcin. PCP2 was also sensitive to tetracycline and Kanamycin. Further, 16S rDNA based identification was carried out because

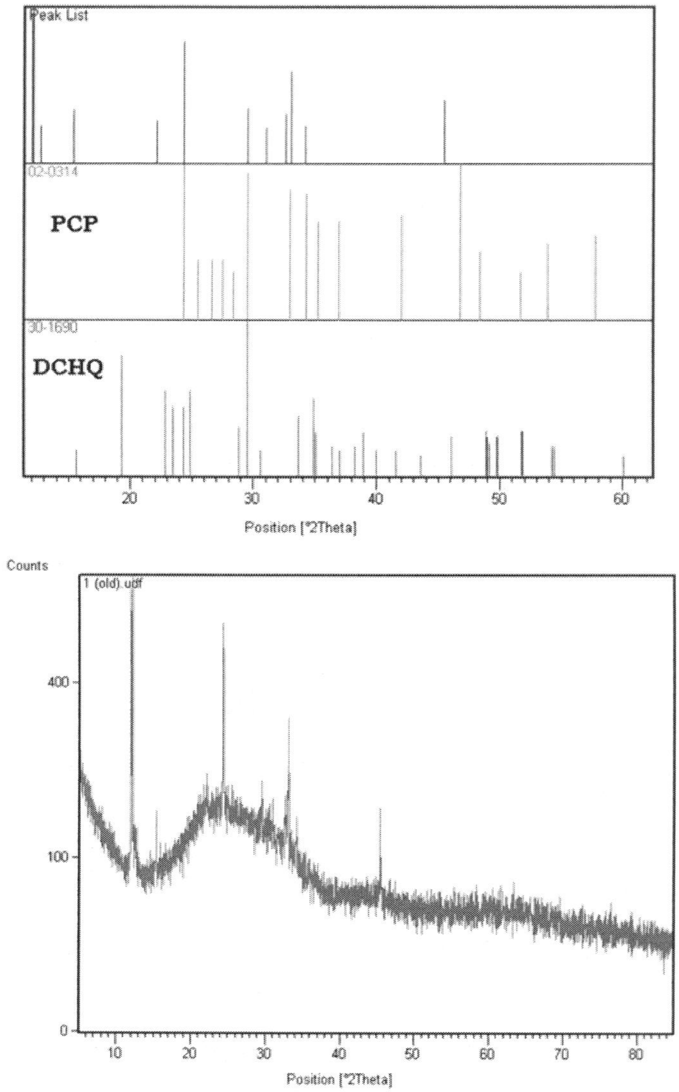

Fig. 3. Diffractogram of metabolites of PCP1 and possible matched peaks. (a.)18 hrs (b.) 72 hrs.

16S rRNA component is very much conserved during course of evolution in prokaryotes and its intermediate size is also reliable over 5S RNA (smaller size) and 23S RNA (larger size). The horizontal transfer of rRNA is believed to rare. The PCR and sequencing strategy used to generate sequence information for a DNA segment comprising partial 16S rDNA for each strain. The DNA were directly sequenced from Microsynth directly by PCR sequencing in which primer P0 (27f) for first set 27f-561r fragment, while P4A (561f) and P6 (1495r) for 561-1495 fragment of 16S rDNA were used for sequencing. Collected sequences were arranged according 5'-3' and attached to find a composite sequence. Sequence comparisons (1487 bp for PCP1 and 1482 for

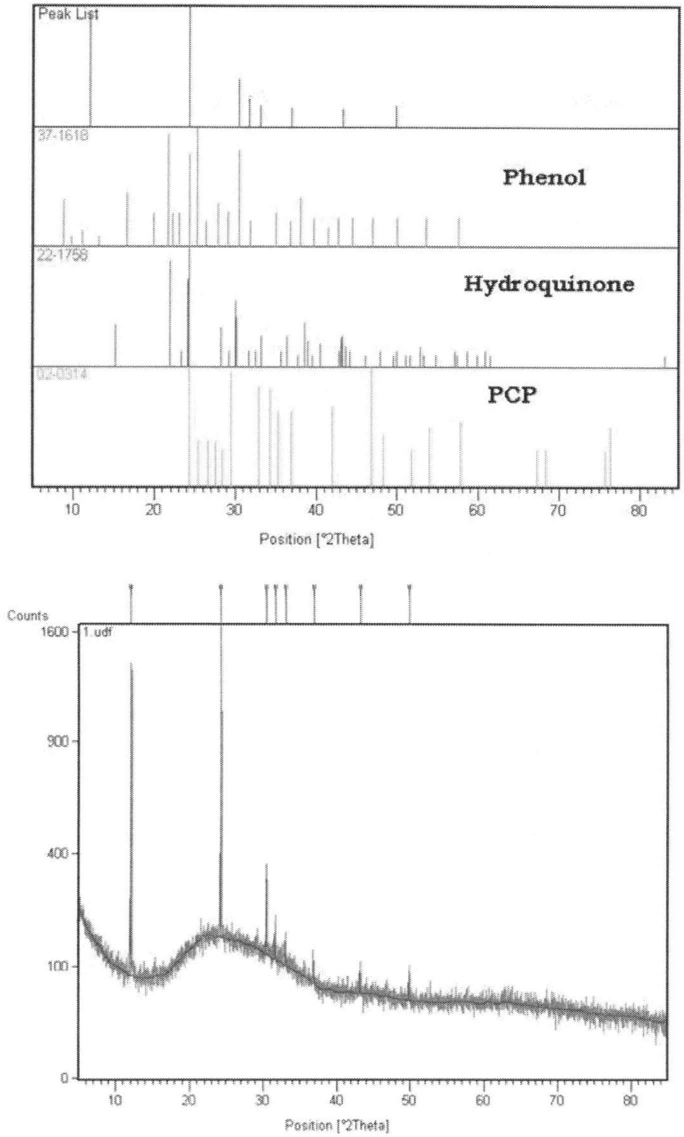

Fig. 4. Diffractogram of metabolites of PCP2 and possible matched peaks.

PCP2) with the 16S rDNA of bacteria were carried out from NCBI public data base. Phylogenetic trees were inferred by the maximum likelihood method using PHYLIP and CLUSTALW with help of tree tool. PCP1 is very close to *Enterobacter* sp. while PCP2 is *Pseudomas.*sp. PCP1 shows closeness to *Enterobacter cloacae*. Genus *Klebsiella* and Enterobacter are very close to each other, but PCP1 is rooted with farther end of both *Klebsiella* and *Enterobacter* separation point, on that basis we can say that PCP1 is *Enterobacter sp.*

Enteric bacteria including genera *Klebsiella*, *Enterobacter*, *Escherichia,* were reported very occasionally for degradation of persistent organics (Hanlon 1995 and Cooper and Skinner 1980). But the potent enteric to degrade PCP is not reported yet, except our finding. While *Pseudomonas* sp. is very much analyzed for PCP degradation (Thakur 2001, 2002 and Lee, 1998).

Gas Chromatography and X-ray Diffraction Analysis. PCP utilization and its bioconversion were clearly indicated by GC. In both strains, the peak of PCP (Retention time 22.9 min) starts shifting and completely replaced by another peak (RT 20.6 min) in PCP1, while in PCP2 two peaks formed stepwise in 72 hrs (RT 20.5 and 19.8 min). Without standard of metabolites, only conversion of PCP was established. In this regard, X-ray diffraction analysis was helpful. The analysis of PCP by X-ray diffraction was reported only for sorption of PCP by clay mineral (Boyd 1988). X-ray diffraction analysis is mainly used for minerals and other compounds have definite crystal structure. Organic compounds also give peaks on diffractogram which is very specific to compound and used for qualitative analysis. The peaks list of a diffractogram is compared to the database of specific peaks of organic compound for identification of organic compound. The software 'X-pert HighScore' suggests the possible compound in sample on the basis of peaks of diffractogram. In PCP1 the suggested metabolites were PCP, Dichlorohydroquinone (DCHQ), Hydroquinone and phenol, while in PCP2 Tetrachloroquinone, DCHQ and 1,4-Benzoquinone were suggested.

In the present investigation of a bacterial consortium containing two isolates belonging to *Enterobactor* sp. and *Pseudomonas* sp. have clearly shown the potency to utilize, as well as degrade PCP. The GC analysis shows the bioconversion of PCP. Finding of X-ray diffraction was very significant and indicates the dechlorination as dechlorination is the major peripheral reaction for chlorinated organics to channel for ring cleavage in case of PCP (Cai and Xun 2002). Both DCHQ and Hydroquinones undergoes ring cleavage catalyze by dioxygenase the product of *pcpA* gene. pcpA gene is evolved from some other catabolic pathway and assembled in *pcp* gene cluster to carry out ring cleavage for mineralization of PCP, in this case also possibility exists for ring cleavage of quinines by *pcpA* gene as DCHQ which is catabolized by *pcpB*, *pcpC* gene of *pcp* gene cluster. *pcpA* gene is also part of *pcp* gene cluster (Cai and Xun 2002). Here is possibility of PCP mineralization by consortium, as quinines are intermediary metabolites reported, that shows the convergence of catabolic pathway at ring cleavage, indicating their dependence on each other (synergistically working).

These two indigenous bacterial strains (NCBI Gene Bank Accession No. DQ873515 and DQ873517) from tannery effluent sediment can be further utilized for pollution abetment as passive tool after optimization at pilot scale then field scale for removal and mineralization of Pentachlorophenol from tannery effluent.

ACKNOWLEDGEMENTS

We thank University Grant Commission, New Delhi for providing funds in the form of research projects and fellowship. We are also thankful to Mr. Venkat, Tech. Assist., Central Instrumental Facility (CIF), School of Environmental Sciences, JNU for helping for X-ray diffraction and analysis of data.

REFERENCES

1. Shiu, W.Y., Ma, K-C., Varhanickova, D. and MacKay, D. (1994). Chlorophenols and alkylphenols: A review and correlation of environmentally relevant properties and fate in an evaluative environment. *Chemosphere* **29(6):** 1155-1224.
2. Thakur, I.S., Verma, P. and Upadhyaya, K.C. (2001). Involvement of plasmid in degradation of pentachlorophenol by *Pseudomonas* sp. from a chemostat. *Biochem. Biophys. Res. Comm.* **286:** 109-113
3. Kringstad, K.P. and Lindstrom, K. (1984). Spent liquors from pulp bleaching. *Environ. Science Technol.* **18:** 236-248.
4. Lovley, D.R. (2003). Cleaning up with genomics: applying molecular biology to bioremediation. *Nat. Rev. Microbial.* 1: 35-44.
5. Neilson, J.W., Josephson, K.L., Pepper, I.L., Arnold, R.B., Diglovanni, G.D. and Sinclair, N.A. (1994). Frequency of horizontal gene transfer of a large catabolic plasmid (pJP4) in soil. *Appl. Environ. Microbial.* **60**: 4053-4058
7. Radehaus, P.M. and Schmidt, S. (1992). Characterization of a novel *Pseudomonas* sp. that mineralizes high concentrations of pentachlorophenol. *Appl. Environ. Microbial.* **58:** 2879-2885.
8. Lee, S.G., Yoon, V.D., Park, Y.H. and On, H.M. (1998). Isolation of nevel Pentachlorophenol degrading Bacterium *Pseudomonas* Sp. BU34. *J. Appl. Microbial.* **85:** 1-8.
9. Haggblom, M.M., Janke, D. and Salkinoja-Salonen, M.S. (1989). Hydroxylation and dechlorination of tetrachlorohydroquinone by *Rhodococcus* sp. strain CP-2 cell extract. *Appl. Environ. Microbial.* **55:** 516-519.
10. Thakur, I.S., Verma, P. and Upadhayaya, K. (2002). Molecular Cloning and Characterization of Pentachlorophenol-Degrading Monooxygenase. *Bioche. Biophys. Re. Comm.* **290:** 770-774.
11. Xun, L., Topp, E. and Orser, C.S. (1992). Purification and characterization of a tetrachloro-*p*-hydroquinone reductive dehalogenase from a *Flavobacterium* sp. *J. Bacteriol.* **174:** 8003-8007.
12. Ohtsubo, Y., Miyauchi, K., Kanda, K., Hatta, T., Kiyohara, H., Senda, T., Nagata, Y., Mitsui, Y. and Takagi, M. (1999). PcpA, which is involved in the degradation of pentachlorophenol in *Sphingomonas chlorophenolica* ATCC 39723, is a novel type of ring-cleavage dioxygenase. *FEBS Lett.* **459:** 395-398.
13. Shah, S. and Thakur, I.S. (2002). Enrichment and characterization of a microbial community from tannery effluent for degradation of pentachlorophenol. *World J. Microbial. Biotechnol.*, **18:** 693-698.
14. Studholme, D.J., Jackson, R.A. and Leak, D.J. (1999). Phylogenetic Analysis of Transformable Strains of Thermophilic *Bacillus* Species. *FEMS Microbio. Lett.* **172:** 85-90.
15. Kuisiene, N., Jomantiene, R., Valiunas, D. and Chitavichius, D. (2002). Characterization of Thermophilic Proteolytic Spore-Forming Bacteria from a Geothermal Site in Lithuania Based on 16S rDNA RFLP and ITS-PCR Analyses. *Microbiology* **71(6):** 712-716.
16. Radehaus, P.M. and Schmidt, S.K. (1992). Characterization of novel *Pseudomonas* sp. that mineralizes high concentration of PCP. *Appl. Environ. Microbial.* **56:** 1392-1397.
17. Edge hill, R. and Finn, R. (1983). Microbial treatment of soil to remove pentachlorophenol. *Appl. Environ. Microbial.* **45:** 1122-1125.
18. Thakur, I.S. (1995). Structural and functional characterization of a stable, 4-cholrosalicylic-acid-degrading bacterial community in a chemostat. *World J. Microbial. Biotechnol.* **11:** 643-645.
19. Bouwer, E.J. and Zehnder, A.J.B. (1993). Bioremediation of organic compounds: microbial metabolism to work. *Trends Biotechnology* **11:** 360-367.
20. Hanlon, S.P., Carpenter, K., Hassan, A. and Cooper, R.A. (1995). Formation *in vitro* of the 3,4,6-trihydroxyphenylalanine quinone cofactor. *Biochem. J.* **306:** 627-630.
21. Cooper, R.A. and Skinner, M.A. (1980). Catabolism of 3- and 4-hydroxyphenylacetate by the 3, 4-dihydroxyphenylacetate pathway in *Escherichia coli. J. Bacteriol.* **143:** 302-306.
22. Lee, S.G., Yoon, V.D., Park, Y.H. and On, H.M. (1998). Isolation of nevel Pentachlorophenol degrading Bacterium Pseudomonas Sp. BU34. *J. Appl. Microbiol.* **85:** 1-8.
23. Boyd, S., Shaobai, S., Lee, J. and Mortland, M.M. (1988). Pentachlorophenol Sorption by Organo-Clays. Clays and Clay Minerals **36:** 125-130.
24. Cai, M. and Xun, L. (2002). Organization and Regulation of Pentachlorophenol-Degrading Genes in *Sphingobium chlorophenolicum* ATCC 39723. *J. Bacterol.* **184:** 4672-4680.

19

Variable Characteristics of Rainwater Status in the Union Territory of Puduchery–A Focus on Puduchery and Karaikal Regions

VARMA, V.N.
Department of Agriculture, Pondicherry
Email: varma–agri@hotmail.com

ABSTRACT

The Union Territory of Puduchery possesses divergent spatial and climatic distributions. The annual seasonal precipitations vary with oceanic influences, vegetation types, cloud types and climatic changes.

The coastal rainwater consists of aerosols particles of sea evaporites, soil and mineral dusts, organic matter, dissolved gases, congested fall out microbes as suspended and dissolved solids in the atmosphere; though in traces, its impact is very dynamic on earth.

In this context, in this paper, we had compared the rainwater characteristics of the coastal Puduchery and Karaikal regions. Our samples analysed data had shown the pH condition varies from acidic to alkaline ranges dominated by the soil dusts and ocean evaporates containing ionic composition of Ca, Fe, S,Cl, Mg, N and C in Puduchery region whereas in Karaikal region, the precipitational water had possessed by the biogenic and evaporite aerosols with pH ranges from acidic to very feeble alkaline conditions dominated by ionic constituents of K, Na, Cl, S, N, C. The rainwater collected for analyses might not be the only inputs from the vertical upper skies but also the lateral layer contacts of the orographic land mass and the ocean evaporites. The organic carbon amounts in the Karaikal region rainwater had been found to consist of slightly higher concentrations than those of the Puduchery region. The acidic rainwater consists of mineral and volatile acids.

The residues of limestones, ocean salts, lateritic and biogenic soil minerals, industrial and garbageous sewage suspensions, fossilated fuel burnings, thunder fall outs gases, lateral passage of ocean evaporites and high altitude contacts may contribute the chief aerosols mixing with the coastal area rainwater. Further in the last three years, abnormal cooling and warming in the less arid climatic areas, bulk precipitaton, sea surface temperature variability were observed, probably linked with climatic changes. The future scope of the study on the effects of vulnerability of microbes and organic biomarkers of the rainwater in the site specific areas will be possible.

Keywords: Rainwater, aerosols, ionic constituents, volatile acids, atmosphere.

INTRODUCTION

The Union Territory of Puduchery had possessed four coastal isolated spatial regions, of which Puduchery and Karaikal regions lie in the coast of Bay of Bengal.

These regions fall under tropical sub-humid, coastal agroclimatic zone with high rainfall intensity during North-East Monsoon. The long range of average temperature is 26.8°C, R.H is 84 % with consistent of hazy weather. For Karaikal region, the temperature is 28.4°C, RH being 71% with hazy weather but the weather shifts from 1970 to 2005 were 10% to Puduchery and 8% to Karaikal region.

Rain is a precious gift by nature and absence of it make the ecosystem lifeless. Puduchery region has the average annual rainfall of 1000 mm and that of Karaikal region has 1200 cm. The rainfall occurred by the N.E. monsoon period during 2006 and 2007 was relatively heavy under the development of the cumulonimbus clouds. Most of the vertical rainfall occurred close to the coastal areas due to supercooling effect close to the freezing condition thereby gaining higher pressurized density. The recent rain temperature had been recorded even below 15°C in Puduchery areas (Table 1).

Table 1: Rainwater samples collection during December 2008.

No.	Sample No.	Place of collection	Temperature	Monsoon
1	PY01	Open air	12°C	North East
2	PY02	Open air	21°C	North East
3	PY03	Open air	18°C	North East
4	PYK4	Open air	21°C	North East
5	PYK5	Roof water fall outs	22°C	North East

The ocean potential temperature, the tidal heights reached at one metre with turbulancy and their gushing at higher intensity were inundating the coasts and raising the water levels of the coastal waterbodies. The coastal floods currents are vigorous. The recent rains had caused the outbreak of contigeous and sporadic diseases. All the factors may contribute the climatic changes. The fossil carbon burning in the north Indian coasts and the garbages cum pollutants in the Southern coasts are felt by the experts for the climate changes which had also bearing effects on the Continental Shelf/margin tidal turbulancy. In the last decade as per the experts findings, the rainfall intensity of the Indian Ocean domain was increasing from India to Sumatra island[1].

SPATIAL RAINWATER CHEMISTRY

In Puduchery, the rainwater samples were collected in the open atmosphere but in Karaikal, the samples were collected from the rooftop pouring. Most of the samples are clear with lesser amount of the suspended solids.

The 4 Rainwater samples thus collected were labelled as flows:

PY01;Kalapeth village-Puduchery region

PY02;Tondamanatham-Puduchery region

PY03;Mettupalayam-Puduchery region

PYK-04;Karaikal–Karaikal region

PYK-05;Vanjore-Karaikal region.

The water samples were analysed for the suspended and dissolved solids, pH, Conductivity, major cations and major anions, Iron and Manganese, organic carbon %, acid and alkaline contents.

Earlier some of the isotopes investigation data were compared with the present data. They consist of radiocarbon, stable oxygen and stable carbon isotope values which were done in the Geophysical and Atomic Mineral institutes at Hyderabad. All the results are documented in Tables 1, 2, 3 and 4.

Table 2a: Physico-chemical data of rainwater in Puduchery and Karaikkal region.

No.	Sample No.	Location	Suspended solids ppm	TDS	NO_3	Fe +Mn	P mg/L	Total acidity mg/L	Volatile acidity mg/L	Organic Carbon %
1	Py01	Kalepath	15	90	Nil	0.1		Nil	Nil	0.15
2	Py02	Tonadamanatham	20	140	Nil	Nil		Nil	Nil	0.11
3	Py03	Metupalayam	42	180	Nil	Nil		15	traces	0.12
4	Pyk 4	Karaikal	38	140	Nil	Nil		Nil	Nil	0.1
5	Pyk5	Vanjore kkl	55	138	Traces	Nil		Nil	Nil	1

Table 2b: Comparison of the rainwater analysis of at other sources.

No	Test	Annamalai Nagar	Hyderabad
	Physical		
1	pH	6.9	5.8
2	TDS	225 mg/L	150
3	TH	95 mg/L	45
	Chemical		
4	Cl ion	28	15
5	Ca/Mg ion	24	12
6	Fe ion	00	0.2
7	NO_3 ion	22	10
8	SO_4 ion	nil	3

Table 3a: Stable isotope data of Puduchery state.

No.	Well location	$\delta^{18}O^{permil}$
I. Karaikal region (1992)[6]		
1	Vizidiyour	-2.0
2	Ambagarathur(Padugai)	-3.1
3	Agalcan	n -3.9
4	Kottuchery	-2.2
5	Poovam (Filter point)	-3.8
6	Peralam (25 feet)	- 2.8
7	Tranquebar (sea intruded)	-1.0
8	Rainwater,Vanjore (Pattinam site)	-3.8

Table 3a: Contd.

No.	Well location	$\delta^{18}O^{permil}$
II. Puduchery region (1983)		
1	Muthirapalayam	-3.7
2	Bahour coast	-2.2
3	Thengaithitoo lagoon padugai	-1.5
4	Rainwater, Puduchery (Bahour site)	-3.3
Rainwater samples in Karaikal and Puduchery $\delta^{13}C$		
	(1983 and 1992)	
1	Rainwater Bahour	-6.1
2	Rainwater-Pattinam	-12.5

Table 3b: Isotopic data of the water samples of marine, coal and petroleum environments.

No.	Water samples	^{14}C as pMC	$\delta^{18}O$ per mil	$\delta^{13}C^{permil}$	^{3}H TU
1	Puduchery-Bahour**	Modern	-3.3	-6.1 } }	20-30[5]
2	Karaikal-Pattinam**	Modern	-3.8	-12.5 }	TU
3	Shallow well (Sakkaramangalam)	Modern Near to & west of Karaikal	-3.0	-4.5	
4	IMine water-NLC, Neyveli	84.3	-6.8	-8	35+/-5
5	Oilfieldwater,Nawagam (Ahamedabad)	2 to 5	NA	–4 to –6	>5
6	Shallow water-Agalancan Karaikal-well water	62.6	-3.9	-6	-

**Rainwater samples,1983 and 1992

Table 4: Chemical analysis of rainwater in Puduchery and Karaikal region.

No.	Sample No.	Location	pH	EC ms/m	HCo_3 Mg/L	CO_3 Mg/L	Cl Mg/L	So_4 Mg/L	Ca Mg/L	Mg Mg/L	Na+K Mg/L
1	Py01	Kalepath	6.1	0.13	Nil	25	22	15	10	4	12
2	Py02	Tonadamanatham	8.1	0.22	2	90	12	3	22	10	10
3	Py03	Metupalayam	5.9	0.3	Nil	15	10	40	6	3	8
4	Pyk 4	Karaikal	8	0.22	Nil	85	18	5	3	3	15
5	Pyk5	Vanjore kkl	8	0.2	Nil	70	25	13	8	12	18

RESULTS AND DISCUSSION

The suspended solids show the moist aerolsols of the soil dusts and the organic suspensions. Karaikal rainwater possessed the black colour settleable matter. The organic carbon contents are found to be even 1% for the Karaikal rainwater (in Vanjore sample) suspended solids but in Puduchery, it is 0.1% which conveys the soil fragments mixing in Puduchery rainwater and the silty organics mixing in the Karaikal rainwater in recent period.

Site specific variations occurred in pH values of Puduchery environs. pH variations from 5.9 to 8.1 may be due to the limestone and gas evaporites mixing with the clouds. Thus, in sample PY02, it was due to the fact of higher concentrations of calcium and Magnesium evaporite mixing; the acidity in sample no. PY03 was due to the effect of acid gaseous emittants from industrial sites in the atmosphere. The contribution of pH and the suspended solids in the rainwater may be from the land inputs. PY01 sample was not affected by acidity in acid soil domain due to orography and vegetation cover.

The salinity data consist of relatively high chloride content along with sulphate contents which support the soluble salts are due to the ocean evaporites mixing with the rainwater as the soils do not contain high salinity but the alkalinity rate is high. Significantly, PY01 area is a small orographic red terrain, the Iron and Manganese value is about 0.1 mg/L in the rainwater. Here, though some areas are acidic in nature, due to the dense vegetation and untapped mineral acidity, the aerosols were not affected by the acidic evaporites. So from the above results, pH condition is related to the land based evaporites and the soluble salts contain mostly ocean moisture salts. The absence of the nitrate contents suggest that there was no intervention of the man-made activities due to the access of the sewage contaminants in the rainwater moisture. There was no phosphorus contaminant in rainwater.

The isotope hydrological investigations on various waterbodies were done during 1983 in Puduchery region and during 1992 in Karaikal region. Isotope studies on radioactive carbon^{14}C, Tritium of radioactive hydrogen ^{3}H, stable oxygen and stable carbon isotope as $\delta^{18}O$ (SMOW) and $\delta^{13}C$ (PDB) were measured earlier in the labs of the National Geophysical Research lab, Hyderabad for carrying out groundwater recharge potential. These variations help us to specify the nature of water body movements and environmental influx in the groundwater and the other water bodies. The rainwater samples were also included for the isotope study.

As observed in the isotope values, all the modern carbon values have 100 PMC (Per cent modern carbon) and above as cited in Table 3. All the rainwaters possess the PMC values above 100 as Modern Carbon in the uncontaminated state. The stable oxygen isotope values range from –2 to –5 per mil. Usually, rainwater values vary from –15 to 0 per mil. The seawater intruded groundwater has –1 per mil. So the oceanwater salts might have interfered the aerosols content of rainwater as observed from the values in Tables 3 and 4. These values are dependent on the origin of moisture, the trajectory of moisture transport and the precipitation history. The higher values of heavy oxygen suggest that the arrival of moisture not from the far off distant areas[4]. The oceanwater intruded salt water in Tranquebar groundwater has the $\delta^{18}O$ value close to -1 per mil and much rainfall moisture may push the heavy oxygen values from –2 to –4 which were observed in Puduchery areas. Further these higher values support the evaporate mixing in the aerosolic moisture at the higher altitude of the clouds with variant temperatures[3]. The stable carbon isotope ($\delta^{13}C$) values fluctutate from –6 to –13 per mil. The low values in Karaikal region confirms the organic moiety aerosols in the rainwater, since the average rainwater consists of $\delta^{13}C$ values of –7 per mil[4]. The organic carbon content of 1% in Vanjore sample confirms the evidence (Table 2). In Table 3, these isotope values of rainwater were compared with other waterbodies including the groundwater for interpretations.

CONCLUSION

The recent rainwater in Puduchery and Karaikal consist of the ocean evaporites and land sediments whence the fall outs occured. The pH variations from alkaline to acidity conditions suggest site specific mineral reactions on the aerosolic adherence moisture. The high rainfall intensity of bulk precipitation as strong vertical downpours due to supercooling effecting pressurized density progressing towards the freezing state was targeting the coastal areas. In acidic rainwater condition in Metupalayam, the gaseous evaporites might have been mixed to cause the low pH with sulphur compounds interference. The man-made activated inducement of nitrates absence may probably be nullified due to the ameoliorates and drainage intensity.

From the stable isotopes measurements, it was observed that the precipitational moisture might not be transported from far away places. The slightly heavy oxygen and higher chloride concentrations compared to bicarbonate values suggest that the ocean salt dominancy was observed in the rainwater soluble salts. The organic carbon values and stable carbon values evince that the Karaikal precipitation consist of the organic suspended solids which was absent in Puduchery rainwater. The relative supercooling, the bulk precipitation and the ocean turbulancy whence on compared with the past 3 decades give clues for the climate changes.

REFERENCES

1. Aragues, A. and Froelich, K. (1998). *J. Geophysical Research*, **103:** 28, 721-728.
2. Gupta, S.K. and Despande, R. (1991). Stable isotope in New Delhi precipitation, **128:** 223-236.
3. Suresh. R. (2007) Application of isotope techniques in conceptualization of multiaquifer system; Workshop on Isotope tracer techniques for the water resources development and Management, Annamalai University, Earth Science Department, 49.
4. Tidan, L. and Yao. T. (2006). Stable isotope variations in Monsoon precipitation, *J. Meteorological Soc. Japan*, **79:** 959-966.
5. Sukhija, B.S., Reddy, D.V., Nagabhushanam, Patil, and Hussain. S. (1990) Development of isotope methodology for the injection water movement studying oilfields; NGRI report, TR-C, 13.
6. Varma, V.N., Thiagarajan, S., Gopalan, S. and Sukhija, B.S. (1991) Mining effects based on the hydrological problems in the coastal groundwater. *J. Indian Soc. Coastal Agric. Research*, **9:** 362.

20

Hydrochemistry and Environmental Status of Pazhayar River, Kanyakumari District, Southern Tamil Nadu

Sabu Joseph*, Byma Benziger and Jobin Thomas
**Department of Environmental Sciences, University of Kerala, Kariavattom, Thiruvananthapuram–695 581, Kerala*
**Email: jsabu2000@yahoo.co.in*

ABSTRACT

For more than a century, rivers have been of interest to hydrologists, geomorphologists, geochemists and civil engineers. Recently, rivers are of great concern to environmentalists as well. Rivers are subjected to various physico-chemical and biological processes which may ultimately result in their environmental deterioration. Acceleration of these processes by human activities, mainly in the form of waste disposal, agricultural run off, industrial pollution, etc. is a matter of great concern. Increased pollution level in river water contributes to health hazards and leads to deterioration of the ecosystem. Hence, a qualitative and quantitative assessment of these pollutants in the riverine system must be seriously looked into.

A systematic investigation was done to elucidate the level of environmental status of a minor river "Pazhayar" (Order = 4th; L= 35 km), originating from the Mahendragiri hills (elevation = 975 m) of Western Ghats in Kanyakumari District. It follows a rectilinear path from NW to SE covering the highland, midland and finally debouches into the Arabian Sea at Manakudi near Kanyakumari. The spatial and temporal variation in the water quality parameters (i.e., physico-chemical and microbial) of water samples (n = 24) from upstream to downstream for three seasons, viz., pre-monsoon-PRM, monsoon-MON and post-monsoon-POM showed that the attributes DO, BOD and P are above permissible limits. Further, the MPN index of faecal coliform was high in the downstream stations. The possible causes of these aspects warrant explanation and are discussed in this paper.

Keywords: Hydrogeochemistry, special temporal variations, Pazhayar, river.

INTRODUCTION

For more than a century, rivers have been of interest to hydrologists, geomorphologists and geochemists and civil engineers. Nowadays, the rivers are of great concern to environmentalists as well. Rivers are subjected to various physico-chemical and biological processes. These processes may ultimately result in their environmental deterioration, if unchecked. Acceleration of these processes may also be caused by human activities mainly in the form of waste disposal, agricultural runoff, industrial pollution, microbial contamination, etc., and is a matter of great concern.

Hydrochemistry of rivers have earned considerable attention during the last decade since they are the signatures of the riverine environments (Tyler, 1995). The types of pollutants, flow rates, dilution capacity and recovery time vary widely with different river basins. Further, river basins differ greatly in size, climate, topography, geology, vegetation, wetlands and land use all of which influence flow rate and pollutant load (Ward and Whiple, 1959).

Continually increasing contamination of surface water in the vicinity of urban centres is a global phenomenon. Point sources discharge pollutants at specific locations through ditches or sewers into bodies of surface water (e.g., a factory). However, non-point sources are sources that cannot be traced to any single discharge (e.g., agricultural runoff). They are usually large, poorly defined areas that pollute water by runoff, subsurface flow and deposition from the atmosphere (Boss, 1956).

Insufficient water to meet the country's needs in the coming years is a ground reality of fast deteriorating water quality. The pollution of rivers and other surface water bodies like lakes, ponds and tanks through discharge of municipal and industrial effluents brings the crisis point a lot closer (Duncan and Rzoska, 1978; Garbow, 1979). Contrary to what critics think, improving river water quality even by marginal levels is a stupendous task given the large habitations on their banks, poor sewage management, agricultural runoff, discharge of untreated industrial effluents and cultural/religious practices of people (WHO, 1984).

Since most of the rivers end up in oceans, the coastal region of India faces a variety of problems due to the contaminated discharges of the rivers flowing through a number of densely populated regions of the country. Many of our coastal areas have problems connected with water quality due to urban waste discharge (Trivedi, et al. 1990).

Hence, the potential of various water resources and their quality (in terms of physical, chemical and biological attributes) should be evaluated thoroughly to generate baseline information for the welfare of society. Further, it is easy for planners and decision-makers to take necessary steps regarding improvement and management of pollution status.

Pazhayar River in Kanyakumari district suffers from a paradoxical situation primarily due to insensible use of land and water resources. Further, no serious studies have so far been made to characterize the quality of water for various purposes. Hence, hydrochemical analysis of surface water is very essential.

STUDY AREA

A minor river called "Pazhayar" (Order = 4th; L= 35 km; N. Lat. 77° 26' to 77° 29' & E. Long. 8° 07' to 8° 15'), originating from Mahendragiri hills (Elevation = 975 m above msl) of Western Ghats in Kanyakumari District, was selected for the present investigation (Fig. 1). It follows a

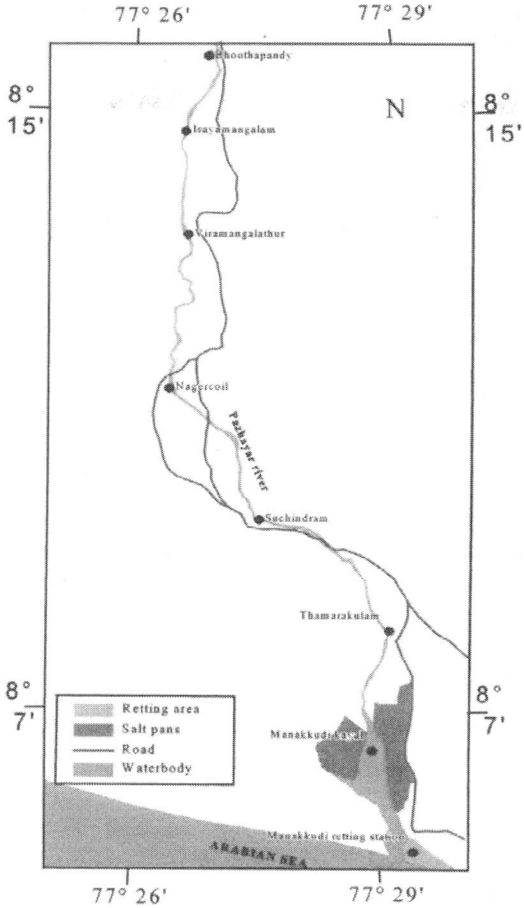

Fig. 1. Map showing sampling stations of Pazhayar river.

rectilinear path from NW to SE, covering the highland, midland and finally debouches into the Arabian Sea at Manakudi near Kanyakumari. The river has eleven check dams for irrigational purposes and the lower most dam is the Mission dam located about 3.5 km landward of the shore. Pazhayar has many tributaries and the major ones include Thodariyar, Ulakkairiver, Alanthuraiyar, Kodyuodai and Poigaiyar. The main irrigation channel is the Nanchilnad Puthanar taking off at the Chattuputhoor Anaikut.

Physiography

Two distinct physiographic regions, i.e. hills and plains, are identifiable throughout the course of the river. The hills are located towards the eastern part with an elevation of 100-600 m above msl and consisting of hills, knolls, valleys and plateaus. The low-lying land between hillocks and knolls are referred to as "elas". It is from this region, the river makes it origin. The plains extend from the foot of the hills in the north to the Arabian Sea in the west. A major portion of river (~ 25 km) stretches along the plains.

The major geomorphological features are less dissected plateaus, sedimentary plains and coastal plains. The general slope of the land is towards west. The river is mainly rainfed and therefore fluctuates in volume during the year.

Geology

The geology of the area indicates the presence of rock types, which includes garnet–sillimanite–graphite–gneiss (i.e., Khondalite), Quaternary fluvio-marine sediment, Quaternary marine sediment and alluvium. In the low flat area lying between the Western Ghats and the Arabian Sea, the gneisses are very much reduced by erosion and appear as subdued hills. Tertiary rocks, mainly of laterite, sandstone and clays occur in a chain of plateau remnants lying parallel to the west coast. The coastal tract contains limestone and calcareous sandstone, which are covered by modern white sand dunes.

Climate

The study area enjoys a semi-arid climate with an annual average rainfall of 800 mm and average annual temperature of 27.97°C. The area receives precipitation mostly from NE monsoon (October to December), which contributes 44% of the total annual precipitation, of which around 90% is in the months of October and November. The SW monsoon (i.e., June to December) contributes about 30% of the total precipitation. The remainder (26%) is sparsely distributed from January to May. The lowest humidity is 65% (in January) and the highest humidity is 84.8% (in June).

Soil

There are three main soil groups present along river basin and surrounding land. Laterite soil predominates in the hilly region or elas. Most of the land cover near the river is of mixed red and alluvial soil. In some places deep red loam is observed. The lateritic and red loam soils are poor to medium in fertility and coastal alluviam is highly fertile. The soils are sandy to sandy loam in texture. Along the coastal line there is a narrow belt of sandy coastal alluvial soil where coconut and rice are mainly cultivated.

Materials and Methods

In order to assess the spatial and temporal variation in the surface water quality parameters (i.e., physico-chemical and microbial), water samples (n = 24) were collected from upstream to downstream (between Bhoothapandi and Manakudi kayal), covering 6 sites (W1-W6) for rivers and one site each for the Manakudi kayal (W7) and a retting station (W8) in the seaward side of the river. The collection was for three seasons, viz., pre-monsoon-PRM (August) 2003, monsoon-MON (December) 2003 and post-monsoon-POM (March, 2004). The parameters include pH, DO, BOD, N, P, K, Na, Ca, Mg, hardness, Cl and salinity along with microbiological attributes (total and faecal coliform) following standard procedures (APHA, 1985). The portraits of the sampling stations are given in Table 1.

Table 1: Portrait of sampling sites, Pazhayar.

Location No.	Sample ID	Name of location	Latitude	Longitude	Remarks
1	W_1	Bhoothapandi	76⁰ 26' 50"	8⁰ 15' 43"	Upstream side, almost clear flowing water
2	W_2	Isanyamangalam	77⁰ 26' 34"	8⁰ 14' 43"	Small tract of paddy field on the banks of river
3	W_3	Viramangalathur	77⁰ 26' 36"	8⁰ 13' 20"	Vast areas of paddy cultivation on the banks
4	W_4	Nagercoil	77⁰ 26' 23"	8⁰ 11' 18"	Urban waste dumping, stagnant, turbid and polluted water
5	W_5	Suchindram	77⁰ 27' 29"	8⁰ 09' 32"	Near to Suchindram Temple. Bathing and washing station
6	W_6	Thamarakulam	77⁰ 29' 05"	8⁰ 08' 05"	Extensive weed growth indicating dangerous eutrophication
7	W_7	Manakudi estuary	77⁰ 28' 53"	8⁰ 06' 28"	Sampling from Manakudi kayal. Saline water is pumped into this kayal from the adjoinging salt plant
8	W_8	Retting Station	77⁰ 29' 24"	8⁰ 05' 09"	Sampling is near to the estuary and located in the retting pit on the eastern side of river

Results and Discussion

The results of the hydrochemical attributes of surface water samples from eight sites along the Pazhayar river for the three sampling seasons, viz., pre-monsoon-PRM, monsoon-MON and post-monsoon-POM, are given in Tables 2 to 4 and Fig. 2. Site-wise seasonal averages are shown in Table 5. Further, the averages of various parameters for the different seasons are given in Table 6.

In the study area, pH oscillated from 5.1 to 8.6. The averages for monsoon (MON), postmonsoon (POM) and premonsoon (PRM) are 7.3, 7.8 and 7.4 respectively. The seasonal average of river is 7.5 and those of Manakudi kayal are 7.7, and are indicative of alkaline condition.

The DO level of river water ranges from 1.42 to 5.67 mg/L. The averages of DO for MON, POM and PRM are 3.73, 4.05 and 3.68 mg/L respectively. Relatively high amount of DO was observed in POM possibly due to the mixing of water during NE monsoon. The seasonal average is 3.82 mg/L and that of Mankudi kayal is 2.18. The BOD recorded a range between 0.43 and 4.62 mg/L. The averages of BOD for MON, POM and PRM are 2.25, 0.98 and 2.85 mg/L respectively. The seasonal average is 2.03 mg/L.

TN value ranges between 0.03 and 2.75 mg/L. The averages of nitrogen for MON, POM and PRM are 0.27, 0.30 and 1.43 mg/L respectively. The seasonal average is 0.67 mg/L. The TP concentration ranged between 0.02 and 1.92 mg/L. The averages of TP for MON, POM and PRM are 0.06, 0.60 and 1.07 mg/L respectively. The seasonal average for river and kayal are 0.58 and 0.6 mg/L respectively. Both TN and TP are relatively high during PRM.

The K content ranged between 2.50 and 13.52 mg/L. The averages for MON, POM and PRM are 3.92, 5.27 and 9.61 mg/L respectively (sea. ave. = 6.27 mg/L). The content of Na oscillated between 8.56 and 70.12 mg/L. The season wise averages are 25.38, 24.46 and 30.77 mg/L respectively (seasonal ave. = 26.8). The average values of K and Na for the kayal are 7.7 and 29.5 respectively. The content of Na is relatively higher than that of K in all seasons. Excess Na enrichment suggests leaching of secondary or saline soil by running water. The silica

Table 2: Hydrochemical attributes of Pazhayar river, pre-monsoon-PRM.

Sample ID	Temp	pH	DO	BOD	N	P	K	Na	Si	Ca	Mg	TH	Cl	Sal
W1	28.70	7.37	4.81	1.52	0.20	0.08	2.5	8.56	13.00	24.04	9.29	33.33	44.32	80.02
W2	27.60	7.56	5.23	1.36	0.03	0.07	4.2	13.28	15.00	16.03	30.63	46.66	60.96	110.06
W3	30.10	6.29	3.90	2.41	0.34	0.07	3.0	25.64	12.00	64.12	9.20	73.33	63.85	115.27
W4	32.20	8.52	1.92	3.86	0.26	0.05	6.0	38.04	10.00	93.46	18.78	112.24	50.27	90.76
W5	20.20	7.17	3.40	2.27	0.18	0.04	2.8	26.58	9.00	64.12	62.53	126.66	41.62	75.15
W6	29.20	6.83	3.10	2.06	0.62	0.06	5.0	40.19	12.00	80.16	107.09	187.25	75.71	136.68
Mean	**28.00**	**7.29**	**3.73**	**2.25**	**0.27**	**0.06**	**3.92**	**25.38**	**11.83**	**56.99**	**39.59**	**96.58**	**56.12**	**101.32**
SD	**4.10**	**0.75**	**1.20**	**0.89**	**0.20**	**0.01**	**1.39**	**12.73**	**2.14**	**30.76**	**38.57**	**57.28**	**13.02**	**23.54**
W7	28.60	8.54	1.23	3.98	0.48	0.09	8.20	28.63	16.00	287.75	132.26	420.00	96.75	174.66
W8	29.70	8.09	Nil	5.16	3.14	0.44	5.28	52.74	5.00	192.06	136.13	328.19	238.04	462.85

Table 3: Hydrochemical attributes of Pazhayar river, monsoon-MON.

Sample ID	Temp	pH	DO	BOD	N	P	K	Na	Si	Ca	Mg	TH	Cl	Sal
W1	29.60	7.65	5.30	1.05	0.10	0.16	3.10	4.52	15.00	6.05	46.23	52.28	34.28	61.90
W2	27.80	6.30	5.67	0.90	0.09	0.09	3.00	12.81	17.00	12.65	55.60	68.25	41.72	77.33
W3	29.20	8.56	4.96	1.43	0.12	0.72	7.60	30.86	13.00	50.32	20.64	70.96	70.53	127.33
W4	30.20	7.94	2.61	0.85	0.52	0.81	6.50	50.02	19.00	71.82	18.70	90.52	45.30	81.79
W5	28.50	8.61	3.50	1.20	0.47	0.02	4.20	32.26	20.00	58.07	40.75	98.82	33.68	60.82
W6	28.80	7.81	2.28	0.43	0.49	1.80	7.20	16.28	17.00	89.95	73.67	160.62	70.92	128.04
Mean	**29.02**	**7.81**	**4.05**	**0.98**	**0.30**	**0.60**	**5.27**	**24.46**	**16.83**	**47.64**	**42.60**	**90.24**	**49.41**	**89.54**
SD	**0.84**	**0.84**	**1.5**	**0.34**	**0.21**	**0.68**	**2.08**	**16.46**	**2.56**	**32.25**	**21.00**	**38.27**	**17.09**	**30.69**
W7	29.50	8.10	2.85	1.23	0.36	1.09	5.80	30.51	16.00	130.19	80.37	210.56	109.86	198.32
W8	30.80	7.91	1.28	4.07	0.03	0.06	2.70	70.43	27.00	176.16	126.02	302.18	292.61	527.09

Table 4: Hydrochemical attributes of Pazhayar river, post-monsoon-POM.

Sample ID	Temp	pH	DO	BOD	N	P	K	Na	Si	Ca	Mg	TH	Cl	Sal
W1	30.20	7.68	3.43	1.56	0.20	1.08	3.50	7.08	18.00	12.36	48.36	60.72	52.58	94.93
W2	31.60	7.45	5.38	0.97	0.51	1.34	12.60	18.19	25.00	20.65	52.16	72.81	76.26	132.26
W3	28.80	5.16	4.61	1.92	0.86	1.92	7.40	26.52	19.00	70.06	20.17	90.23	72.63	131.12
W4	29.10	8.45	1.42	3.89	1.86	0.29	10.78	70.12	21.00	68.26	32.70	100.96	62.74	113.27
W5	28.40	7.98	2.54	4.62	2.75	0.56	13.52	30.24	22.00	90.62	39.66	130.28	40.70	73.49
W6	32.40	7.61	2.42	4.12	2.40	1.20	9.86	32.47	26.00	95.89	84.36	180.25	82.62	149.15
Mean	**30.08**	**7.39**	**3.30**	**2.85**	**1.43**	**1.07**	**9.61**	**30.77**	**21.83**	**59.64**	**46.24**	**105.88**	**64.09**	**115.70**
SD	**1.62**	**1.15**	**1.48**	**1.54**	**1.05**	**0.58**	**3.68**	**21.40**	**3.19**	**35.25**	**21.90**	**43.65**	**15.37**	**27.75**
W7	34.10	6.42	2.48	3.98	0.96	0.87	9.12	29.64	28.00	232.52	98.24	330.76	112.31	202.74
W8	31.20	8.26	0.28	4.48	0.99	1.52	11.04	67.42	12.00	178.10	103.88	281.18	238.04	429.62

Table 5: Site-wise seasonal average of hydrochemical attributes of Pazhayar river.

Sample ID	Temp	pH	DO	BOD	N	P	K	Na	Si	Ca	Mg	TH	Cl	Sal
W1	29.50	7.57	4.51	1.38	0.17	0.44	3.03	6.72	15.33	14.15	34.63	48.78	43.73	78.95
W2	29.00	7.10	5.43	1.08	0.21	0.50	6.60	14.76	19.00	16.44	46.13	62.57	58.65	106.55
W3	29.37	6.67	4.49	1.92	0.44	0.90	6.00	27.76	14.67	61.65	16.67	78.17	69.00	124.50
W4	30.50	8.30	1.98	2.87	0.88	0.38	7.76	52.73	16.67	77.85	23.39	101.24	52.77	95.28
W5	25.70	7.92	3.15	2.70	1.13	0.21	6.84	29.69	17.00	70.94	47.65	118.59	38.67	69.82
W6	30.17	7.42	2.60	2.20	1.17	1.02	7.35	29.55	18.33	87.67	88.37	176.04	76.42	137.96
W7	30.73	7.67	2.19	3.06	0.60	0.68	7.71	29.59	20.00	216.82	103.62	320.44	106.31	161.91
W8	30.57	8.09	0.78	4.57	1.39	0.67	6.34	65.20	14.67	182.11	122.01	303.85	256.23	473.19

Table 6: Seasonal averages of hydrochemical attributes of Pazhayar river and Manakudi estuary.

	Temp	pH	DO	BOD	N	P	K	Na	Si	Ca	Mg	TH	Cl	Sal
Pazhayar River	29.03	7.50	3.82	2.03	0.67	0.58	6.27	26.87	16.83	54.76	42.81	97.57	56.54	102.19
Manakudi estuary	30.73	7.69	2.18	3.06	0.68	0.60	7.70	29.51	20.00	216.81	103.62	320.44	106.30	191.90
Retting station	30.56	8.08	0.52	4.57	1.38	0.67	6.34	63.53	14.66	182.10	122.01	303.85	256.23	473.18

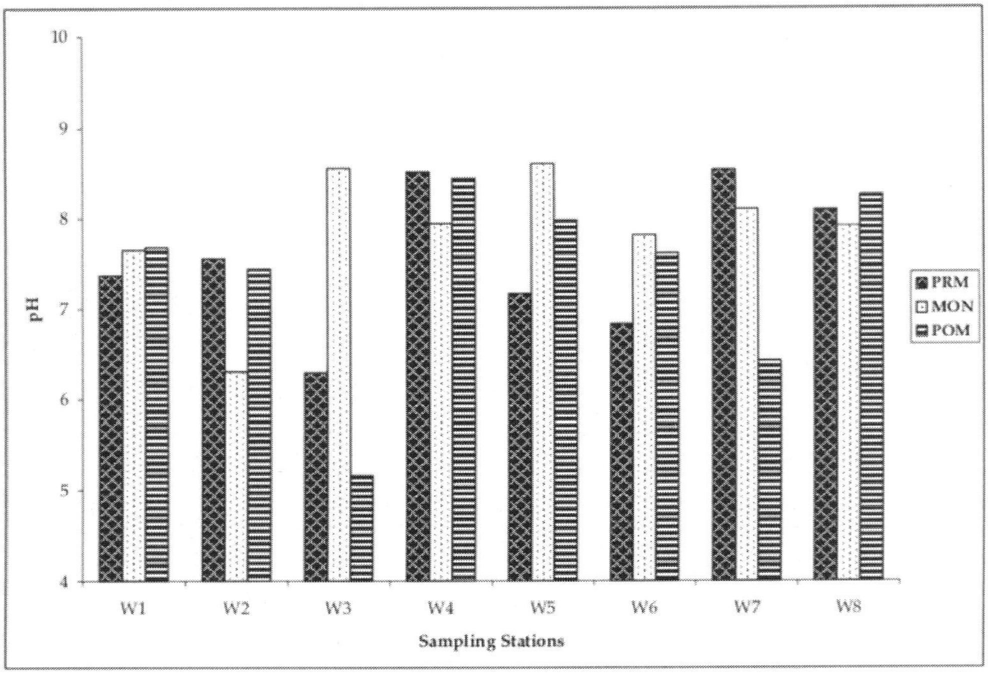

Fig. 2. Spatio-temporal variation of pH.

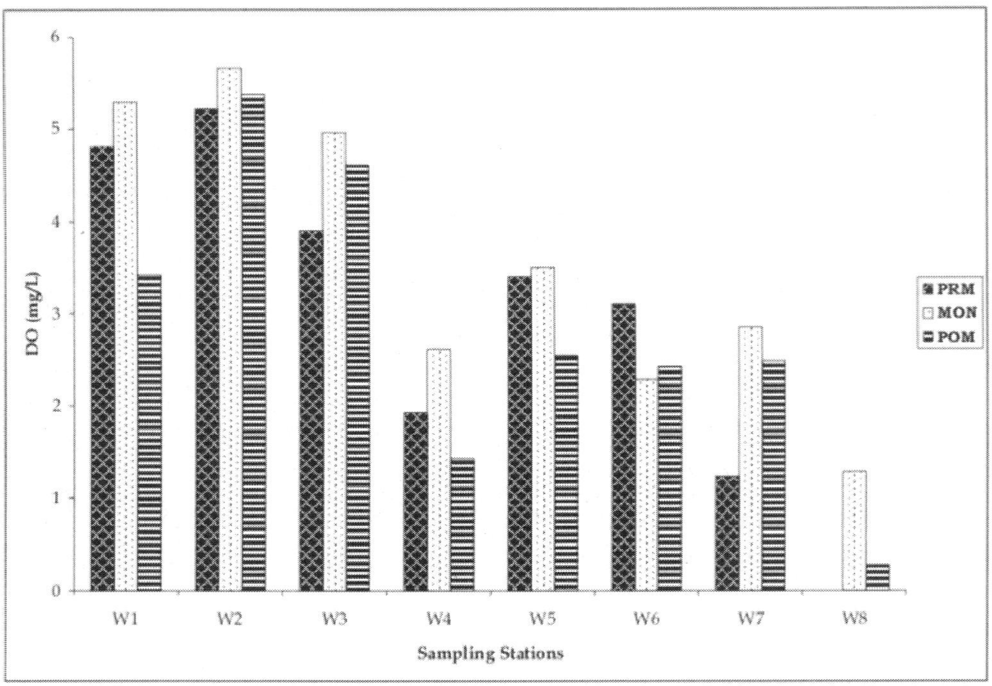

Fig. 3. Spatio-temporal variation of dissolved oxygen.

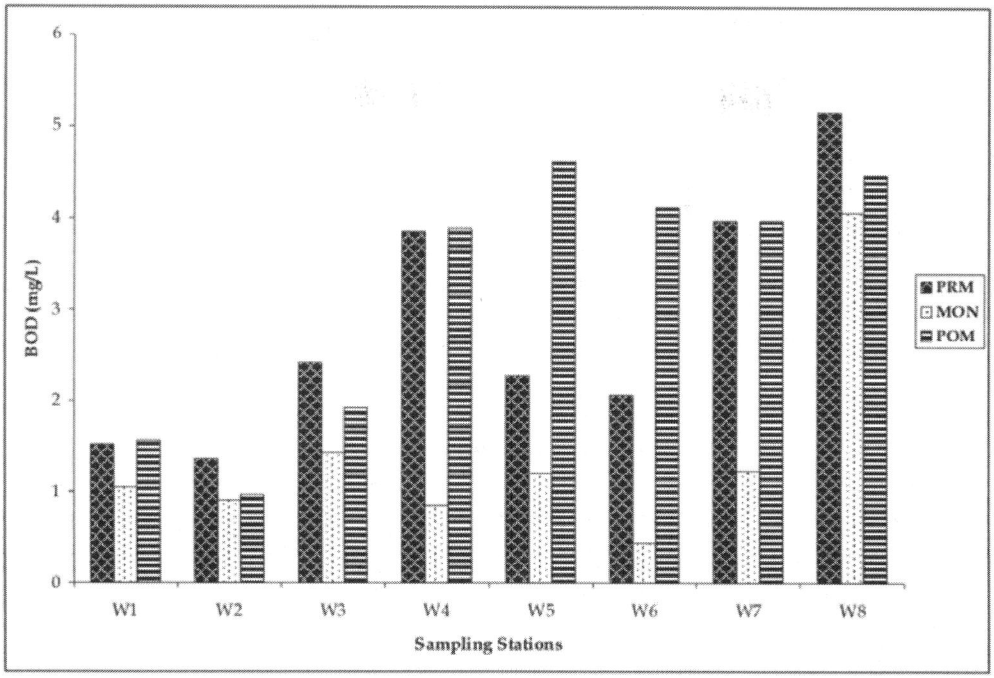

Fig. 4. Spatio-temporal variation of biochemical oxygen demand.

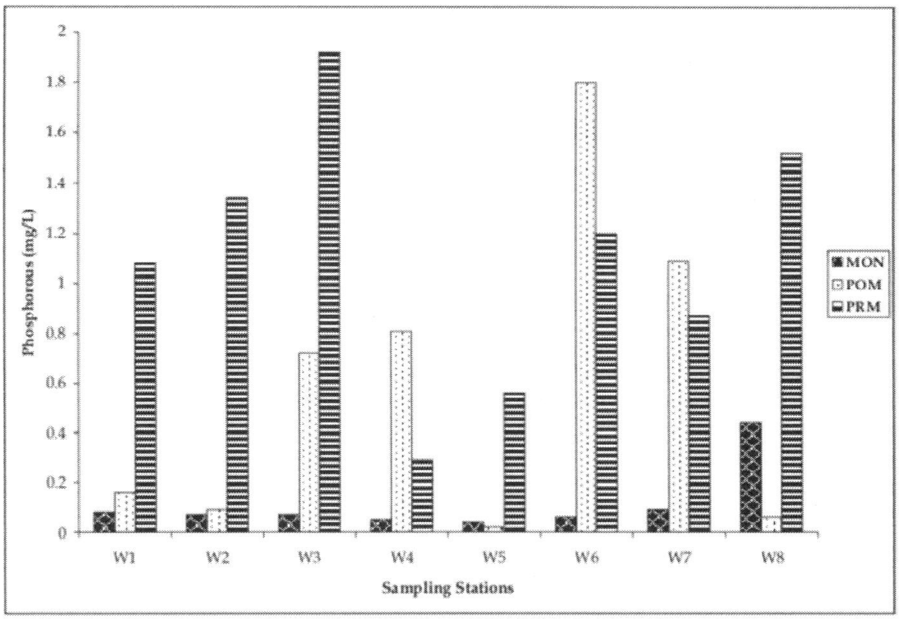

Fig. 5. Spatio-temporal variation of phosphorous.

values obtained fluctuate from 9 to 26 mg/L. In Manakudi kayal, the seasonal average of silica is 20 mg/L.

Hardness of water is caused by dissolved solid present in water especially calcium and magnesium. Hardness ranges between 33.33 and 187.25 mg/L. The averages for MON, POM and PRM are 96.58, 90.24 and 105.88 mg/L respectively (seas. ave. = 97.57 mg/L). Site-wise variation shows an increasing trend from W1 to W6. Further, sample W6 records high value for Ca and Mg in all seasons. Low Ca (47.64) and Mg (42.60 mg/L) contents are seen in POM. It was noticed that as the content of Ca and Mg increases, the content of hardness also shoots up. The maximum permissible limit of hardness is 500 mg/L (WHO, 1984). It is seen that samples of both river and kayal in all the seasons fall below this limit.

The chloride ion is the most predominant natural form of element chlorine and is extremely stable in water. The chloride value oscillates between 33.68 and 109.86 mg/L. The same stations again record the lowest (60.82 mg/L) and highest (198.32 mg/L) salinity values and indicate that a positive correlation exists between chloride and salinity. The averages of chloride for MON, POM and PRM are 56.12, 49.41 and 64.09 mg/L and that of salinity are 101.32, 89.54 and 115.70 mg/L respectively. Their seasonal averages are 56.54 and 102.19 mg/L respectively. River water records least chloride and salinity contents during POM due to high rainfall and highest during PRM. According to WHO standards, maximum recommended concentration of chloride is 250 mg/L. It is seen that samples from both river and kayal in all the seasons fall well within this limit.

From the foregoing study, it has been found that water quality parameters like N, K, Na, Si, Ca, Mg, Hardness, Cl and salinity are within the permissible limit during the study period. However, parameters like pH, BOD and P of river water exhibit higher values above permissible limit and DO show lower values. There is no significant seasonal variation in the water quality of the river. However, it has been found that some of the parameters like DO, BOD and P show relatively high value during PRM.

Nevertheless, no significant site-wise variation (except W4 and W6) is seen for the hydrochemical attributes in the river. The W4 station is an urban centre in Nagercoil and it shows relatively high pH (seas. ave. = 8.3), low DO (= 1.98), high BOD (= 2.87). This location is under increasing eco-degradation due to anthropogenic stress in the form of discharge of sewage from urban centres and households. The station W6 is from Thamarakulam, located in the downstream side of the river and near to the confluence of river with kayal. It shows relatively high content of P (seas. ave. = 1.02), Ca (= 87.6), Mg (= 88.3), hardness (= 176.0), chloride (= 76) and salinity (= 137). The P content in W6 is above the permissible limit and is responsible for the eutrophication as evidenced by dense mat of water hyacinths there. The high content of the above attributes in W6 is due to the influence of the kayal.

Even though the water sample (W7) from the Manakudi kayal shows relatively high content of BOD, Si, Ca, Mg, hardness, chloride and salinity than the river samples, the quality of water in the kayal is deteriorated by the ret liquor from the retting station adjacent to the kayal. The various hydrochemical attributes of the retting liquor like pH, BOD, N, Na, Ca, Mg, hardness, chloride and salinity are higher than that of river and kayal samples, and DO lower than that of river and kayal (Tables 2-4).

From the study it is evident that chloride (ave. = 56.54 mg/L) is the dominant anion in the area and is followed by nitrate (= 0.67) and phosphate (= 0.58). The dominant cation in the water samples is Ca (ave. = 54.76 mg/L) and is followed by Mg (= 42.81), Na (= 26.87) and K (= 6.27 mg/L).

Sodium Absorption Ratio, SAR

This is an important character for assessing the suitability of water for irrigation purpose after considering the likely sodium hazards. It is calculated by using the formula:

$$SAR = \frac{Na^{+}}{\sqrt{\frac{Ca^{2+} + Mg^{2+}}{2}}}$$

The average SAR values for MON, POM and PRM seasons are 2.82, 1.37 and 3.60 respectively. The seasonal average of SAR for the entire river is 2.59. Moreover, the site-wise variation of SAR ratio shows a range between 1.06 and 3.84. The SAR values for three seasons and site-wise values fall well within the 0 – 10 category which implies that the water is good for all types of crops and all types of soil except for those crops which are highly sensitive to Na.

Microbial Analysis

The analysis of coliform (total and faecal) from seven stations of river for the post-monsoon (POM) and pre-monsoon (PRM) shows that the total coliform ranges from 420/100 mL (W1) to 18000/100 mL (W5) during POM (ave.= 8408), while during PRM it ranges from 780/100 mL (W1) to 26000/100 mL (W7) with an average of 12476 (Table 7). However, the faecal coliform ranges between 0/100 mL (W1) and 625/100 mL (W6) in POM (ave.= 260) and between 6/100 mL (W1) and 1588/100 mL (W7) during PRM (= 749).

Table 7: Total and faecal coliform count (in MPN/100 ml) of Pazhayar.

Sample ID	Total Coliform Count (MPN/100 mL)			Faecal Coliform Count (MPN/100 mL)		
	POM	PRM	Average	POM	PRM	Average
W1	420	780	600	0	6	3
W2	765	1600	1183	30	22	26
W3	3258	2850	3054	68	12	40
W4	7600	18000	12800	520	850	685
W5	18000	22100	20050	272	575	424
W6	12850	16000	14425	625	900	763
W7	15960	26000	20980	300	2876	1588

The site-wise variation of coliform count follows a gradual increase from W1 to W5 with a sudden fall at W6 and again rising up in W7. A variety of environmental factors such as organic matter present in sediment and temperature of water influence the count (Clark and Pagel, 1977). As organic matter increases count increases and vice versa for the temperature.

Based on some selected water quality parameters (viz., pH, DO, BOD, P, total coliform count), has classified water standards into five broad categories, i.e., A, B, C, D & E. The categories A and C are for drinking purposes, B for bathing, D propagation of wildlife and fisheries and E for irrigation.

The site-wise seasonal average of the hydrochemical parameters shows that the attributes like pH, DO, BOD and P of stations W1, W2, W3 are falling within the recommended limit of the suitability of water for category B, C & E. Further, the total coliform and faecal coliform count also falls within the recommended limit for category B, C& E. Hence, it is obvious that water from the upstream site like W1, W2, W3 can be used for drinking purpose after conventional treatment and disinfection. However, the pH, DO, BOD of samples from W4, W5 & W6 show values which are beyond the recommended limit for use as B & C and hence, are unfit for drinking and domestic use.

However, the above parameters of water for the entire river (W1 – W6) fall within the limit recommended for use as E category (i.e., irrigation purpose). The seasonal and site-wise average of SAR ratio corroborates this statement. Further, the above parameters of Manakudi kayal show that this water is also fit for use as category E (i.e., irrigation).

SUMMARY AND CONCLUSION

Water quality parameters of river like N, K, Na, Si, Ca, Mg, Hardness, Cl and Salinity are within the permissible limit during the study period. However, parameters like pH, BOD and P of river water exhibit values above permissible limit and DO show lower values.

There is no significant seasonal variation in the water quality of the river. But, it has been found that some of the parameters like DO, BOD and P show relatively high value during PRM. No significant site-wise variation (except W4 and W6) is seen for the hydrochemical attributes of the river. The W4 is an urban centre in Nagercoil and due to anthropogenic activities and sewage dumping.

Microbial analysis shows an alarming rate of total coliform and faecal coliform count with an average of 10442 and 504/100 mL for December 2003 and March 2004 respectively. All the river samples are contaminated with microbes and the content increases in the downstream direction.

According to it is obvious that water from the upstream sites like W1, W2, W3 (i.e., Bhoothapandy to Veeramangalathoor, L = 10 km) can be used for drinking purpose after conventional treatment and disinfection. However, the parameters (pH, DO, BOD, P and Total Coliform) of water from sites like W4, W5 and W6 show beyond the recommended limit for use of samples from and are, hence, unfit for drinking and domestic use. However, based on the SAR ratio, the water from the entire river course is suitable for irrigation.

ACKNOWLEDGEMENTS

The second author expresses her deep gratitude to Fr. Raju Kallumpurath and nuns of Stella Mary's Convent, Kanyakumari for the most wanted help rendered during the field work.

REFERENCES

APHA, (1985) Standard method for the examination of water and wastewater, American Public Health Association, 19th edition, New York.

Boss, B.B. (1956) Observations on the hydrology of the Hoogly estuary. *Ind. Jour. Fish.*, 101 pp.

Clarke, J.A. and Pagel, J.E. (1977) Pollution indicator bacteria dissociated with municipal drain and drinking water supplies. *Can. Jour. Microbiol.*, 23, 465-470 pp.

Duncan, N. and Rzoska, J. (1978) Landuse impact on lake and reservoir ecosystem. Prc. MAB project 5 workshop, Warsaw, Poland, facultas Veriag, Wein.

Garbow, W.O.K. (1979) Disinfection of water: Pros and cons. *Water SA*, 5(2), 98-108 pp.

Trivedi, R.K. Khatavkar, A.Y. Kulkarni and Shrotri, A.C. (1990) Ecology and pollution of River Krishna in Maharashtra, Ministry of Environment and Forests, Government of India.

Tyller, M.G. (1995) Living in the environment: Principles, connections and solutions. 5th edition, Wadsworth Publishing Co., California.

Ward, H.B. and Whipple, G.C. (1959) Freshwater biology. John Wiley & Sons Inc., 1248 pp.

WHO, (1984) Guidelines for drinking water quality. World Health Organization, Geneva, 1-3 pp.

21

Groundwater Quality Assessment of Kammapuram Block, Cuddalore District, Tamil Nadu

Aruljoicy, C., Manivel, T., Sivaraman, K. and Lawrence, J.F.*
Department of Geology Presidency College, Chennai
**Email: jorlaw@gmail.com*

ABSTRACT

Water is essential for both animal and plant kingdom and also for industrial and agricultural needs. Nowaday, the available surface water is not fulfilling the needs of human beings. Hence, to meet the demand of water the subsurface resources are tapped. Kammapuram block is one of the major panchayat union with 54 villages and has been located in centre part of Cuddalore district. It lies in between 79°19' and 79°33' East longitude and 11°24' - 11°34' North latitude. In Kammapuram block people mainly depend on the groundwater for their domestic and agricultural needs. The quality of groundwater is as important as its quantity. In the present study, a part of Kammapuram block has been selected and 30 groundwater samples have been collected in a systematic way from the existing bore well. The water samples have been analyzed for major cations and anions. With the analyzed results, the water has been classified using HYCH programme. The following thematic maps have been prepared in GIS environment, viz:- TDS, Total Hardness, Corrosivity ratio, Gibb's plot and Groundwater classification. From the above study, it is understood that the groundwater of the study area is of moderate quality. Hydrogeologically groundwater occurs under unconfined to confined conditions in the study area. Most part of the study area, the TDS under permissible limit which can be used for drinking and domestic purposes. The area is mostly occupied by non-corrosive water which can be transported through metal pipes. From this study, it is inferred that the groundwater occurs under water table condition in Kammapuram and soft in quality. The groundwater quality and quantity of this region can be improved further by adopting artificial recharge method such as rainwater harvesting.

Keywords: Groundwater, water quality, classification, utility.

INTRODUCTION

Groundwater is a most important natural resource required for drinking irrigation and industrialization. The resource can be optimally used and sustained only when quantity and quality of groundwater is assessed. Groundwater is nothing but the water held in the rock underground by certain forces replenished by natural according to the climate and local geology and consequently variable in both amount and quantity. The condensation of cloud in the form of rainfall is the main source of groundwater. The process of water sinking into the soil is called infiltration capacity of the soil which helps the rainwater (Carrold,1962) to enter into the soil. Capillarity and gravity are the two main forces involve in the downward movement of water in a soil through pore space, holes, cracks and joints that are present near surface rock.

STUDY AREA

Kammapuram is one of the major panchayat union with 54 villages and located in centre part of the Cuddalore district. It lies between 79°19 and 79 °33 'East longitude and 11°24' and 11°34' Northern latitude occurring in 58M/6 Survey of India toposheet (Fig. 1). The climate of the study area is hot and dry for the most of the months of the year. In the months of June, July and August the area receives scattered rainfall from south-west monsoon. During the months of October, November and December, this area receives nearly 70% of annual rain fall through north-east monsoon. The area experiences an arid to semi-arid climate with an average annual rainfall of 1208 mm most of which is received through NE monsoon prevail from October to December.

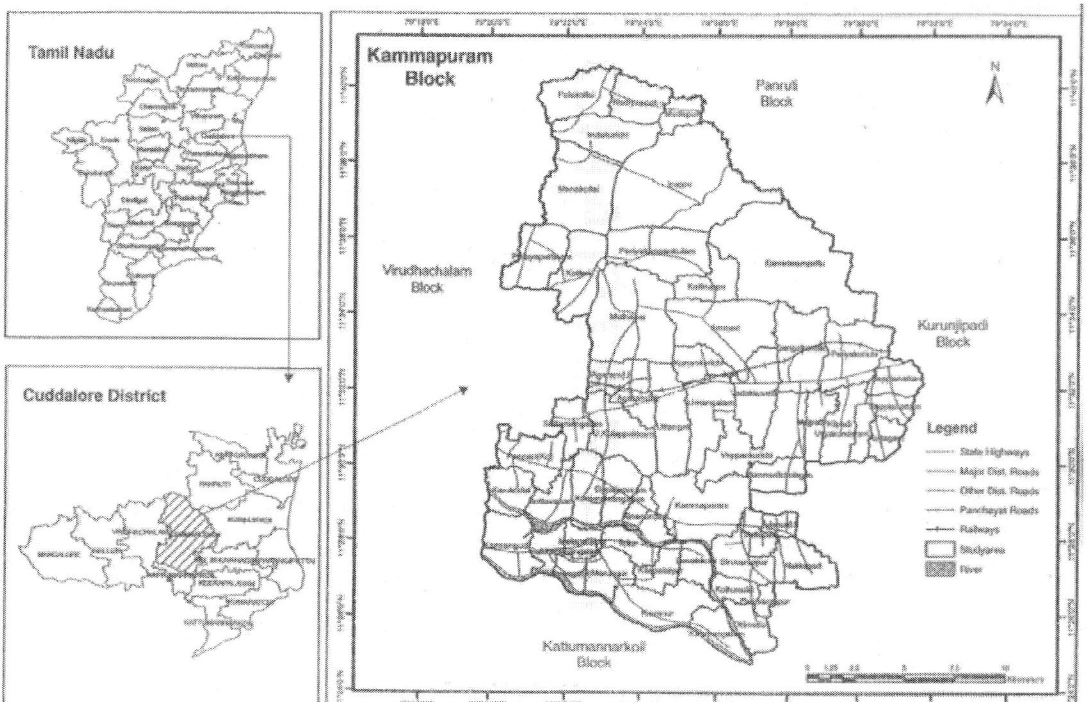

Fig. 1.

GEOLOGY

The study area is underlain by various geological formations ranging in age from Achaean through semi-consolidate formation of Mesozoic and alluvium formation of Quaternary age. Most of the area is underlained by Achaean rocks. The fossiliferous limestone and calcareous sandstone also cover considerable area which is of Ariyalur formation. The alluvium with sand and sandy clay mixed is seen along the river courses. The Cuddalore sandstone covers the large area along the western region where Neyveli lignite has been mined.

HYDROGEOLOGY

Groundwater constitutes one position of the hydrologic cycle as even it is to be regarded as the renewable natural resource. Groundwater occurrence in unconsolidated rocks strikingly different from that of in consolidated sediments (Todd, 1980, Karanth, 1993). The lithology is one of the parameters that control the groundwater quality and other factors include evaporation at the surface prior to infiltration, evapotranspiration and nature of aquifer and residence time.

METHODOLOGY

Quality of groundwater is an important parameter. The quantity and suitability of groundwater for human consumption and irrigation are determined by its physical, chemical and bacteriological parameters. Rainwater which is pure in nature which, fall on the ground and dissolves the stable minerals before it infiltrates to the subsurface formations. Hence, the quality of groundwater water depends on the nature of the formation and time duration it stays in that formation. Depending upon various physical and chemical parameters, the groundwater can be utilitised for different purpose. To analyse the physical and chemical parameter to evolve the utilization and to take certain preventive measure, water sample have been collected from 30 locations, which are evenly distributed in the entire part of the study area. The availability groundwater in the study area is extensively tapped through open wells and bore wells for drinking and agriculture purposes.

RESULTS AND DISCUSSION

Total dissolved solids (TDS) of the groundwater is above 1000 mg/L in U.Adhanur (Loc.20). Fresh to moderate quality water is seen southern portion of the study area, while entire northern part is occupied by freshwater (TDS $<$ 500 mg/L). This fresh water can be utilized for drinking and domestic purposes (Ryzner, 1994) (Fig. 2). With reference to total hardness, hard water exists in Uthangal (Loc. 10), U. Mangalam (Loc. 11), Kammapuram (Loc. 5), Santhamangalam (Loc. 5) and moderate hardness exist in Pallakollai (Loc. 1), Periyakurichi (Loc. 9) and Sepplanatham (Loc. 8) (Fig. 3). U. Agaram (Loc. 7) Uthangal (Loc. 10) area groundwater found to have low sodium and salinity content that can be used for irrigation of any type of crops. Groundwater around Palayapattinam (Loc. 2), Kotteri (Loc. 4), Kammapuram (Loc. 6), Periyakurichi (Loc. 9), U.Mangalam (Loc. 11), Melphati (Loc. 12), Sathamangalam (Loc. 15) and Vepankurichi (Loc. 29) have medium saline and low sodium water which can be used for plant with moderate salt tolerance (Richards, 1954). The groundwater in Ammeri (Loc. 5) has very high saline water, which is not suitable for irrigation (Fig. 4).

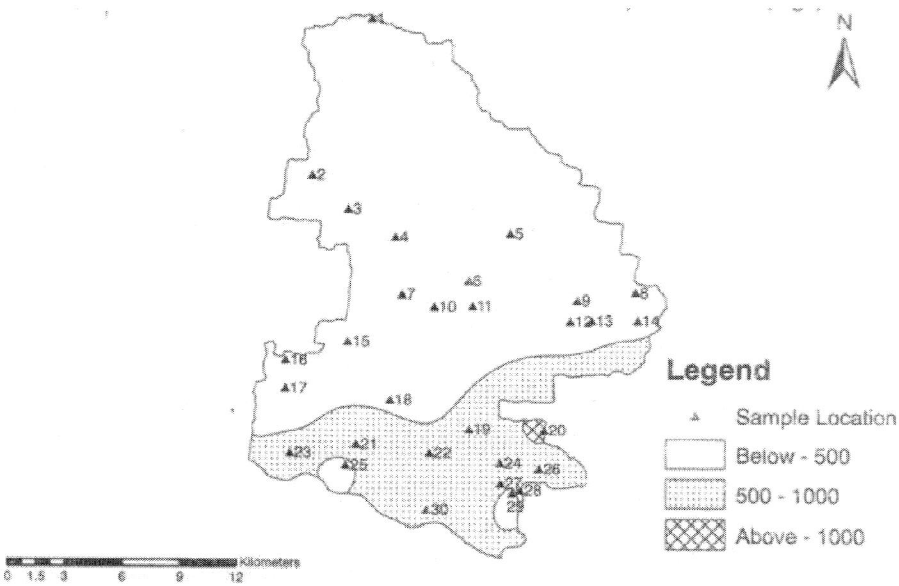

Fig. 2. Total dissolved solids in groundwater (mg/L).

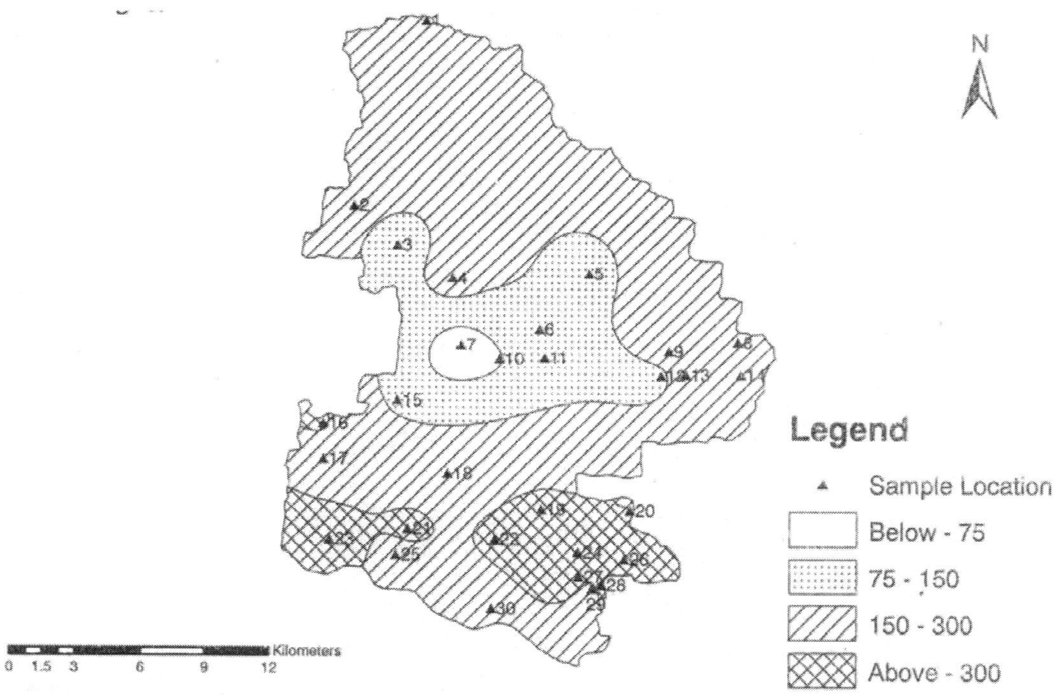

Fig. 3. Total hardness in groundwater (mg/L).

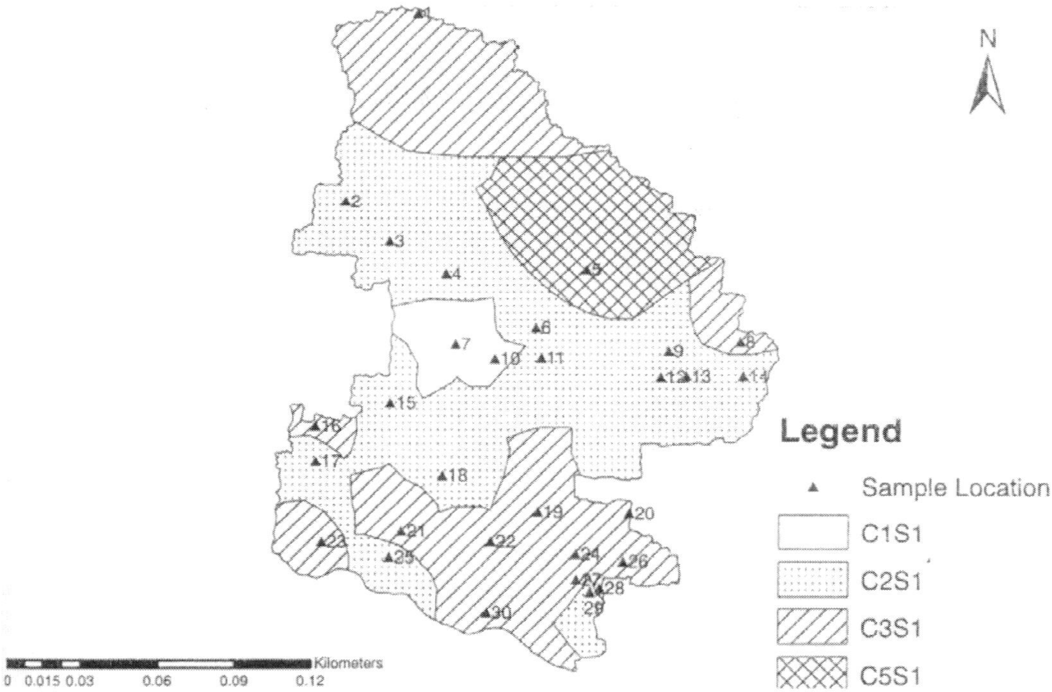

Fig. 4. USSL classification of groundwater.

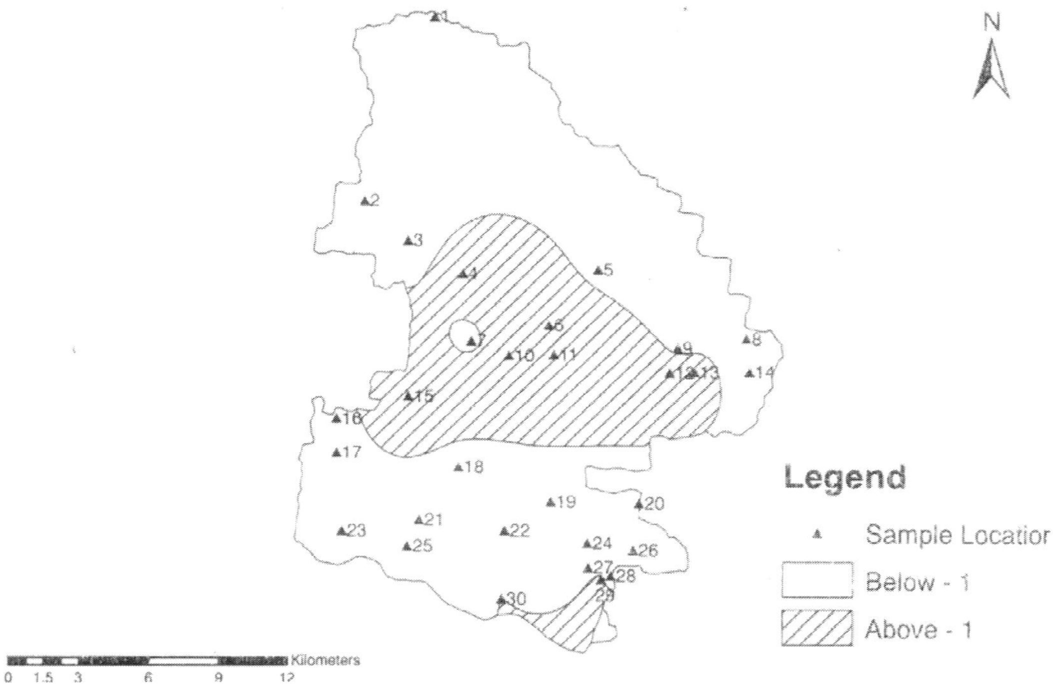

Fig. 5. Corrosivity ratio in groundwater.

The groundwater of following region Mudanai (Loc. 4) Kunankurichi (Loc. 6), Uthangal (Loc. 10), Melpathi (Loc. 12), Sathamangalam (Loc. 15) and Kavanur (Loc. 30) falls under corrosive type of water and it can only be transported through PVC pipes. The other parts of the study area found to be non-corrosive, which can be easily transported through metal pipes.

SUMMARY AND CONCLUSION

Geologically the Kammapuram is underline by various geological formation ranging in age from archean to quaternary age from the different sedimentary rock. Hydrogeologically groundwater occurs under unconfined to confined condition in the study area. Most part of the study area the TDS value found to be below 1000 mg/L indicating the extend of freshwater and they are potable water and it can also be used for drinking and domestics purpose. From this study, it is inferred that the groundwater occur under water table condition in Kammapuram and soft in quality.

REFERENCES

Carrol, D. (1962). Rainwater as a chemical agent of geologic processes. A review U.S. Geological survey water supply paper 520-f pp. 97-104.

Karanth, K.R. (1993). Groundwater assessment Development and management Tata McGraw-Hill Publishing Company Ltd. p. 715.

Richards, (1954). Diagnoses and improvement of saline alkali soils USDA hand book, **60:** p. 160.

Ryzner, J.W. (1994). A New index for determining amount of calcium carbonate scale formed by water, J, Ames W, Wass. 36, pp. 472-486.

Todd D.K. (1980). Groundwater hydrology John Wiley, pp. 535.

22

Hydrogeochemical Study of Groundwater in Gummudipoondi Taluk, Thiruvalluvar District, Tamil Nadu

Ganesh, R*., Sathishkumar, P., Nancy, S., Rajeswari, P. and Lawrence, J.F.

Department of Geology Presidency College, Chennai

**Email: ganeshgeo83@yahoo.com*

ABSTRACT

Groundwater is a precious and the most widely distributed resources of the Earth. Groundwater is the water held in the pore space of rocks beneath the ground and replenished by nature and consequently variable in both quality and quantity. The study area forms a part of Gummudipoondi taluk which is located between 79°59' – 80°13' East longitude and 13°22' – 13°32' North latitude. The areal extent of the study area is 175 sq kms. In this area people mainly depend on the ground water for their domestic and other uses. The quality of groundwater is as important as its quantity. For the present study, a part of Gummudipoondi area has been selected and 28 groundwater samples have been collected in a systematic way from the exiting bore well. The water samples have been analyzed using APHA (1984) procedure for major cations and anions. With the analyzed results the water has been classified using HYCH programme. The following thematic maps have been prepared, viz., Total Dissolved Solids (TDS), Total Hardness (TH), Cl/HCO_3, Corrosivity Ratio, Water Type, Groundwater Classification and Gibb's plot. Trace elements study also has been carried out for Fe, Pb, Zn and Cd. From the study, it is inferred that the groundwater occurs underground in water table condition and are good to moderate in quality. Cadmium in groundwater is found to be more than permissible limit in the northern region of the study. Consuming this water for longer duration may lead to health hazard. The groundwater quality and quantity of this region can be further improved by adopting artificial recharge methods such as roof top harvesting, percolation pond and check dam.

Keywords: Hydrogeochemistry, factors, classification, utility.

INTRODUCTION

Groundwater is the most important natural resource required for drinking, irrigation and industrialization purposes. The resource can be optimally used and sustained only when quantity and quality of groundwater are assessed. Hydrogeochemistry of groundwater is a dynamic process, which undergoes constant changes with reference to space and time. The geological quality of groundwater is also related to nature of host rock as well as the overlying rock types. Geochemical characteristics of groundwater are essential component of scientific management of existing groundwater resource. The quality of groundwater is determined by its physical, chemical and biological parameters.

The main objective of this study is to evaluate the available groundwater resources using its physical and chemical parameters and classify according to usage and to delineate freshwater zones (Carrold, 1962). The study also suggests the management strategy so as optimize the utility of groundwater.

STUDY AREA

Gummudipoondi taluk is one of the important taluk in Thiruvalluvar district. The study area forms a part of Gummudipoondi taluk which includes 28 villages in it. It is geographically located between 79°59' and 80°13'E longitude and 13°22' to 13°32' north latitude. The area extend is well connected with other parts of the state by state and national highways. The other major towns are connected through district roads. It lines in the Chennai – New Delhi Railway lines. This study area annual rainfall is 750 mm and most of which is contributed by north-east monsoon.

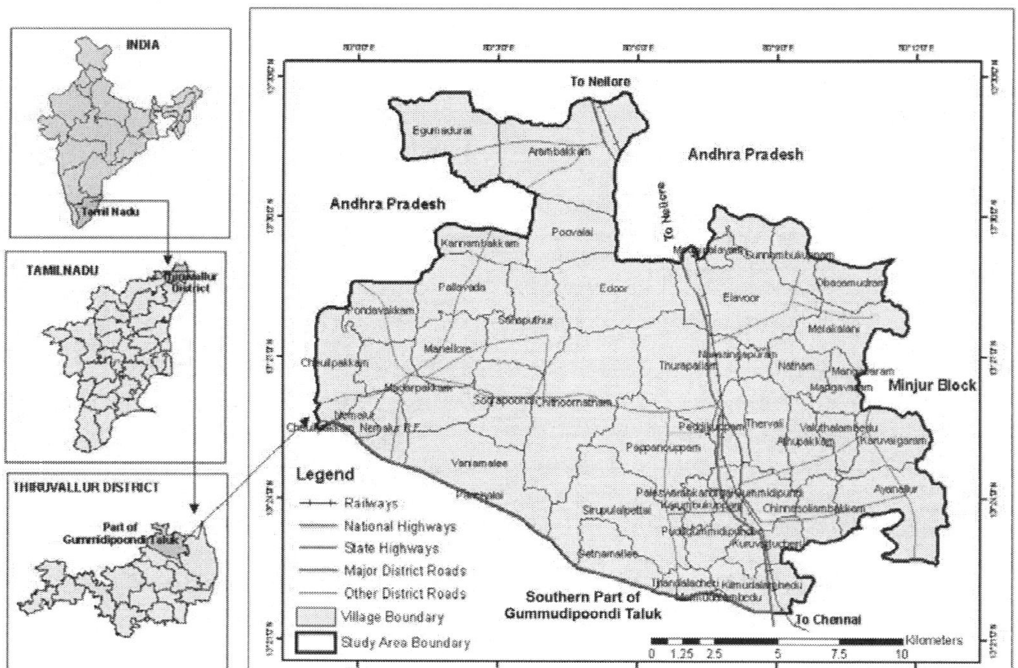

Fig. 1. Part of Gummudipoondi Taluk - Location.

GEOLOGY

The study area is underlain by various geological formations from Achaean to the Recent Alluvium. It is mainly covered with large amount of laterite and with a few out crops of Alluvium and coastal Alluvium with an unconformity. The geology map of the study area has been prepared from the satellite imageries IRS 1C, LISS, FCC Geo-coded data on 1:50,000 scale.

METHODOLOGY

An integrated approach is very much required for any groundwater studies due to the complex nature of land water. To study hydrogeochemical characters of the groundwater along the study area, 28 locations have been collected and the water samples have been analyzed to obtain various physical chemical parameters (cations & anions) analyzed using APHA (1984) procedure. The analyzed results have been processed using HYCH programme to classify the groundwater. From HYCH output, the thematic maps have been prepared in GIS environment.

RESULTS AND DISCUSSION

Groundwater TDS above 1500 mg/L exists in north-eastern part of study area at Obasamudram (Loc.6) which is brackish water and is not potable for domestic use. Freshwater having TDS less than 500 mg/L exist in Egumadurai (Loc. 1), Arampakkam (Loc. 2), Naidukuppam (Loc. 3), Edur (Loc. 5), Pallavada (Loc. 7), Sanaputtur (Loc. 8), Chedulpakkam (Loc. 11), Sanaputtur (Loc. 13), Periaobulapuram (Loc. 18), Eguvarpalayam (Loc. 17), Soorapoondi (Loc. 20), Nemallur (Loc. 21) and Eguvarpalayam (Loc. 23). TDS in groundwater between 500 and 1000 mg/L occupies most part of the study area indicating freshwater horizons and may be used for drinking purpose. TDS value between 1000-1500 mg/L occur as patches in north and central region.

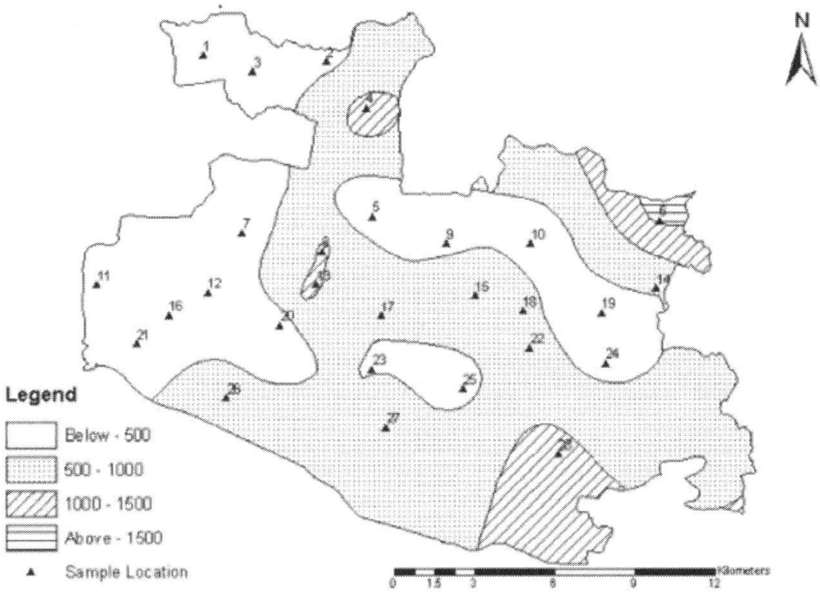

Fig. 2. Total dissolved solids in groundwater (mg/L).

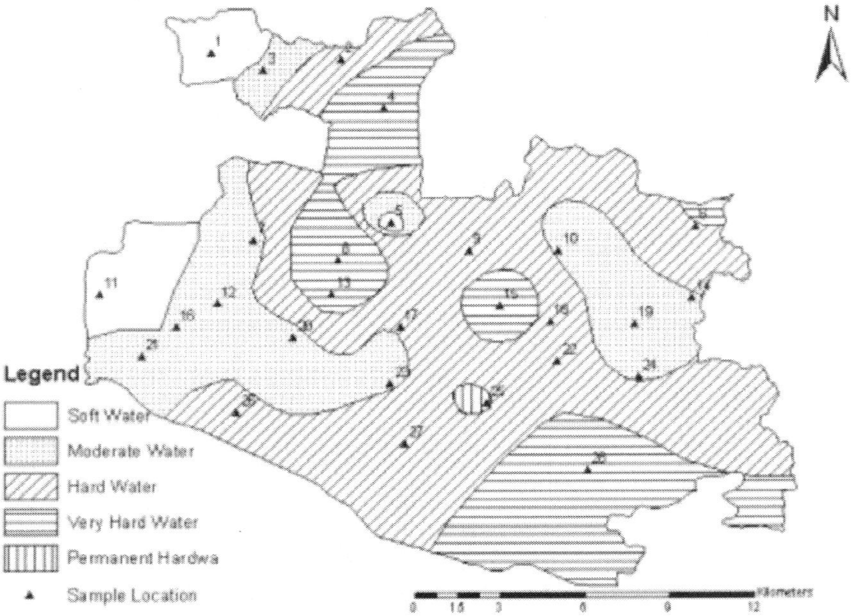

Fig. 3. Total hardness in groundwater (mg/L).

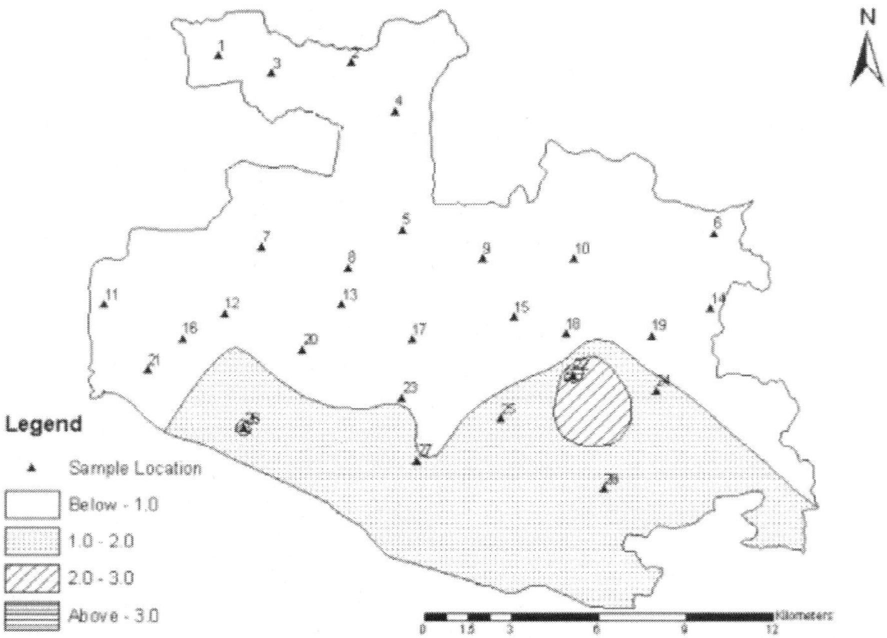

Fig. 4. Cl/HCO_3 ratio in groundwater.

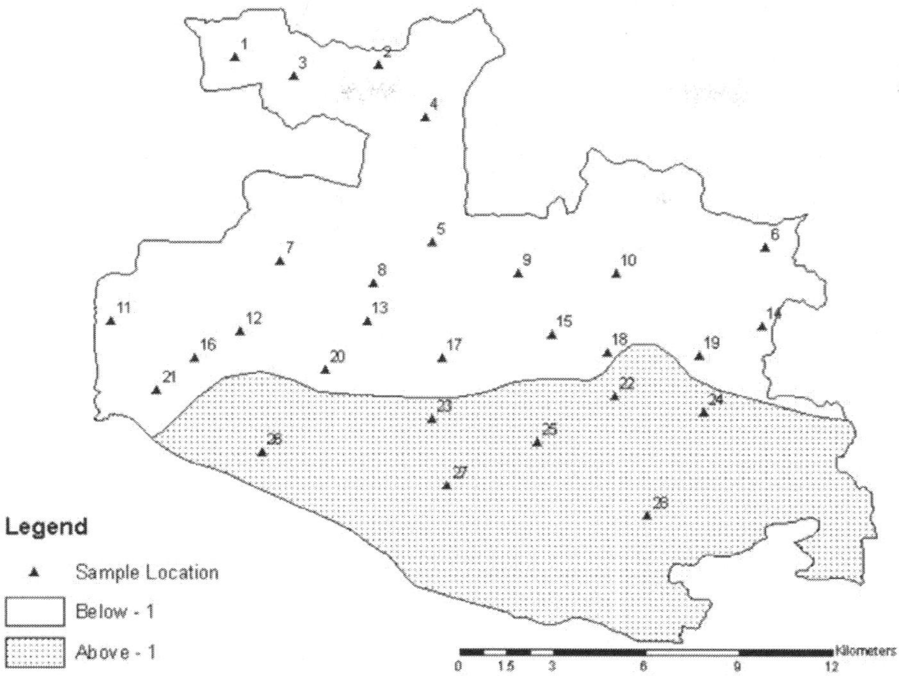

Fig. 5. Corrosivity ratio in groundwater.

Total hardness in groundwater less than 75 mg/L which is soft water (Richards, 1954), exist in Egumadurai (Loc. 1), Edur (Loc. 5), and Chedulpakkam (Loc. 11). Moderate hard water occur in Naidukuppam (Loc. 3), Pallavada (Loc. 7), Sanaputtur (Loc. 8), Menallure (Loc. 12), Periaobulapuram (Loc. 18) Eguvarpalaym (Loc. 17), Soorapoondi (Loc. 20), Nemallur (Loc. 21), Athupakkam (Loc. 24), and Eguvarpalayam (Loc. 23). Hard water (TH ranging 150 – 300 mg/L) occur in Arampakkam (Loc. 2) Obasamudram (Loc. 6), Elavoor (Loc. 10), Natham (Loc. 19), Madharpakkam (Loc. 16), Pedikuppam (Loc. 22), Vaniamallai (Loc. 26), and Sirupuzhalpettai (Loc. 27). Very hard water (TH < 300) occurs in Arampakkam (Loc. 4), Edur (Loc. 9), Thurapallam (Loc. 15) and Gummudipoondi (Loc. 28).

The chloride/bicarbonate ratio in groundwater will indicate the quantum of seawater intruded in any region (Ryzner, 1944). In the study area southern part is an indication of sea water intrusion (Fig. 4). This may be due to over exploitation of groundwater in this area. The corrosivity ratio (Fig. 5) in groundwater less than one are considered as non corrosive and exist in Egumadurai (Loc. 1), Arampakkam (Loc. 20), Naidukuppam (Loc. 3), Arampakkam (Loc. 4), Edur (Loc. 5) Obasamudram (Loc. 6) Pallavada (Loc. 7), Edur (Loc. 9), Elavoor (Loc. 10), Sanaputtur (Loc. 8), Chedulpakkam (Loc. 11) Sanaputtur (Loc. 13), Melakkalanai (Loc. 14) Manellore (Loc. 12) Thurapallam (Loc. 15), Periabobulapuram (Loc. 180, Natham (Loc. 19), Madharapakkam (Loc. 16), Eguvarpalayam (Loc. 17), Soorapoondi (Loc. 20) and Nemallur (Loc. 21). The Corrosivity ratios is above than one are considered as corrosive and occur at Pedikuppam (Loc. 22), Athupakkam (Loc. 24), Eguvarpalayam (Loc. 23), Pappankuppam (Loc. 25), Vaniamalli (Loc. 26), Sirupuzhalpettai (Loc. 27) and Gummudipoondi (Loc. 28).

From HYCH Program output, USSL Classification (Karanth 1993) map has been prepared for types of water (Fig. 6). C_1S_1 and C_2S_1 type of water occurs in Egumadurai (Loc. 1), Naidukuppam (Loc. 3), Edur (Loc. 5), and Chedulpakkam (Loc. 11), Arampakkam (Loc. 20), pallavada (Loc. 7), Elavoor (Loc. 10), Sanaputtur (Loc.8), Sanaputtur (Loc. 13), Manellore (Loc. 12) < Periaobulapuram (Loc. 18), Madharpakkam (Loc. 16), Soorapoondi (Loc. 20), Nemallur (Loc. 21), Eguvarpalayam (Loc. 23), and Pappankuppam (Loc. 25). This type of water is the most preferable one for all types of crops and recommended for plantation.

C_3 S type of water occurs in Arampakam (Loc. 4), Edur (Loc. 9), Thurapallam (Loc. 15), Natham (Loc. 19), Eguvarpalayam (Loc. 17), Peddikuppam (Loc. 22), Athupakkam (Loc. 24) and sirupxhulpetai (Loc. 27). This water is slightly higher saline water and salt tolerant crops can be cultivated. C_3S_2 type of water occurs in Melakkalani (Loc. 14) and Gummudipoondi (Loc. 28). Water of this type can be used only for more salt tolerant crops. C_4S_2 type of water occurs only in Obasamudram (Loc. 6) which is highly saline and high sodium water and cannot be used for irrigational purposes.

The rock water interaction is the dominant mechanism than evaporation and precipitation that controls the water quality of this study area (Fig. 7). Groundwater classification map using Stuyfzand's classification is made from the HYCH output results (Fig. 8). Oligohaline water to Freshwater occurs in Egumadurai (Loc. 1), Arampakkam (Loc. 2), Aaidukuppam (Loc. 3), Edur (Loc. 5), Pallavada (Loc. 7), Elavoor (Loc. 10), Chedulpakkam (Loc. 11), Sanaputtur (Loc. 13) Periaobulapuram (Loc. 18) Soorapoondi (Loc. 20) and Nemaluur (Loc. 21), Edur (Loc. 9), Obasamudram (Loc. 6), Melakkalani (Loc. 14), Natham (Loc. 19), Madharapakkam (Loc. 16),

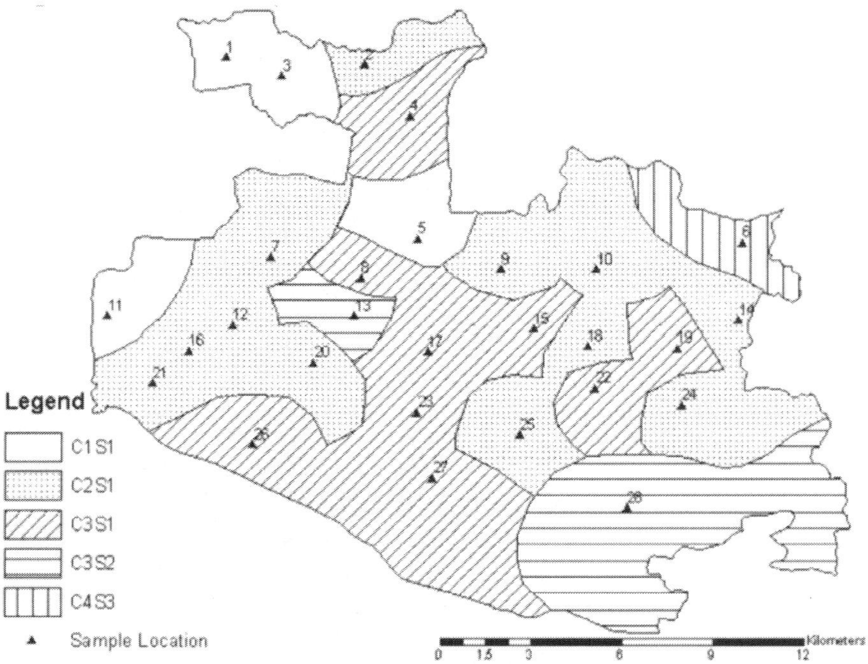

Fig. 6. USSL classification of groundwater.

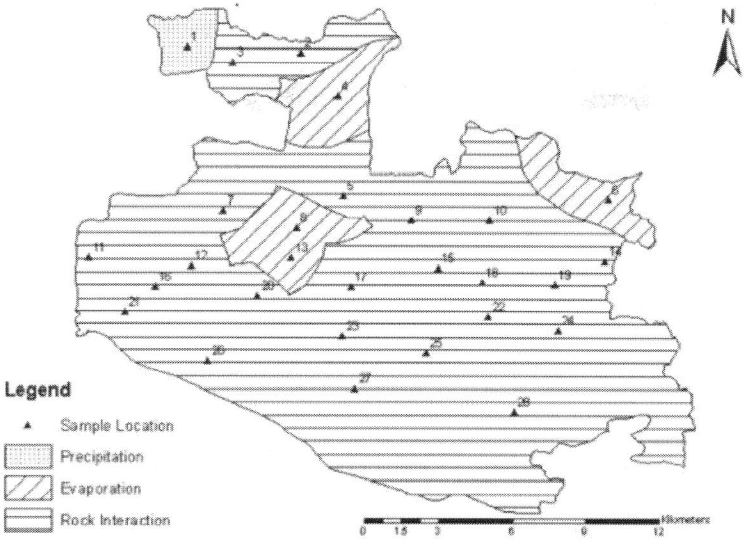

Fig. 7. Gibbs Plot.

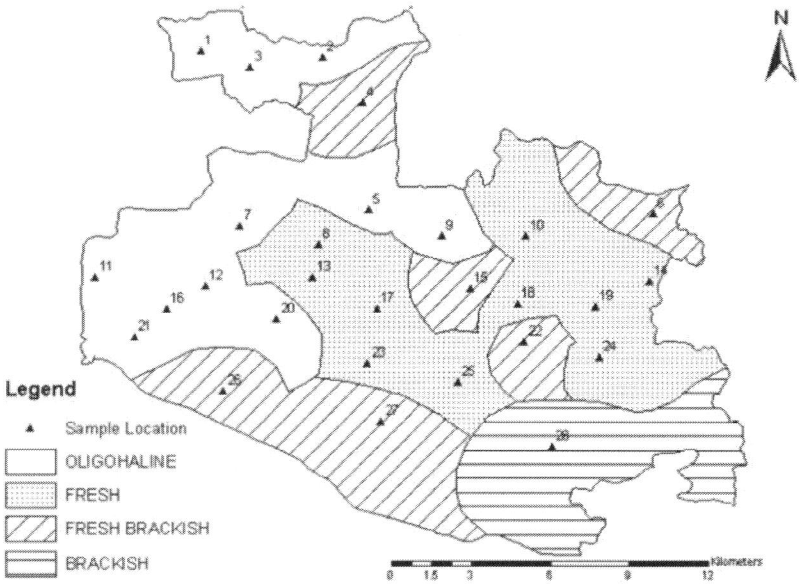

Fig. 8. Stuyfzand's classification of groundwater.

Eguvarpalayam (Loc. 17), Athupakkam (Loc. 24), Eguvarpalayam (Loc. 23) and Pappankuppam (Loc. 25). Fresh–brackish water exist in Arampakkam (Loc. 4), Obasamudram (Loc. 6), Thurapallam (Loc. 15), Peddikuppam (Loc. 22), Vaniamalli (Loc. 26), and Sirupuzhalpettai (Loc. 27) while brackish water is seen only in Gummudipoodi (Loc. 28).

SUMMARY AND CONCLUSION

Geologically the study area is part of Gummudipoondi taluk and is underlain by various geological formations ranging in age from Achaean to Quaternary age. Hydrologically groundwater occurs under unconfined to confined condition in the study area. Most part of the study area, the groundwater TDS is found to be below 1000 mg/L indicating the existing of freshwater can be used for domestic and agriculture purposes. North-eastern part of study area is occupied by high TDS and permanent hard water. Saline water intrusion is also identified in this zone which may be of over-exploitation. Consuming this water for longer duration may lead to health hazard. The groundwater quality of this region can be further improved by adopting artificial recharge method such as rooftop harvesting percolation pond and check dams.

REFERENCES

APHA_AWWA and WPCF (1984). Standard methods for examination of water and waste water,16th Ed. Washington.

Carrold, D. (1962). DC, APHA,1268P. Rain water as a chemical agent geological process. A view, U.S Geological survey water- supply paper 520f pp. 97-104

Karanth, K.R. (1993). Groundwater assessment development and management, Tata McGraw-Hill Publishing Company Ltd., p. 715.

Richard, C.A. (1954). Diagnosis and improvement of saline alkali soils USDA handbook 60: p. 160.

Ryzner, J.W. (1944). A New index for determining amont of calcium carbonate scale formed by water, J. Amer W.W. Assn, 36, pp. 472-486.

23

Trend Analysis of Groundwater Levels in Stressed Aquifers of Tiruvannamalai District, Tamil Nadu

Setty, G.Y.
Scientist-B, CGWB, Chennai
Email: sujathasetty@yahoo.com

ABSTRACT

Depletion and contamination of groundwater resources, either by natural or anthropogenic factors has assumed serious proportions in recent decades in several countries of the world. In a country like India, where vast areas are underlain by hard rocks with very limited groundwater potential, groundwater depiction has proved to be a major constraint in sustainable management of water resources. Groundwater level trend zonation map of study area has been prepared based on long-term trend analysis of groundwater levels and demarcated stress and moderately stress areas in terms of groundwater extraction.

Keywords: Groundwater level, water table, fluctuation, long-term trend analysis, hardrock aquifer.

INTRODUCTION

Increasing demand for freshwater resources coupled with diminishing supplies has been putting a severe stress on the limited available freshwater resources, especially in the case of hard rock areas. The available resources are also being increasingly threatened by pollution from various sources. Conservation, judicious management of supply and demand and prevention of pollution have become imperative to avoid a water crisis. Increasing demands for safe drinking water and requirements to maintain healthy ecosystems are leading policy makers all over the world to ask complex social and scientific questions about ways to assess and manage our precious water resources.

Groundwater is basically a dynamic resource, expressed as the quantity of water measured in terms of difference between maximum and minimum water table fluctuation within the aquifer.

The annual periodic fluctuations of groundwater table result from changes in recharge and discharge parameters of the hydrologic cycle. The aquifers are normally recharged in monsoon period leaving behind a usable dynamic groundwater resource for the rest of the year. Withdrawal of groundwater from the aquifers creates stress on groundwater regime. When the withdrawal exceeds recharge for a number of years, a continuous decline in water table may occur. It is essential to study the long-term groundwater table fluctuations for a proper planning of groundwater development and management of a particular region. In this study, an attempt has been made to study the long-term groundwater level trend using statistical techniques.

Data Source, Material and Methodology

The database for the analysis has been derived from the periodic and routine monitoring of depth to water levels of observation wells being monitored by State Groundwater Department, Govt. of Tamil Nadu and CGWB, Govt. of India, Chennai. The database and pre- and post-monsoon water levels of last 20 years have been considered for the study.

The database of the depth to water level data is converted with respect to mean sea level to prepare the depth to water table counter map, which depicts flow direction of groundwater. The IMD rainfall data has been considered for analysis. The statistical technique of Simple Linear Regression (SLR) was used for regression lines to determine the long-term trend of water table for the period of study. The softwares Map-info and GEMS have been used for preparation of groundwater trend maps and generation of hydrographs.

General Features and Hydrogeology

The study area lies between N. Latitudes 12°00'00" and 12°45'00" and between East Longitudes 78°43'00" and 79°15'00" and falls in Survey of India Toposheets 57L and 57 P. A map showing location of the area investigated and its administrative subdivisions, a part of Tiruvannamalai district, is shown in Fig. 1.

The area under study consists of high-grade metamorphic rocks of the peninsular gneissic complex intruded by acid, basic and ultra basic dykes. The rocks are extensively weathered. The general geology of the area is shown in Fig. 2. The area experiences subtropical climatic conditions with moderate humidity and temperature. The normal annual rainfall varies from 980 mm to 1000 mm.

Groundwater occurs under phreatic conditions in the weathered mantle of crystalline rocks and semi-confined conditions in the fractured zones. Productive fracture zones have been encountered in the depth range of 11.00 to 128.00 mbgl. Depth to water level is 1.5 to 16.4 mbgl during pre-monsoon and 1.6 to 14.90 mbgl in post-monsoon periods.

Groundwater Table Elevation and Flow Net Analysis

A map showing the elevation of groundwater levels during pre-monsoon period is presented in Fig. 3. The water table contours follow generally the topographical configuration. The water table elevation various between 130 and 220 m above msl. The highest elevation of water table over –220 m above msl is at the SW part of the area in Tandarampattu block. The general gradient is towards south-eastern direction in major part of the area. The hydraulic gradient is

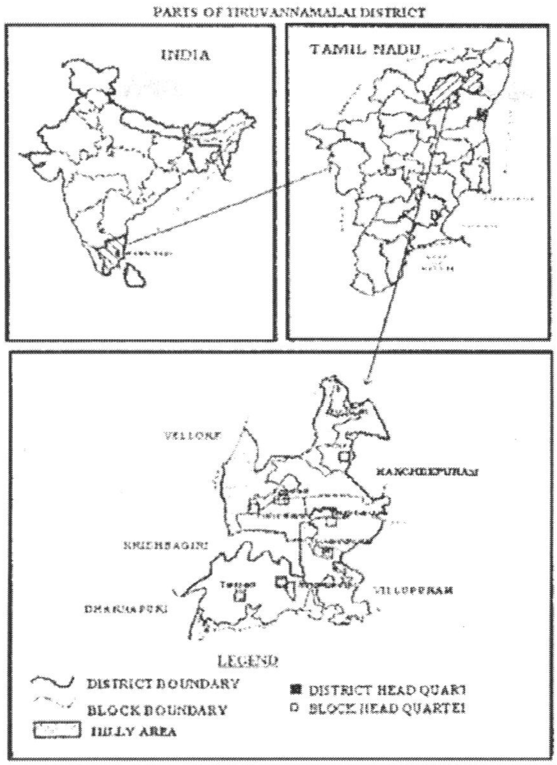

Fig. 1. Location of study area.

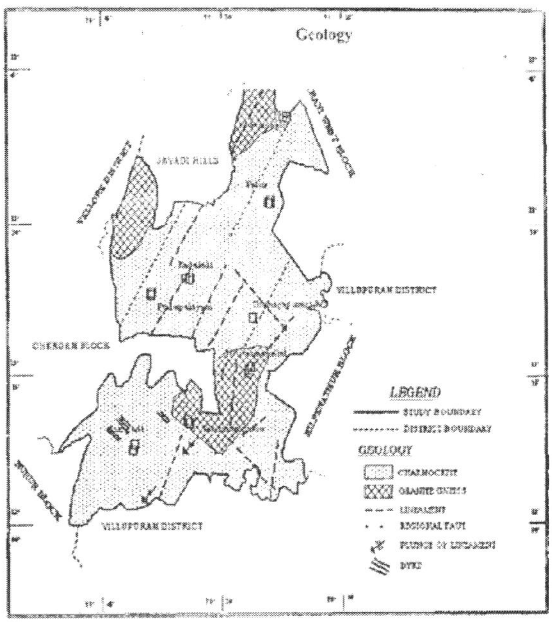

Fig. 2. Parts of Tiruvannamalai district, Tamil Nadu.

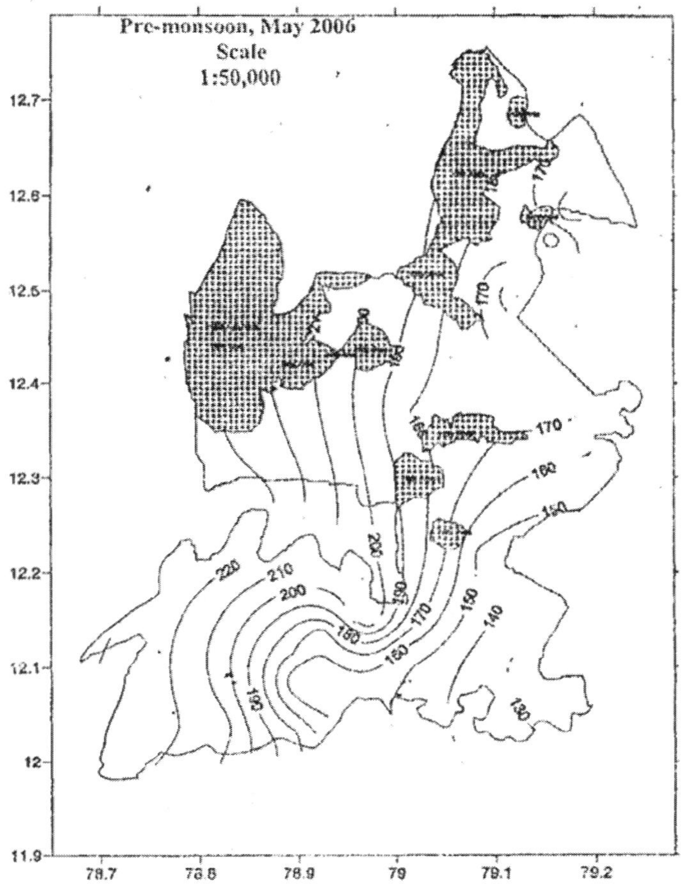

Fig. 3. Groundwater table contour map parts of Tiruvannamalai district.

steep to moderately steep in the area indicating low permeability. The amount of gradient varies depending on the aquifer material. The values of gradients for different areas are given in the Table 1.

Table1: Groundwater table gradient at select locations.

Locations	Gradient in/km)
Tiruvannamalai	4.0
Kalasapakkam	4.3

LONG–TERM WATER LEVEL FLUCTUATIONS

The long-term water level functions in the shallow zone have been studied using the historical water level data of network hydrograph stations of central groundwater board and state ground and surface water data resource centre of public works department Govt. of Tamil Nadu. Two approaches/techniques have been used for studying the long-term functions of groundwater levels shown in Table 2.

Table 2: Study of long-term water level fluctuations.

	Technique – 1	Comparison of pre- and post-monsoon water levels (May 2006 and Jan. 2007) with mean water levels for the last decade.
Methodology	Technique – II	Simple linear regression (SLT) analysis of water levels: Computation of annual pre-monsoon trends for each well to get to get an idea about long-term behaviour of groundwater levels using simple linear regression (SLR) analsyis.

Comparison of Water Levels with Decadal Mean: Technique

The depth to water level during June 2006 and January 2007, compared with the mean water levels of corresponding seasons for the one decade. Both rise and fall have been observed in groundwater levels when compared with the decade mean in both the seasons.

In the pre-monsoon period (June 2006), water levels have risen by 0.400 to 3.490 m in comparison with the decade mean in 75% of wells analysed. The rise in water levels are in excess of 2.0 m in the observation wells at Periyakulam (Pudupalayam block), Varagur in (Tandarampattu block), Mallavadi and Thachampatuu in Tiruvannamalai Block. Groundwater levels have shown a decline during the period in 25% of by amounts ranging from –0.222 to –0.915 m. The decline is most predominant in wells located Pudupalayam, Rayandapuram in Pudupalayam block at Gancshpuram in Polur Block. During the post-monsoon period (January 2007), 57% of wells analysed have shown a rise in groundwater levels ranging from 0.065 to 3.460 m in comparison with the mean water levels for 1997-2006. 43% of wells shows decline in groundwater levels by amounts ranging from –0.080 to –9.130 m.

Long-term Water Level Fluctuations: Technique – II

Hydrograph Analysis

To elucidate the general trend and fluctuation of water level characteristics, hydrographs of observation wells representing pre- and post-monsoon water levels have been generated. Best fit straight lines (regression lines) using least square technique were computed for pre- and post-monsoon data sets for all the stations. The slope of each line i.e., dy/dx, gives an effective resultant trend of pre- and post-monsoon groundwater level for the specific period. A positive gradient indicate a rising trend of water level, whereas a negative gradient implies a falling trend of water level. Zero gradients, on the other hand, represent a static situation where there is no change in net storage of groundwater. The range of variation of gradient lie between –0.002 and 0.625 m. Long-term annual trend equations which indicate a fall or rise in water level occur in the order of 0.002 to 0.625 m per year.

Map Analysis and Interpretation

Maps depicting spatial distribution of groundwater level trends have been generated using Map-info and Surfer software's for annual, pre- and post-monsoon periods for the study area.

Pre-Monsoon Water Level Trend — Analysis and Interpretation

It has been observed from the pre-monsoon water level trend map (Fig. 4) that the declining trend of water levels ranging from 0.012 to 0.651 m/year in and around Athuvambadi, Santhavasal, Murugapadi and Siruvallur in Polur block and as localized phenomenon at the villages Nayudumangalam, Mangalam, Melmudiyanur. Sirugalambadi, Mallapambadi and Kalaspakkam. Positive gradient has been noticed in the order of 0.002 to 0.224 m/year during pre – monsoon period in major part of the area.

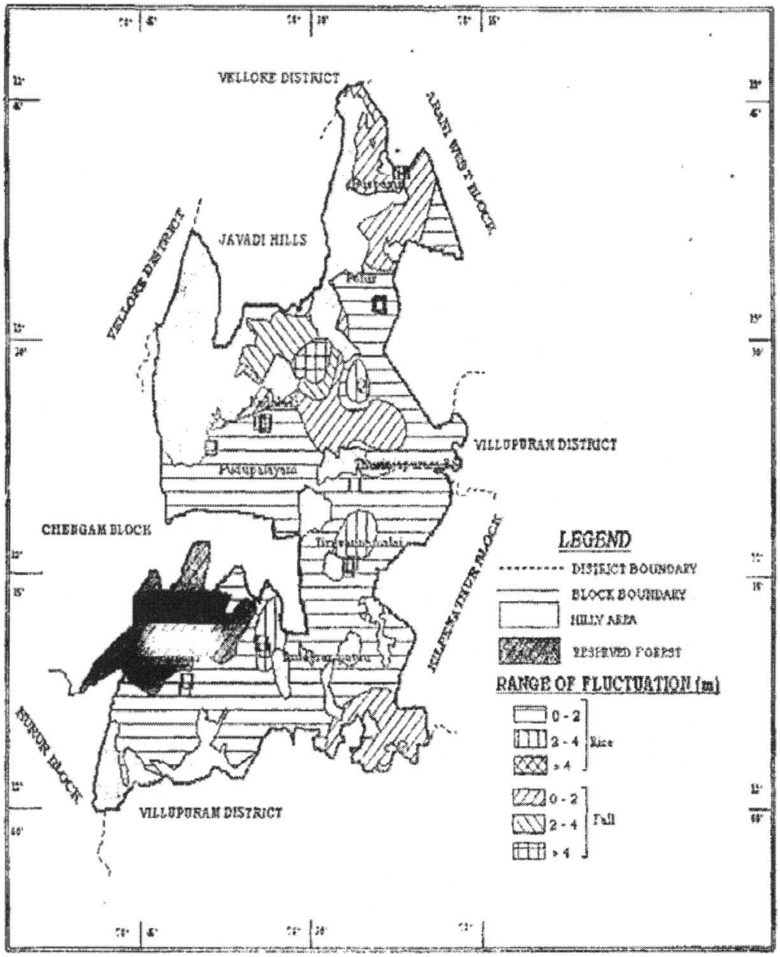

Fig. 4. Part of Tiruvannamalai district, Tamil Nadu. Groundwater level trend (pre-monsoon - MAT - 2007).

Post-monsoon Water Level Trend — Analysis and Interpretation

Post-monsoon water level trend map. (Fig. 5) reveals that the declining trend of water levels, ranging from –0.002 to –0.520 m/year, lies been observed mostly in part in parts of Polur, Thurinjapuram, and Pudupalayam blocks. Positive gradient of 0.010 to 0.475 m/year has been observed in the rest of the area during post-monsoon.

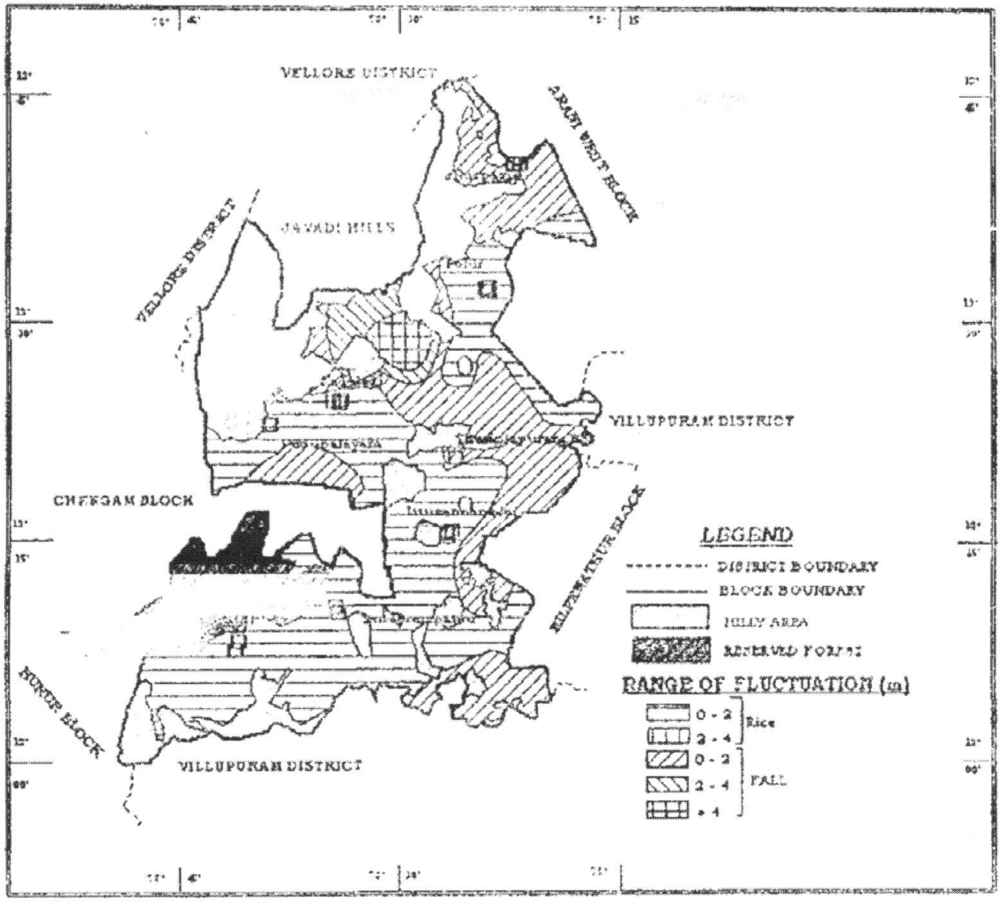

Fig. 5. Annual water level trend–analysis and interpretation. Part of Tiruvannamalai district, Tamil Nadu. Groundwater level trend (post-monsoon - JAN - 2008).

Annual water level trend map (Fig. 6) reveals that the declining trend of water levels, ranging from –0.002 to –0.520 m/year, has been observed in parts of Polur, Thurinjapuram, and Pudupalayam blocks. Positive gradient of 0.010 to 0.475 m/year has been observed in rest of the area.

Composite Pre- and Post-monsoon Water Level Trend Map—Analysis and Interpretation

A composite map of water level trends shown in Fig. 7 has been prepared using Map-info software by integration of both pre- and post-monsoon water level trend layers. It depicts four distinct trend zones as shown in the Table 3. An illustration of hydrographs of observation wells representing each trend zone are presented in Figs. 8 to 12.

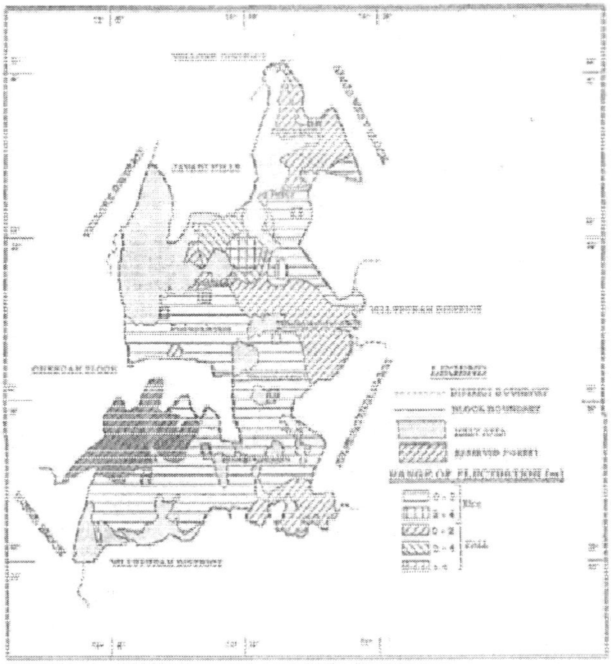

Fig. 6. Part of Tiruvannamalai district, Tamil Nadu. Groundwater level trend (Annual - JAN - 2007).

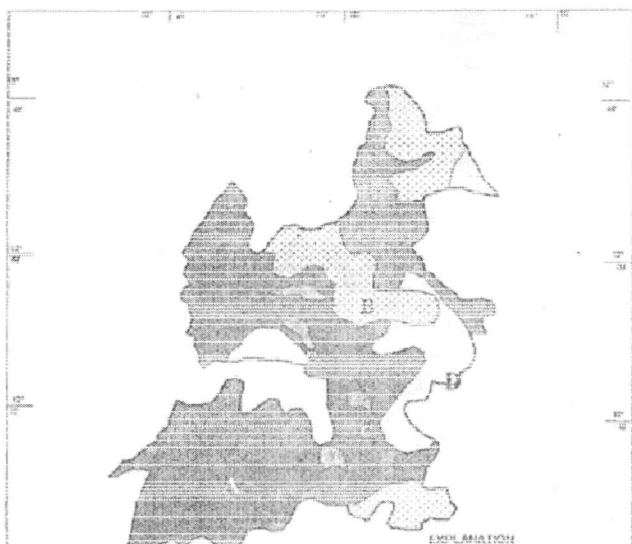

Fig. 7. Groundwater level trend zone. Part of Tiruvannamalai district.

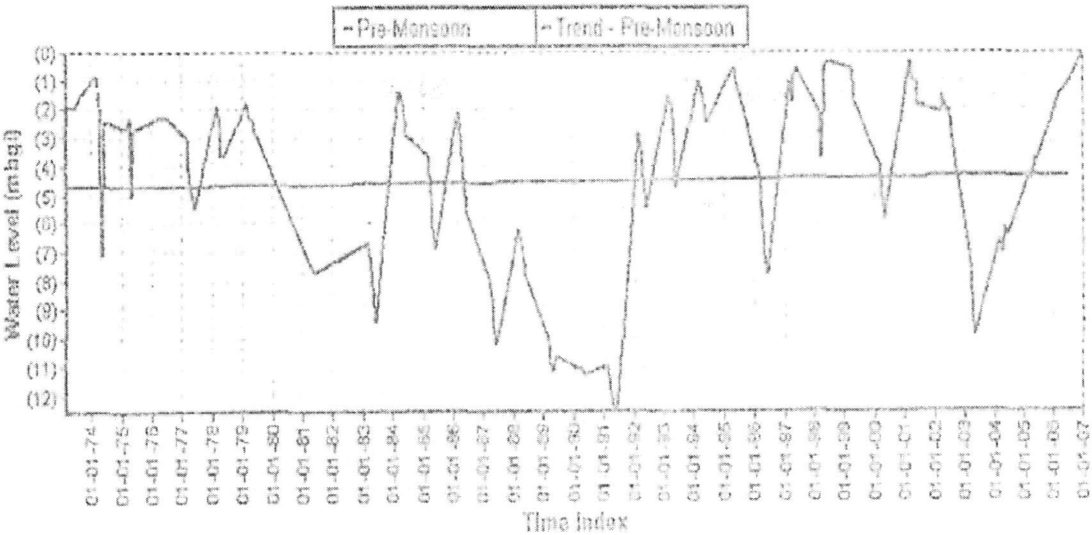

Pre-monsoon : Slope= 0.012 Intercept= -4.69

Fig. 8. Hydrograph of dugwell at Thatchampattu, Tiruvannamalai block, Tiruvannamalai district long-term pre-monsoon water level trend (1973-07)

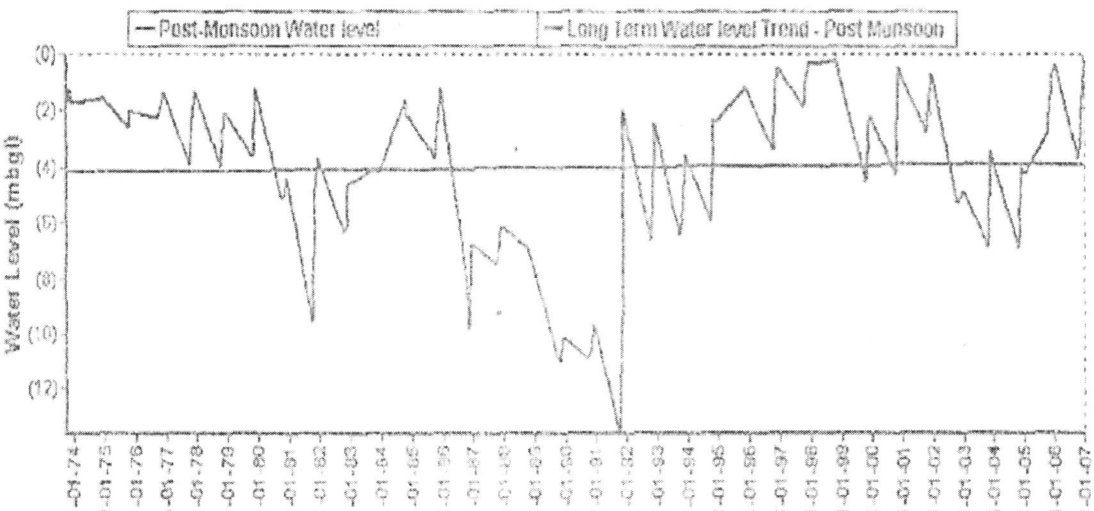

Post-monsoon : Slope= 0.010 Intercept= -4.13

Fig. 8a. Hydrograph of dugwell at Thatchampattu, Triruvannamalai block, Tiruvannamalai district long-term post-monsoon water level trend (1973-07).

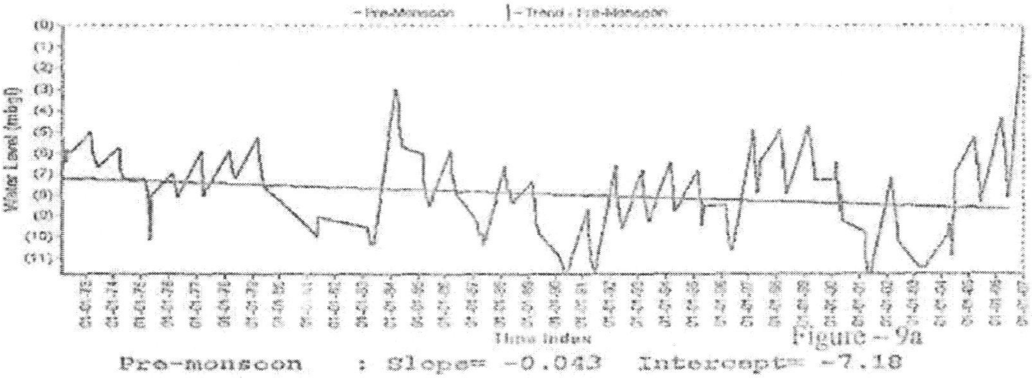

Fig. 9. Hydrograph of dugwell at Nayadimangalam Thurinjapuram block, Tiruvannamalai district, long-term pre-monsoon water level trend (1972-07).

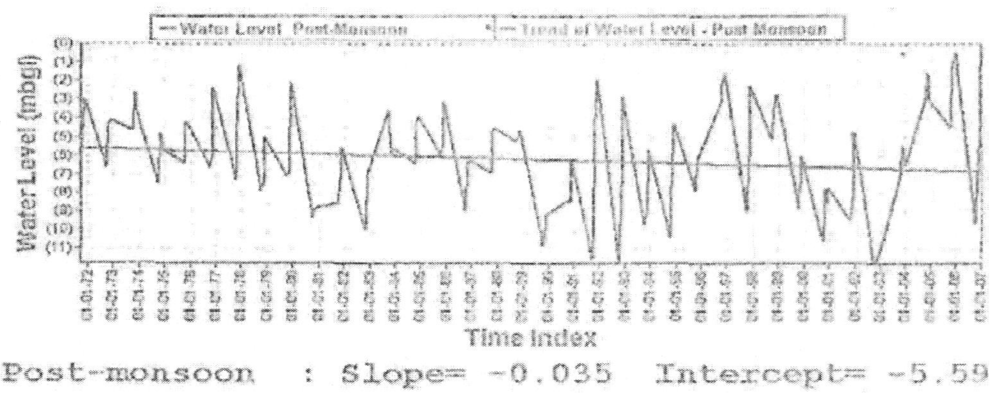

Fig. 9a. Hydrograph of dugwell at Nayadimangalam Thurinjapuram block, Tiruvannamalai district, long-term post-monsoon water level trend (1972-07).

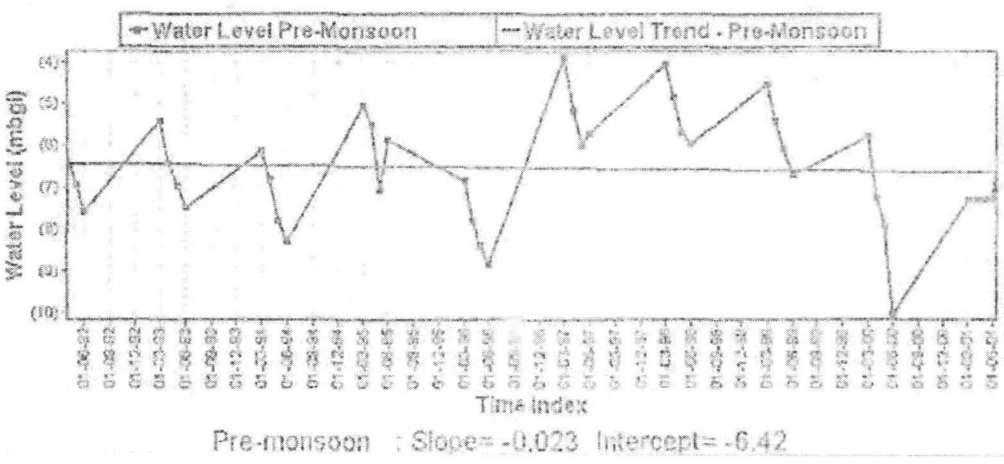

Fig. 10. Hydrograph of Dugwell at Kalasapakkam, Kalasapakkam block, Tiruvannamalai district, long-term pre-monsoon water level trend (1992-07).

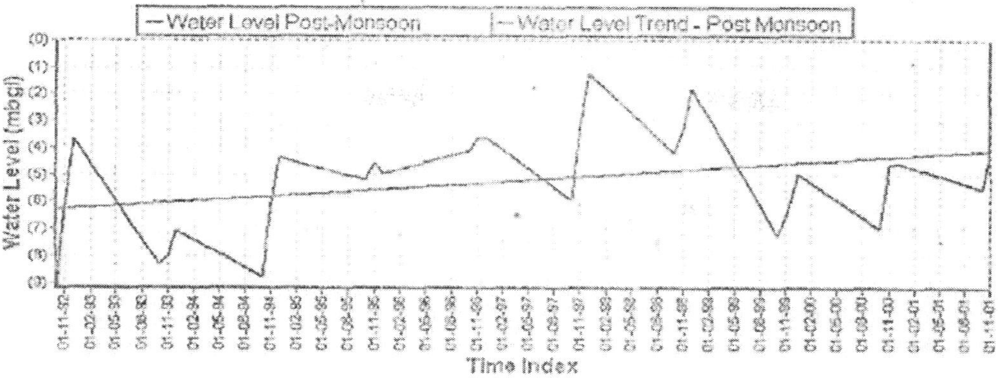

Post-monsoon : Slope= 0.243 Intercept= -6.51

Fig. 10a. Hydrograph of Dugwell at Kalasapakkam, Kalasapakkam block, Tiruvannamalai district, long-term post-monsoon water level trend (1992-01).

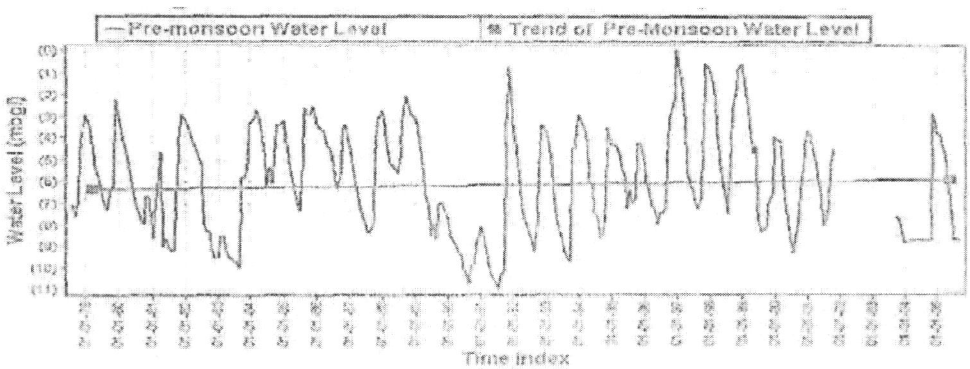

Pre-monsoon : Slope= 0.017 Intercept= -6.33

Fig. 11. Hydrograph of Dugwell at Adivaram, Pudupalayam block, Tiruvannamalai district, long-term pre-monsoon water level trend (1980-2005).

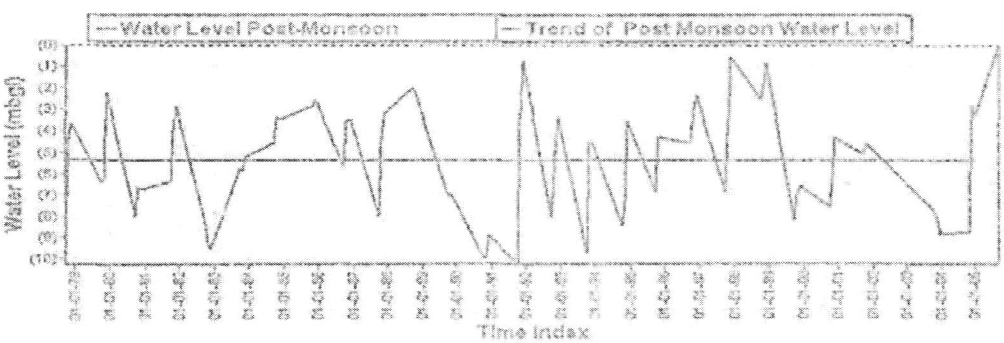

Post-monsoon : Slope= -0.002 Intercept= -5.34

Fig. 11a. Hydrograph of Dugwell at Adivaram, Pudupalayam block, Tiruvannamalai district, long-term post-monsoon water level trend (1979-2005).

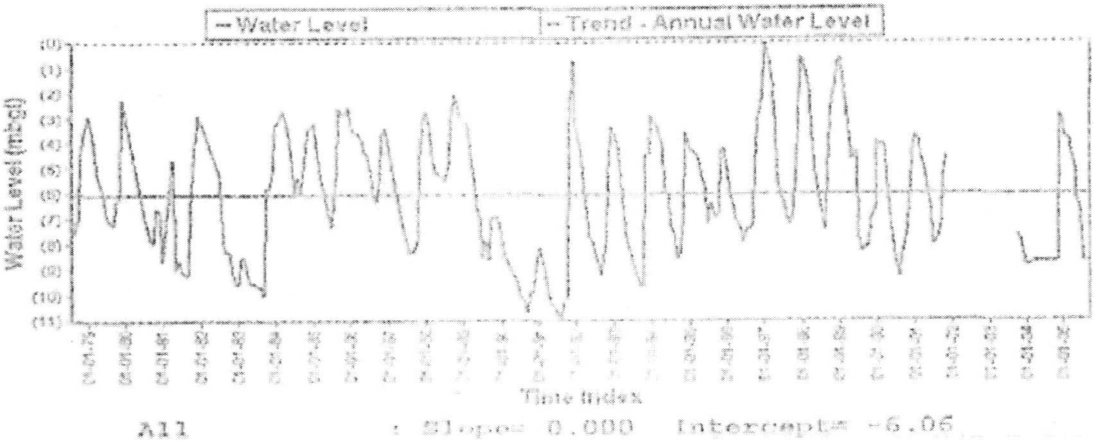

Fig. 12. Hydrograph of Dugwell at Adivaram, Pudupalayam block, Tiruvannamalai district, long-term annual water level trend (1972-2006).

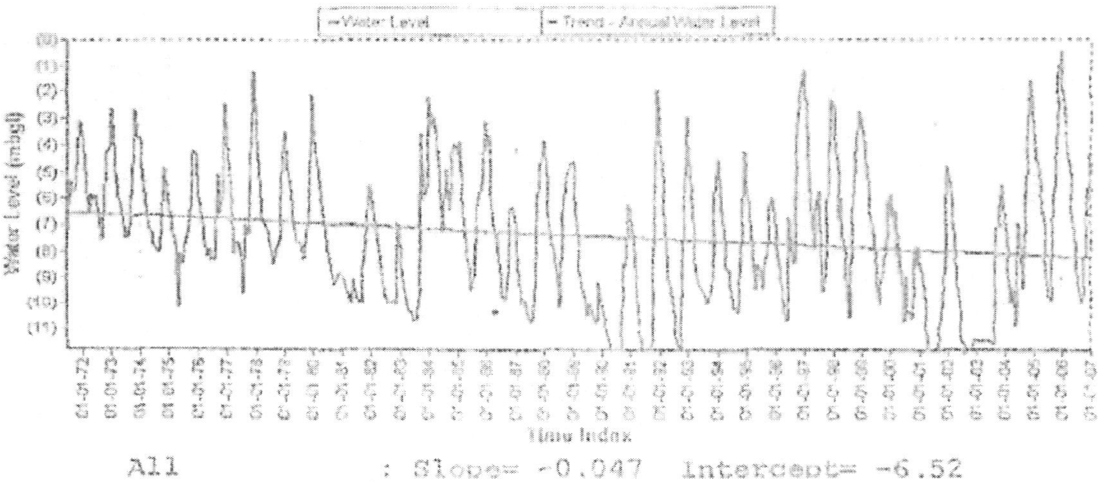

Fig. 12a. Hydrograph of Dugwell at Adivaram, Pudupalayam block, Tiruvannamalai district, long-term annual water level trend (1972-2007).

Table 3: Discrimination of different trend zones and characteristic features.

Sl. No.	Trend Zones	Trend of water level Pre-monsoon	Post-monsoon	Remarks on the over all long-term resultant trend
1.	A	Positive (+ve)	Positive (+ve)	Not under stress very good
2.	B	Negative (-ve)	Negative (-ve)	Highly stressed condition
3.	C	Negative (-ve)	Positive (+ve)	Moderately stressed recovering condition
4	D	Positive (+ve)	Negative (-ve)	Stressed condition

Table 4: Groundwater level Trend Zone—A Rising trend of long-term groundwater levels in both pre- and post-monsoon

Block	Location	Long-term Trend of Groundwater Levels			Period Considered
		Pre-monsoon	Post-monsoon	Annual	
Tiruvauanialai	Thandarai	+0.043	+0.029	+ 0.029	1972-'07
	Thalchampattu	+0.012	+0.010	+0.060	1973-'07
Thurinjapuram	Kunnumunjeri	+0.228	+0. 279	+0.245	1979-'04
Tandarampattu	Kilriandal	+0.008	+0.185	+0.086	1979-'04
	Pithandal	+0.068	+ 0.082	+0.067	1979-'04
	Kootaru	+0.065	+0.060	+0.049	1979-'04
	Malayanur Chakkadi	+0.091	+ 0.017	+0.036	1979-'04
	Olagalapadi	+0.070	+ 0.074	+0.056	1972-MM
	Perungalathur	+0.081	+0.077	+0.069	1973-'04
	Puduchakkadi	+0.104	+0.104	+0.123	iy/3-'04
	Rayandapuram	+0.11	+0.137	+0.100	1980-'0(5
	Tandarampattu	+0.123	+0.246	+0.177	I973-'07
	Tanipadi	+0.123	+0.130	+0.124	1973-*07
	Varagur	+0.070	+0.041	+0.048	1972-'07
Pudupalayam	Periyakulam	+ 0.004	+0.046	+0.012	1979-'07
	Pudupalayam	+0.027	4 0.034	+0.023	1972-*07
Polur	Mambattu	+0.081	+0.101	+0.079	1978-4Q7
	Gancspunuri	+0.017	1 +0.062	+0.030	1979-'07
	Padavecdu	+ 0.026	+0.043	+0.029	1975-'07
	Pillur	+ 0.249	+0.437	+0.304	1992-'97
Kalasapakkam	Kadaladi	+ 0.063	f 0.063	+0.056	1975+07
	Padayam	+0.052	+0.034	+0.047	1975-*04

Table 5: Groundwater level Trend Zone—B falling trend of long-term groundwater levels in both pre- and post-monsoon

Block	Location	Long-term Trend of water Levels			Period Considered
		Pre-monsoon	Post-monsoon	Annual	
Polur	Athuvambadi	-0.010	-0.040	-0.024	1975-'05
	Murugapadi	-0.059	-0.065	-0.069	1972-'07
	Siruvallur	-0.651	-0.521	-0.625	1992-'07
	Santhavasal	-0.005	-0.001	-0.011	1975-'07
Thurinjapuram	Nayudumangalam	-0.043	-0.035	-0.047	1972-'07
	Sirugalambadi	-0.112	-0.105	-0.123	1972-'07

Table 6: Groundwater level Trend Zone—C falling trend of long-term groundwater levels la pre-monsoon and rising trend in post-monsoon period.

Block	Location	Long-term Trend of Groundwater Levels			Period Considered
		Pre-monsoon	Post-monsoon	Annual	
Kalasapakkam	Kalasapakkam	–0.023	+0.234	–0.037	**I972-'07**
Tiruvannamalai	Mallapambadi	–0.129	+0.078	–0.012	1972-'07
Pudupalayam	Melmudiayarmr	–0,026	+0.036	–0.002	**1979-'07**
Tandarampattu	Mangalam	**–0.017**	+0.031	–0.038	I972-'07

Table 7: Groundwater level Trend Zone—D Rising *trend* of long-term groundwater levels in pre-monsoon *and* falling trend in post-monsoon period.

Block	Location	Long-term Trend of Groundwater Levels			Period Considered
		Pre-monsoon	Post-monsoon	Annual	
Pudupalayam	Adivaram	+0.017	-0.002	0.000	1979-'05

Long-term Groundwater Level Trend Zones

Trend Zone A: (Pre and *Post-monsoon both Positive Trend)*

The area characterised by positive groundwater level trend (Table 4) both in pre- and and post-monsoon periods comes under the Trend Zone A category. The areas falling under this category are predominately in the blocks of Tiruvannamalai Thurinjapuram, Tandarampaltu, Pudupalayam, Polur and Kalasapakkam. Since the groundwater levels shows rising trend both in pre- and post-monsoon periods, despite of the increasing demand and withdrawal and the zone may be considered as adequately recharged zone. Hydrographs of observation well at Thachampattu representing the Trend Zone A is shown in Fig. 8 and 8a. It is observed from the hydrographs that the groundwater levels shows increasing trend of 0.012 m and 0.010 m/year in pre- and post-monsoon periods respectively. Trend Zone B: (Pre- and Post-monsoon both Negative 'Trend).

The area characterised by negative groundwater level trend (Tabie 5) both in pre- and post-monsoon periods classified as Trend Zone B. The area falling under this category is predominately in parts of Polur and Thurinjapuram blocks. Hydrographs of observation well at Nayadimangalam representing the water level Trend Zone B is shown in Fig. 9 and 9a.

Falling trend of 0.043 m and –0.035 m/year in pre- and post-monsoon periods respectively has been noticed at Nayadimangalam observation well. This situation indicates a highly stressed condition where as the recharge is not able to adequately compensate the ground water withdrawal, thereby results in, declining ground water trend. Hence, the zone can be considered as highly stressed zone.

The area characterised by negative ground water level trend in pre-monsoon and positive trend (Table 6) in post-monsoon periods is designated by the Trend Zone C category. This phenomenon has observed predominantly in parts of Kalasapakkam, Tiruvannamalai, Pudupalayam and Tandarampattu. Hydrographs of Observation well at Kalasapakkam representing the Water level Trend Zone C is shown in Fig. 10 and 10a. Falling trend of water levels at a rate of –0.023 m/year in pre-monsoon and 0.243 m/year in post-monsoon period has been noticed at this observation well. The declining trend during pre-monsoon period indicates the greater groundwater withdrawal while the rising water level trend during post-monsoon period indicates that the recharge is able to compensate the groundwater extraction. Hence, the zone may be considered as moderately stressed zone.

Trend Zone D: (Pre-monsoon Positive and Post-monsoon Negative Trend)

The area characterised by positive ground water level trend in pre-monsoon negative trend (Table 7) in post-monsoon periods comes under the Trend Zone D. This trend zone is identified predominantly in and around Adivaram area in Pudupaiayani block. Hydrographs of Observation well at Adivarani representing the water level Trend Zone D is shown Fig. 11 and 11a. Rising trend of water levels at a rate of 0.017 in/year in pre-monsoon and declining trend in the order of –0.002 m/year in post-monsoon period has been noticed. Although water level shows rising trend during pre-monsoon lean period, it shows declining during post-monsoon period indicating that recharge is not adequately compensating the groundwater extraction. Consequent to which ground water extraction may also get reduced which might be the reason for rising trend of ground water level during pre-monsoon period. This zone can be treated as stressed zone.

Annual Long-term Groundwater Levels Trend

The analysis carried out with the historical water level data for the period 1990-07 indicate that groundwater levels are showing a long-term (annual) rising trend in the major part of the area. The falling trend is confined mostly to localized patches and can be attributed to localized overdraft for irrigation. Representative hydrographs of select wells, along with annual trend lines are shown in Fig. 12 and 12a. The rise in water levels in the major part of the district is an indicative of the effect of the good monsoons during (1994-98 and 2006) in the study area on the groundwater regime.

CONCLUSION

As far as the groundwater development is concerned, the area is classified as over exploited. The situation become bad to worse in the areas comes under Trend Zones B and D due to increasing demand of groundwater and the zones can be considered as highly stressed zone. The declining trend during pre-monsoon period in the zone C indicates the greater groundwater withdrawal while the rising water level trend during post-monsoon period indicates that the recharge is able to compensate the ground water extraction. Hence, the zone is considered as moderately stressed zone and the condition is in recovering stage. The groundwater situation in the area falling under Trend Zone A is good and shows rising trend though it is classified as overexploited area as on Jan. 2004 as per GEC 1997 methodology and the situation might have recovered *due* to the implementation of artificial recharge structures in the area. Implementation of artificial recharge programme and rainwater harvesting on scientific lines will bring back the situation at least to desirable level.

ACKNOWLEDGEMENTS

The authors are grateful to Sri B.M. Jha, Chairman, CGWB, Faridabad Dr. N. Varadaraj, Regional Director, CGWB, and Chennai for constant encouragement in preparation and permitting to present the paper. The authors also thankful to Dr. S. Suresh Scientist D of CGWB, Chennai for valuable suggestions and guidance.

REFERENCES

Ground Water Exploration in Tamil Nadu and U.T of Pondicherry (as on 1996), CGWB, SECR, Chennai

Ground Water Resources and Development prospects in Tiruvannmalai district, Tamil Nadu-Unpublished report, CGWB, SECR, Chennai.

Ground Water Perspectives, A Profile of Tiruvannamaiai District, Tamil Nadu, State Ground and Surface Water resources data centre PWD Govt, of Tamil Nadu, Chennai 2000.

Ground Year Book 2005, CGWB, SECR, Chennai.

Index

Author Index